Gas-Phase Pyrolytic Reactions

Gas-Phase Pyrolytic Reactions

Synthesis, Mechanisms, and Kinetics

Nouria A. Al-Awadi
Professor of Chemistry
Chemistry Department
Kuwait University
State of Kuwait

Registered Office
John Wiley & Sons, Inc., 111 River Street, Hoboken, NJ 07030, USA

Editorial Office
111 River Street, Hoboken, NJ 07030, USA

For details of our global editorial offices, customer services, and more information about Wiley products visit us at www.wiley.com.

Library of Congress Cataloging-in-Publication data applied for

ISBN: 9781118057476

Cover Design: Wiley
Cover Image: © ririro/Shutterstock

Set in 11/13pt WarnockPro by SPi Global, Chennai, India

Printed in the United States of America

V10014729_101619

To the memory of my beloved parents

Zamzam Al-Awadi
Abdulkareem Al-Awadi

Contents

Preface

Pyrolysis is a classic example of a field of science in which developments in one area of the subject prompt the emergence of a closely related area, ultimately leading to a broad-based and wide-encompassing discipline.

Pyrolysis has moved from the experimental method used to provide analytical information on the behavior of chemical substances under the action of heat to the current state, providing in addition an alternative route to organic synthesis and industrial processes. The history of the subject is also a source of information on how the development of ever-more-sophisticated analytical tools and physical methods, reactor design, and technology breakthroughs have given practitioners in the field the insight and know-how to venture into novel terrains, which in turn served to extend and enrich the science.

Gas-phase pyrolysis has come of age. What once started as a purely analytical technique has now matured into a fully fledged analytical, synthetic, and preparative technology encompassing an astonishing mass of processes and materials. Gas-phase pyrolysis has also emerged as a science in which reaction pathways and mechanisms can be postulated based on the analysis of the kinetics and products of reaction and often also supported by empirical and theoretical investigations. A book on up-to-date discourse on gas-phase pyrolysis is needed now to follow on earlier references that are a few decades in the past. This book, as the titles of its seven chapters indicate, covers aspects relating to gas-phase pyrolysis that have not so far been brought together between two covers.

I am grateful to Prof. Curt Wentrup from the University of Queensland, Australia, for reviewing parts of the manuscript and for his valuable comments.

List of Abbreviations

ab initio SCF	*Ab initio* self-consistent field
B3LYP/6-31+G**	Becke, 3-parameter, Lee–Yang–Parr/6-31+G** basic set
CAMS	Collisional activation mass spectrometry
DFT	Density functional theory
DNA	Deoxyribonucleic acid
ESR	Electron spin resonance
FID	Flame ionization detector
FS-FVP	Falling solid–flash vacuum pyrolysis
FVP	Flash vacuum pyrolysis
GC	Gas chromatography
GC/MS	Gas chromatography–mass spectrometry
GPPy	Gas-phase pyrolysis
hPa	Hectopascal
HPLC	High-performance liquid chromatography
ID	Internal diameter
IR	Infrared
LCMS	Liquid chromatography–mass spectrometry
MP2/ 6-31+G**	Second-order Møller–Plesset perturbation theory/6-31+G** basic set
nPa	Nanopascal
ONIOM	Our own n-layered integrated molecular orbital and molecular mechanics
PAHs	Polycyclic aromatic hydrocarbons
Py-GC	Pyrolysis gas chromatography

Py-GC/MS	Pyrolysis gas chromatography–mass spectrometry
SS-FVP	Solution spray–flash vacuum pyrolysis
STP	Static pyrolysis
UV	Ultraviolet

About the Author

 **Nouria A. Al-Awadi** is professor of chemistry in the Department of Chemistry, Faculty of Sciences, Kuwait University. She obtained her B.Sc. from Kuwait University and received her Ph.D. in gas-phase kinetics under the supervision of Prof. David Bigley from the University of Kent-Canterbury, UK. In addition, she earned a postdoctoral fellowship at the School of Chemistry and Molecular Sciences, Sussex University, working in the laboratories of the late Dr. Roger Taylor.

On her return to Kuwait, she was appointed to the staff of the Chemistry Department at Kuwait University where she has established a strong independent research group.

As a fully fledged faculty member at Kuwait University since 1979, her leadership roles are extensive. She served as Head of the Department of Chemistry and as Dean of the Faculty of Sciences. She also served as Provost for Academic Affairs.

She has been recognized for her role in establishing Analab, an analytical service laboratory, which is superbly equipped, managed, and supported. This was expanded faculty-wide during the period when she was dean through SAF (Science Analytical Facilities), which transmitted the benefits of its resources and capabilities to meet the wide-ranging and diversified requirements of other departments.

A dynamic researcher with world-wide collaborations, she is credited with over 150 publications. As her publication record demonstrates, she has worked with prominent researchers in the field in the UK. She is in contact with all of the significant gas-phase kinetic work worldwide.

Among her main hobbies are collecting unique tea and coffee mugs from all around the world (her collection currently exceeds 1000 pieces) and arboriculture; she currently cares for 35 date palm trees and a large number of jasmine shrubs.

1

Methodologies and Reactors

This chapter presents an overview of pyrolytic reactions, which may be carried out either in sealed tubes (static reactors) or in flow systems, including flash vacuum pyrolysis (FVP) reactors, the use of static pyrolysis (STP) in kinetic investigation, and why flow systems were used in organic synthesis. It also covers the combination of the various pyrolytic reactors and online systems with advanced physiochemical techniques. A comparison of each type of pyrolytical methodology has also been given.

1.1 Static Pyrolysis

In a static reaction, the substrate is heated continuously in a solid phase, in solution, or in a gas phase in a sealed vessel. This type of pyrolysis is

Gas-Phase Pyrolytic Reactions: Synthesis, Mechanisms, and Kinetics,
First Edition. Nouria A. Al-Awadi.
© 2020 John Wiley & Sons, Inc. Published 2020 by John Wiley & Sons, Inc.

performed in furnace pyrolyzers. The sample is heated for a relatively long period of time, generally at a relatively low temperature (below 450 °C). The pyrolysis products are further analyzed, commonly by an offline analytical technique such as HPLC, GC, GC/MS, IR, or LCMS. The residence time of the substrate in the hot zone plays a crucial role in determining the nature of the products formed. The longer the contact time, the higher the probability that the primary products will undergo secondary reactions. This technique has long been used for a large number of classical thermal reactions: for example, eliminations, fragmentations, and rearrangements, including pericyclic processes. Thermally labile products, however, cannot be isolated using STP, as these may undergo further inter- and intramolecular reactions. STP is the right choice for pyrolyzing substrates with low volatility or which involve intermolecular reactions or reactive intermediates. Nevertheless, static reactors are widely used in the study of gas kinetics [1–4].

We, in our laboratory, have successfully used this technique to study gas-phase kinetics, where the ultimate product formation was not disturbed by secondary reactions due to a relatively long residence time of the substrate in the hot zone. The two types of static reactors that were used for kinetic studies are discussed next.

1.1.1 Sealed-Tube Reactor

This system consists of two parts: the oven pyrolyzer and the reaction tube.

1.1.1.1 Pyrolyzer
The pyrolyzer is a custom-made unit made from a cylinder of an insulated aluminum block, which can be heated to any preselected temperature up to 530 °C. Aluminum is chosen for this purpose because of its high thermal conductivity, which ensures an exceptionally low temperature gradient throughout the block. The temperature is controlled by a precision temperature regulator set to provide a 0.1 °C incremental change achieved by a digital switch, which gives an overall temperature output with an accuracy of ±0.5 °C. The actual pyrolysis temperature [5a, b, 6a, b] is measured by a platinum resistance thermocouple within the pyrolyzer unit, very close to the reaction vessel, which is connected to a microprocessor thermometer.

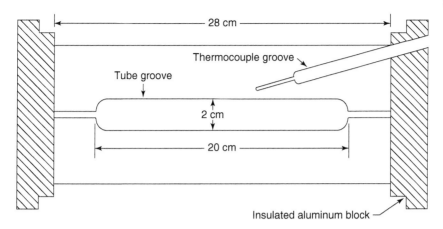

Figure 1.1 Schematic diagram of the pyrolyzer.

The block was hollowed where necessary to fit the pyrex reaction vessel and the tip of the platinum resistance thermocouple for actual reaction temperature read-out; the latter was fitted in a hole drilled diagonally along the cylindrical axis (Figure 1.1).

1.1.1.2 Reaction Tube

The pyrex reaction tube shown in Figure 1.2 is used for both kinetic studies and product analysis. Samples of the starting material in very dilute solution together with an internal standard are introduced into the reaction tube, which is placed in liquid nitrogen in order to freeze the contents of the tube. The tube is then sealed under vacuum to elimi-nate the possibility of combustion reactions and ensure unimolecularity and conversion of the substrate into vapor prior to reaction. The sealed tube with the sample is then placed in the niche in the pyrolysis unit set to the preselected temperature.

1.1.1.3 Kinetic Studies

A stock solution (7 ml) is prepared by dissolving 6–10 mg of the substrate in acetonitrile to give a concentration of 1000–2000 ppm.

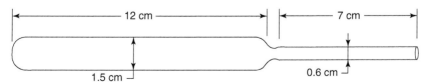

Figure 1.2 Schematic diagram of the pyrolysis tube.

An internal standard is then added, the amount of which is adjusted to give the desired peak area ratio of substrate to standard (2.5,1) in a HPLC/GC analysis. The solvent and the internal standard are selected so that both are stable under the conditions of pyrolysis, and so that they do not react with either the substrate or the products. Generally, the compounds used as internal standards are chlorobenzene, 1,3-dichlorobenzene, and 1,2,4-trichlorobenzene. Each solution is filtered to ensure that a homogeneous solution is obtained.

The reaction rate is obtained by tracing the rate of disappearance of the substrate with respect to the internal standard as follows:

An aliquot (0.2 ml) of each solution containing the substrate and the internal standard is pipetted into the reaction tube, which is then placed in the pyrolyzer for six minutes under non-thermal conditions. The sample is then analyzed using the HPLC/GC probe, and the standardization value (A_o) is calculated.

Several HPLC/GC measurements are obtained with a consistency $\geq 98\%$. The temperature of the pyrolysis block is then raised until approximately 10% pyrolysis is deemed to have occurred over a given period of residence time. This process is repeated after each $10-15\,^\circ C$ rise in temperature of the pyrolyzer until $\geq 84\%$ pyrolysis has taken place. The relative ratios of the integration values of the sample and the internal standard (A) at the pyrolysis temperature are then calculated. A minimum of two kinetic runs are carried out at each $10-15\,^\circ C$ rise in the temperature of the pyrolyzer to ensure reproducible values of (A). Kinetic runs are also repeated in the presence of a radical inhibitor, such as cyclohexene or toluene, depending on the temperature range of pyrolysis, to check the possibility of radical mechanisms, and by using a pyrolysis tube packed with helices to increase the surface area (which helps to ascertain that there are no surface reactions).

1.1.1.4 Treatment of Kinetic Results

For the first-order reaction at a given temperature, the integrated rate equation is given by Eq. (1.1):

$$k = \frac{2.303}{t} \log \frac{A_0}{A} \tag{1.1}$$

The variation of the rate constant (k) with temperature can be expressed satisfactorily by the logarithmic form of the Arrhenius equation Eq. (1.2) to give Eq. (1.3):

$$k = A\, e^{\frac{-E_a}{RT}} \tag{1.2}$$

$$\log k = \log A - \frac{E_a}{2.303\,RT} \qquad (1.3)$$

where:

k = rate constant for first-order reaction (s^{-1})
A = frequency or pre-exponential factor (s^{-1})
E_a = energy of activation $(kcal\,mol^{-1})$
R = gas constant $(1.98722\,cal\,K^{-1}\,mol^{-1})$
T = absolute temperature (K)

The plot of log k versus $1/T$ gives a straight line with a slope equal to $(-E_a/2.303\,R)$ and intercept equal to (log A), from which the energy of activation (E_a) and the frequency factor (A) can be calculated.

The entropy of activation is temperature-dependent [7] and is determined from the log A factor at the appropriate temperature using Eq. (1.4):

$$\Delta S^{\#} = 2.303\,R\left(\log A - \log \frac{KT}{h}\right) \qquad (1.4)$$

where:

K = Boltzman constant
h = Plank constant

1.1.2 Static Apparatus

The apparatus shown in Figure 1.3 used for determining the reaction kinetics in the static system consists of a cylindrical reaction vessel of approximately 185 ml volume with two chambers divided by a very thin stainless steel diaphragm equipped with a platinum contact on the outside, and sealed with a copper gasket crushed between knife edges, to give more reliable seal; the diaphragm acts as a null-point gauge [8].

The reaction vessel is made of stainless steel, for its good resistance to hot acids and hardness to metal seals, which is of prime importance. To avoid leaks, parts are machined from a single piece of steel or by electron-beam welding of the joints. On one end of the cylindrical reactor are two valves, the injection and evacuation valves, each being perpendicular to the cylinder end, and screwed in to finger tightness to provide a metal-to-metal seal after injection. The valve shafts are fitted with rubber ring seals, and the valve extremities with water cooling coils. The injection port is sealed with a silicon rubber septum and fitted

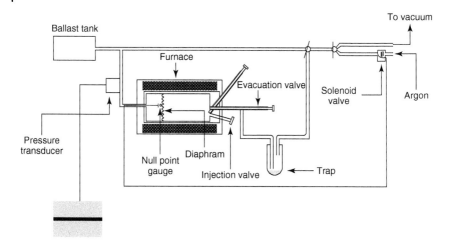

Figure 1.3 Schematic diagram of the static pyrolysis (STP) apparatus.

with a Perspex dome with a hole in the center to allow the passage of the hypodermic needle and with argon admitted through the side. This protects the reactor surface from any possible leakage through the silicon rubber septum during the injection [9–11].

1.1.2.1 Reaction Chamber

The reaction chamber is inside a cylindrical, well-insulated aluminum block with a capacity of $\sim 1\,\text{ft}^3$, which is fitted with a temperature controller, dual thermocouples to record the temperature, and stainless steel heater rods operated by a platinum-resistance thermometer. Temperature stability with an accuracy of $\pm 0.1\,°C$ is achieved over several days, and the temperature can be preset to within $0.3\,°C$ or better.

The balancing gas is 99.99% argon, which is regulated by a solenoid valve controlled by the null-point gauge. Pressure changes are monitored by a pressure transducer and recorded by a strip-chart recorder. Also, the apparatus is fitted with a mercury manometer. All connections in the gas line (mainly glass) are restricted to flexible metal couplings with neoprene O-rings. By inclusion of 2.5 l ballast in the external system, the pressure can be increased in very small increments.

The sample to be analyzed is injected into the reactor by using a Hamilton gas-tight hypodermic syringe (Teflon plunger) modified by a rigid nylon cap fitted to the upper end, through which argon can pass. This prevents any air from entering the reactor during the injection, as oxygen can activate the reactor surface.

1.1.2.2 The Pyrolysis Method

In the kinetic method, the reactor is evacuated. The sample to be analyzed (ca. 100 µl) is then injected into the evacuated reaction chamber through a silicon rubber septum using a micro-syringe with a 5 in. needle. The hypodermic needle is partly withdrawn, and the injection valve is moved to the closed position. Samples are injected either as liquids or solids, or as solution in chlorobenzene. Prior to injection, the sample is degassed by a sharp withdrawal of the plunger in order to eliminate any air bubbles. During vaporization, the pressure buildup in the reactor causes the diaphragm to deflect, as it is sensitive to approximately 0.02 mm pressure.

This activates a solenoid valve, which admits oxygen-free argon to the balancing line. Pressure in the balancing gas line is monitored by a pressure transducer coupled to a strip-chart recorder; it draws the first-order kinetic plot from the monitored pressure and time intervals (measured as a distance in inches), which are taken for increments between the pressure at the initial time (P_0) and at 10 half-lives (P_∞), where the reaction is stopped. On completion of the reaction, the products are pumped out through a cold trap, so that they may be examined as necessary. This method gives linear first-order kinetic plots to more than 95% of the reaction. The method is rapid, and more than 20 kinetic runs can be made per day under favorable conditions.

The main disadvantage of pyrolysis kinetics is the possibility of a surface catalyzed reaction. This can be minimized by pre-pyrolysis of 3-butenoic acid at high temperature or by coating the reactor surface with iridium-free gold [11]. A smooth linear Arrhenius plot (to >99%) shows the absence of surface catalysis. If it is present, scattered or curved plots result; the scatter is worse at low temperatures, where the surface catalysis is more obvious. The kinetics, therefore, should be carried out at a minimum temperature range of at least 50 °C or until a first-order linear plot is obtained.

1.1.2.3 Treatment of the Results

As the reaction takes place in the gas phase, the pressure is equivalent to the concentration. The initial concentration (a_0) is proportional to (P_∞), and (a) is proportional to (P_t). A plot of log $(P_\infty - P_t)$ versus time shows a linear relationship of slope (m), from which the reaction rate constant for the first-order reaction (Eq. 1.1) is determined.

The frequency or pre-exponential factors (log A s^{-1}) and activation energies (E_a) are obtained from measurement of the rate constant at a series of temperatures. The variation of the rate constant (k) with temperature can be expressed satisfactorily by the logarithmic form of the Arrhenius Eq. (1.3).

A plot of log k against $1/T$ gives a straight line with slope ($-E_a/2.303\,R$) and intercept log A. The temperature-dependent entropy of activation is determined from log A at the appropriate temperature using Eq. (1.4).

1.2 Dynamic Flow Pyrolysis

In dynamic flow processes, the substrate is transported in the gaseous state through a heated zone followed by rapid cooling of the products formed. Compared to static procedures, higher reaction temperatures are usually employed. As is the case with any method that depends on gas flow, the residence time of substrate molecules in the hot reaction zone is a key parameter [12], and this in turn depends on a number of variables including pressure, temperature, and the dimensions of the reaction tube. The heating time in the reactor has to be very short, and the products have to be cooled down very rapidly, so that the critical temperature range is passed quickly, and the reaction stops suddenly [13]. By far the most representative technique of dynamic flow pyrolysis in the gas phase is flash vacuum pyrolysis.

1.2.1 Flash Vacuum Pyrolysis

The technique of flash vacuum pyrolysis (FVP) experimentally involves the vacuum sublimation or distillation of a substrate through a hot tube (generally at temperatures between 300 and 1000 °C) to induce chemical change. After passing through the tube, the products are quenched at low temperatures in a trap placed at the exit of the furnace tube [3, 12]. The pyrolysis is usually carried out under dilute, short-contact-time conditions in the gas phase such that individual molecules react intramolecularly in the effective absence of other molecules of the substrate, products, or a substantial amount of oxygen. The technique therefore produces much cleaner results than other forms of pyrolysis. FVP is a robust, highly reproducible method and has found widespread use in applications ranging from matrix isolation and mechanistic studies to preparative organic chemistry [14]. As the heating time is very short and by rapid cooling, substances that are not stable at room temperature can be synthesized. FVP, therefore, permits straightforward synthesis of many sensitive and highly reactive organic compounds that are not readily obtained from reactions in solutions [13].

Pyrolysis products can be used to prepare certain compounds that are not accessible by other methods or only with difficulty. These reactions can be performed on the micro scale, the products being identified by analytical or spectroscopic methods; or, if the reaction is successful, the scale can be increased for the practical preparation of the substance [15].

In FVP, the substrate is sublimed or distilled in vacuo (10^{-1} to 10^{-3} torr) or sometimes in a reduced-pressure nitrogen flow system using a sublimation oven through a horizontal fused quartz tube, which is heated externally to the desired pyrolysis temperature and monitored by a thermocouple situated at the center of the furnace. The substrate can be subjected to the requisite amount of thermal energy required to undergo the desired conversion by careful selection of the dimensions of the hot zone (length and width), and by choosing the appropriate temperature, flow rate, and pressure [16]. Thermal excitation of the substrate molecules occurs mainly by collision with the hot wall of the reactor. The products are collected in a U-shaped trap cooled in liquid nitrogen. The pressure is maintained and is measured with a Pirani gauge situated between the cold trap and the pump. Under these conditions, the typical contact time in the hot zone is estimated to be 10^{-3} to 10^{-1} seconds. The shorter the contact time, the smaller the probability of the primary product molecules undergoing secondary reactions. In FVP, the reactive intermediates are generated under unimolecular conditions in the gas phase, and the low-pressure flow conditions ensure that the individual molecules spend only a short time (on the order of milliseconds) in the reaction zone, so that even thermally unstable products can often be quenched without decomposition. Since molecules acquire energy by collision with the hot wall, wall-catalyzed (surface) reactions are also possible. This is particularly prevalent for tautomerizable molecules. Different zones of the products, ideally free from contaminants, are collected in the U-shaped trap without further experimental handling. They are then analyzed and characterized by different analytical instruments such as ^{1}H NMR, ^{13}C NMR, LCMS, and GC–MS [17]. For many gas-phase flow pyrolyses, simple equipment is usually adequate to obtain the desired preparative results. Figure 1.4 shows a systematic diagram of the FVP apparatus used in our laboratory, and a photograph of it is shown in Figure 1.5.

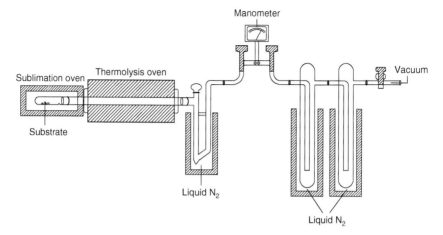

Figure 1.4 Schematic diagram of the FVP apparatus.

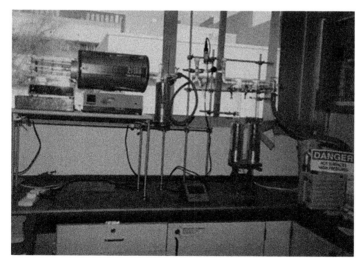

Figure 1.5 FVP apparatus.

1.2.2 Synthetic Applications of FVP

The synthetic application of gas-phase pyrolysis reactions using FVP conditions has been reviewed by McNab [18a]. The main problem with condensed-phase methods is that reactive intermediates are generated in the presence of a precursor, products, and (usually) solvent, so many unwanted secondary reactions can take place; there are added difficulties if the required product is itself thermally unstable. This method is therefore ideal for unimolecular reactions but is often inefficient for gas-phase bimolecular reactions. However, reactive molecules that have sufficient lifetime to survive to the cold trap can be trapped chemically in the condensed phase by added reagents, and this can be a useful synthetic procedure. For example, ketenes and thioketenes are usually too reactive for isolation at room temperature, but these are readily trapped with added alcohols or amines to form (thio)esters or (thio)amides [14b, 18b]. Similarly, nitrile imines are usually reactive intermediates, which can nevertheless in some cases be isolated at liquid nitrogen temperature and trapped in 1,3-dipolar cycloaddition reactions [18c].

Although almost any organic molecule can be thermally decomposed under FVP conditions if a high enough furnace temperature is used, the most constructive use of the technique for preparative purposes involves the intentional use of a reaction with low activation energy. This can either generate a product directly or generate a reactive intermediate in an appropriate environment for intramolecular trapping. Such reactions fall into three categories:

- *Pericyclic reactions.* Sigmatropic shifts, electrocyclization, Claisen and Cope rearrangements, cheletropic reactions, *retro*-ene and *retro*-Diels-Alder–type processes [14]. The latter are favored in the "reverse" direction because of entropy factors. The "forward" direction is relatively rare under FVP conditions because of the tight transition states. Similar factors can favor radical processes over *retro*-ene reactions [18a].
- *Extrusion of small molecules* [14] (e.g. N_2, CO, CO_2). The thermodynamic stability of the extruded molecules is the driving force for this type of reactions and consequently can lead to the formation of high-energy intermediates such as carbenes, nitrenes, or diradicals, as well as control the direction of fragmentation of heterocyclic systems [18a]. The extrusion reactions are often symmetry-allowed cheletropic reactions.
- *Radical reactions.* These are initiated by cleavage of the weakest single bond in the substrate. This type is conceptually the simplest; many direct synthetic applications have been reported, and it is mechanistically important [18d].

Usually, the cleavage of a second bond leads to the products, as for example in the pyrolysis of isoxazole derivatives, which starts with the cleavage of the weak N—O bond but proceeds with elimination of CO_2 and a nitrile [18e]. The pyrolysis of *N-(tert-*butyl) imines proceeds by homolysis of a C—C bond in the *tert-*butyl group [18f, g].

Because these processes are dominated by cleavage, it follows that most FVP reactions are formally oxidative rather than reductive, and these often involve the creation of unsaturated centers [18a].

Discrete ionic intermediates are never encountered under the vibrational activation conditions of the FVP experiment, because in the absence of solvation, ionization energies are very high. However, zwitterionic or mesoionic compounds may be formed.

The use of FVP in the total synthesis of natural products can be classified into three different approaches that can be distinguished according to the application:

- Generation of reactive intermediates, which are used as synthons to build up a structure.
- Generation of a desired functionality via a thermal reaction principle.
- Modification of a molecule with a thermolabile group to enable certain synthetic transformations, after which the modified system is regenerated by removing the protective group.

In order to assess whether a substrate is likely to undergo fragmentation or a clean rearrangement in FVP, the following generalizations can be drawn [19]:

a. Multiple unsaturated and/or polycyclic structures undergo either concerted or homolytic rearrangements, or *retro-*fragmentations. Such systems containing heteroatoms (nitrogen, oxygen, silicon, phosphorus, sulfur, and selenium) and polyhalogen compounds often engage in parallel, orbital symmetry–controlled reaction patterns.
b. Polycyclic and/or unsaturated systems that can lose small gaseous fragments, such as N_2, CO, CO_2, CS_2, S, SO, SO_2, CF_2, CH_2, C_2H_4, CH_2O, and $(CH_3)_2CO$, are often chelotropic in nature but may also proceed via radical and carbene intermediates that will produce the end products by rearrangement.
c. Thermal elimination reactions (e.g. loss of HCl, H_2O, HCN, ROH and/or RCOOH, dehydrogenation, dealkylation, dechlorination, and ester as well as sulfoxide pyrolysis) proceed easily under FVP conditions, usually as 1,2- or 1,4-eliminations.

1.2.3 Gas-Flow Pyrolysis vs. STP

The most impressive advantage of gas-flow pyrolysis over a static process is the ability to produce highly reactive compounds, which can be isolated as such or trapped in a matrix at extremely low temperature and then identified by spectroscopic or chemical analyses. The use of moderate or high vacuum is essential here, in order to isolate the individual product molecule from the surroundings and therefore block the intermolecular secondary reaction, which otherwise could take place in the hot zone. Also, for the synthesis of more stable compounds, gas-flow pyrolysis is generally the technique of choice, when relatively high temperature is required [14].

Other major advantages of using a dynamic rather than a static system are:

i) The absence of any solvent, which may interfere in the reaction or complicate workup; and
ii) The continuous nature of the process, which not only allows easy scaling up, but also makes the technique suitable for substrates with relatively low volatility, needing more time to be processed.

The preparative pyrolysis of organic compounds is best run in a flow system in which the primary products are removed from the hot zone as quickly as possible. The isolation of these products with avoidance of secondary and intermolecular reactions is often readily achieved by FVP [20, 21].

1.2.4 Limitations of FVP

There are also restrictions on the dynamic pyrolysis technique. In order to make gas-flow pyrolysis a practical synthetic method, several conditions must be met:

- The substrate must be volatile at low pressure. Poorly volatile precursors decompose on heating the inlet tube in the sublimation oven. It may also be very difficult to scale up if the substrate is of low or poor volatility. However, this limitation is less important if small quantities of the substrate are used.
- FVP requires a higher temperature than static reactors to compensate for the short residence time in the hot zone. Generally, a reaction that occurs in solution at 180–200 °C may require temperatures in excess of 750 °C under FVP conditions.

- Under these conditions of high temperature, degradation of some functional groups can take place. Most functional groups are stable under mild FVP conditions (up to about 650 °C). In practice, efficient FVP generally is not successful for substrates with multiple functional groups.
- Although the FVP technique is highly reproducible using one set of apparatus, the variables often are not reported accurately in publications of FVP applications, thus making it difficult to reproduce the conditions in other laboratories without carrying out a series of trial experiments [22].

A most promising solution to these problems has been introduced by Meth-Cohn et al., who developed a simple, preparative spray pyrolysis technique.

1.2.5 Spray Pyrolysis

Figure 1.6 shows the apparatus used by Meth-Cohn and his group for preparative spray pyrolysis [23]. The substrate is introduced into the furnace through a finely controlled Teflon tap. A gentle stream of nitrogen injects the sample as a fine spray into the electrically heated furnace. Using this equipment, samples of 5–10 g can be introduced at a rate of 0.5–1.0 g h^{-1}; higher rates of sample introduction are also effective, but at the expense of yield.

Different methods of handling solid substrates based on their melting points were used. For less stable solids, a mixture of the solid with azobenzene, which is fairly inert, was used to depress the melting

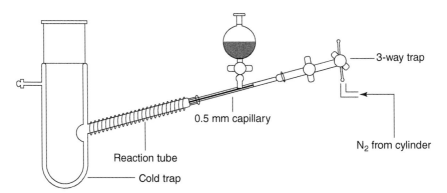

Figure 1.6 Spray pyrolysis equipment.

point. This method was successfully and safely applied in pyrolyzing various azides in preparative scale with a high yield of products [24] (Scheme 1.1). However, for a larger amount of products, two pyrolysis tubes in tandem with condensation at the same cold finger can be used.

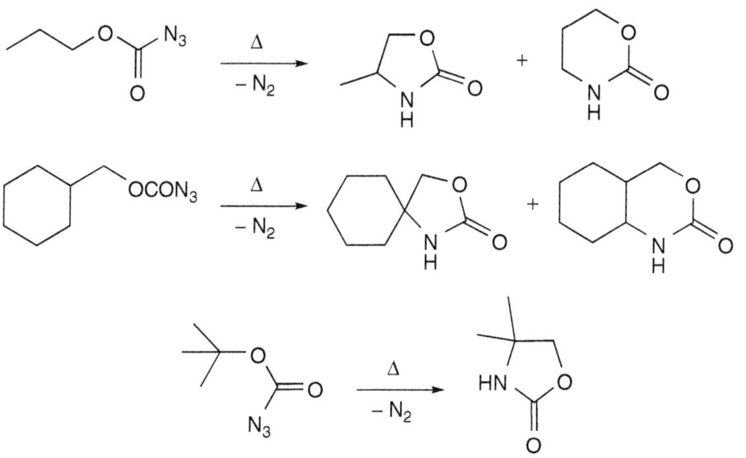

Scheme 1.1

This apparatus were successfully used [24, 25] to overcome the problem encountered in nitrene chemistry, where hot azides have a tendency to explode. The efficiency of this method was tested by Diederich et al. [26] for the synthesis of linear poly-ynes **1a-j** from substituted 3,4-Dialkylnyl-3-cyclobutene-1,2-diones **2a-j** (Scheme 1.2). However, they were not successful under these conditions of modified technique of spray pyrolysis. Olefines polymerized to a great extent without subliming under vacuum of 2×10^{-5} torr, 25–100 °C. Diederich observed that the direct introduction of a solution of the substrate in benzene as a sprayed aerosol inside a hot quartz tube maintained under vacuum of 1–2 torr and filled with quartz rings (Figure 1.7) has achieved the purpose of converting the diones **2a-j** to alkynes **1a-j** in 30–90% yield.

Diederich concluded that this method, which is based on solution spray flash vacuum pyrolysis (SS-FVP), is the method of choice for the elimination of two carbonyl groups from substituted cyclobutenedione. This shows that there is no single method or equipment that works for all purposes.

Wentrup has successfully used this technique in pyrolysis of the involatile di-1-napthyldiazomethane in benzene solution at 660 °C,

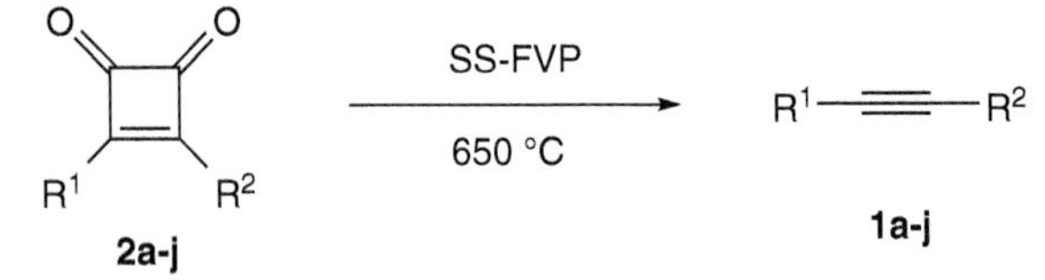

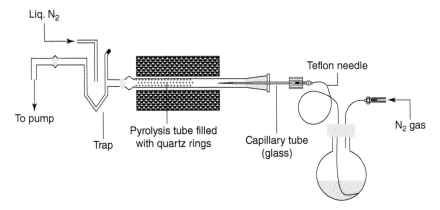

Dione	R^1	R^2	Yield
2a	Ph—	Ph—	1a (98%)
2b	$PhC \equiv C —$	$PhC \equiv C —$	1b (97%)
2c	$PhC \equiv CC \equiv C$	$PhC \equiv CC \equiv C —$	1c (59%)
2d	$n\text{-}PrC \equiv C —$	$n\text{-}PrC \equiv C —$	1d (78%)
2e	$Me_3SiC \equiv C —$	$Me_3SiC \equiv C —$	1e (99%)
2f	$t\text{-}BuMe_2SiC \equiv C —$	$t\text{-}BuMe_2SiC \equiv C —$	1f (99%)
2g	$i\text{-}Pr_3SiC \equiv C —$	$i\text{-}Pr_3SiC \equiv C —$	1g (95%)
2h	$i\text{-}Pr_3SiC \equiv CC \equiv C —$	$i\text{-}Pr_3SiC \equiv CC \equiv C —$	1h (42%)
2i	$i\text{-}Pr_3SiC \equiv C —$	$Me_3SiC \equiv C —$	1i (71%)

Scheme 1.2

Figure 1.7 Solution-spray flash vacuum pyrolysis (SS-FVP) apparatus.

which gave dibenzofluorene (Scheme 1.3) by carbene-carbene rearrangement [27].

Wentrup et al. also realized another major problem with SS-FVP: the amount of solvent used is much higher than the amount of substrate, which usually results in blockage of the cold trap when the apparatus is used for preparative scale. To solve this problem, they developed a falling-solid flash vacuum pyrolysis technique (FS-FVP).

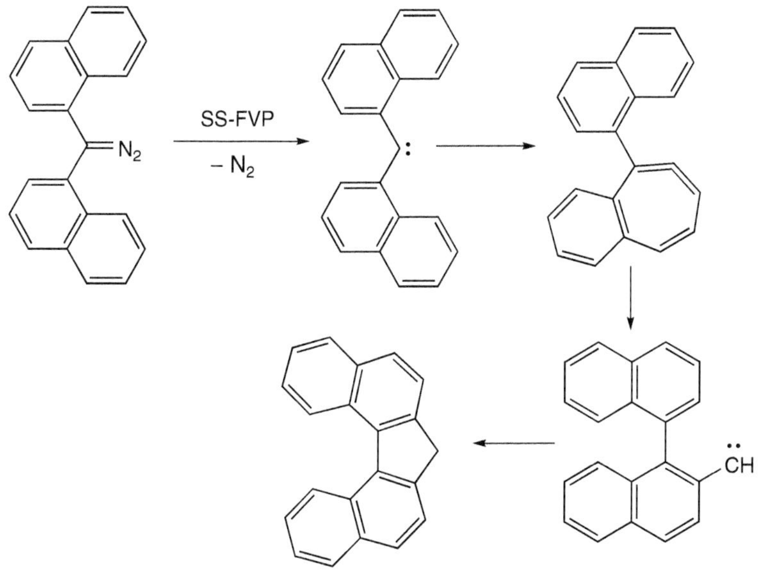

Scheme 1.3

1.2.6 Falling-Solid Pyrolysis

The involatile compound is allowed into the pyrolysis zone to let it fall as a powdered solid. Various methods have been deployed to let the involatile compound fall, using a spinning-band motor that drives a dense brush fitted in a vertical sample-delivery tube. The sample to be pyrolyzed is filled on top of the brush [28a]. The spinning brush acts as a vertical conveyer, delivering the solid as a solid spray into the vertical pyrolysis tube. To avoid the solid falling unchanged through the pyrolysis tube, a loose plug of quartz wool is fitted in the center of the tube (Figure 1.8).

Evaporation takes place on the hot quartz wool, and the vapor so formed then undergoes normal FVP. The method has been used to pyrolyze involatile isoxazolones, which produce aryl- and heteroarylacetylenes in good yields on a multi-gram scale [28a, b] (Scheme 1.4). The method is also useful for the generation of carbenes by pyrolysis of involatile tetrazoles or tosylhydrazone salts [29a, b] (Scheme 1.5).

It is noteworthy that in conventional FVP, involatile substrates often decompose to a tarry material in the solid state. The advantage of the FS-FVP technique is in bringing the solid substrate directly (in the

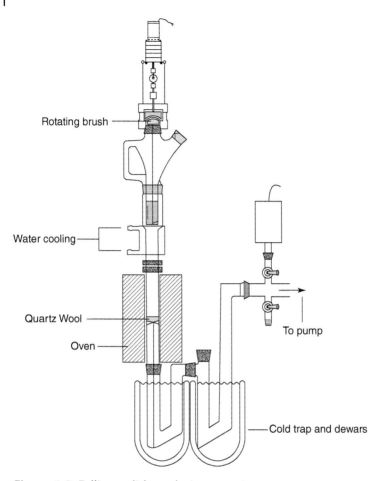

Figure 1.8 Falling-solid pyrolysis apparatus.

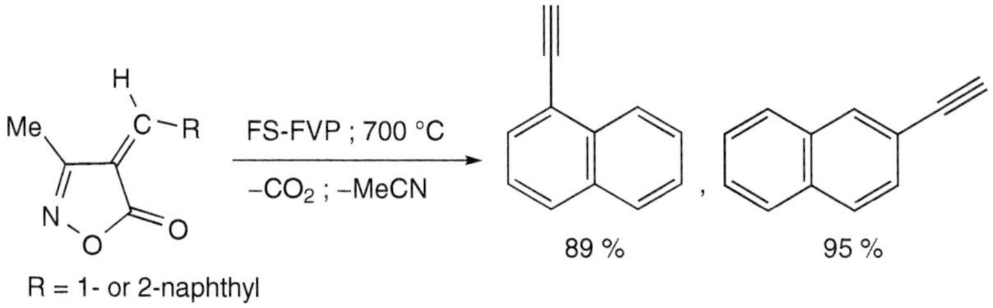

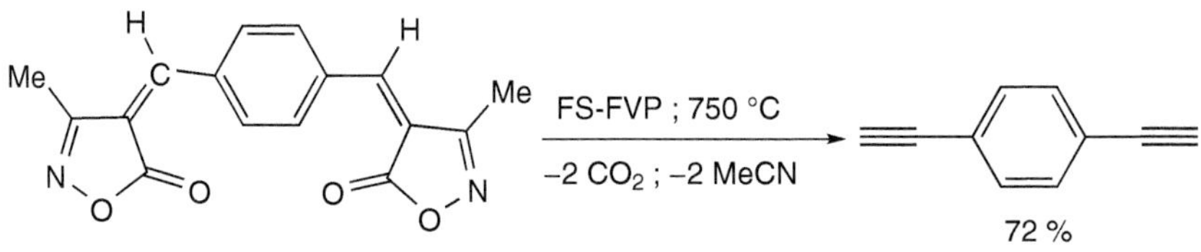

Scheme 1.4

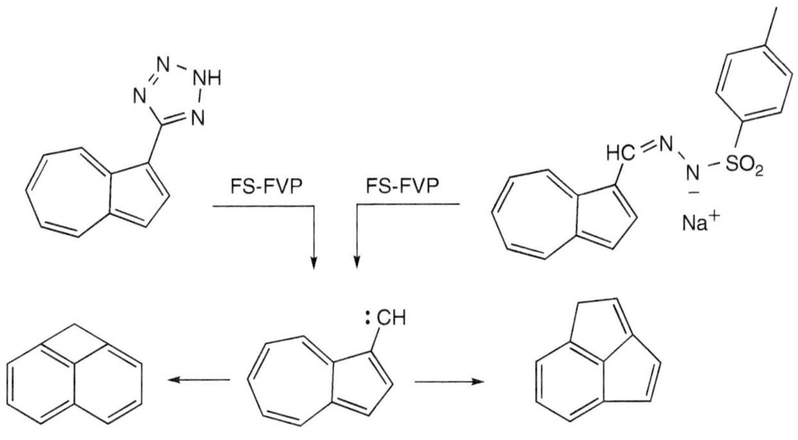

Scheme 1.5

absence of a solvent) into the hot zone, so a rapid FVP reaction is possible.

1.3 Analytical Pyrolysis

A significant step in the evolution of analytical pyrolysis, i.e. for the efficient separation and/or identification of the products, is the combination of pyrolysis with physiochemical techniques.

The essential requirement of the pyrolysis unit in analytical pyrolysis is reproducibility. Selection of the analytical instrument for the analysis of the pyrolysate is an important step for obtaining the desired results. Several techniques and procedures have been reported for the transfer of the pyrolysate to the analytical instrument. These may depend on the pyrolysis unit and also on the procedures utilized for the analysis of the pyrolysate online or offline.

1.3.1 Pyrolysis Gas Chromatography (Py-GC)

One of the most common analytical techniques for investigation of the products of a pyrolysis reaction is gas chromatography. The clear advantages of these techniques, such as sensitivity and the ability to identify compounds, explain their popularity. The 1960s was a period of intense interest in Py-GC. This ease with which a pyrolyzer can be interfaced with a chromatograph led to the introduction of many different designs.

The pyrolysate is injected onto the GC column, onto which ideally the products elute one by one at different times.

1.3.1.1 Online Py-GC

In online Py-GC, the pyrolyzer is directly interfaced with the GC. A simplified diagram of an online (Py-GC) apparatus in the author's laboratory is shown in Figure 1.9, where a gas chromatograph interfaced with flame ionization detector (FID).

This setup allows the pyrolysis products to be analyzed directly, so that transfer losses and secondary decomposition products are minimized. In Py-GC, a mixture of the substrate and an internal standard dissolved in acetonitrile or dichloromethane (1 µl) is injected manually into the pyrolyzer using a gas-tight syringe. The pyrolyzer is set to a given temperature determined through trial and error. The temperature is increased gradually for each successive injection to cover the temperature range of substrate pyrolysis up to more than 90% conversion. The flow rate of the carrier gas and the temperature of the oven of the gas

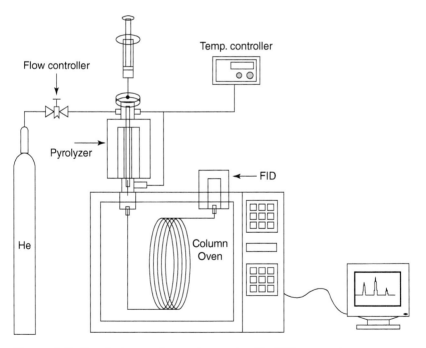

Figure 1.9 Pyrolysis gas chromatography (Py-GC) apparatus.

chromatograph are adjusted in order to give optimum separation of the reaction products relative to the internal standard, with the use of the proper capillary column and a programmed GC oven temperature. The pyrolysis products and the internal standard are detected (FID) and analyzed by the GC software program. The resulting GC peaks may then be identified by reference to authentic samples.

The pyrolysis experiments are carried out using the continuous mode with a temperature range of 0–900 °C under non-oxidizing conditions.

The main features of the pyrolyzer include the following:

- The hot zone is a quartz glass liner, 60 mm in length and with a 2 mm internal diameter.
- Temperature is controlled by a microprocessor digital temperature controller. The "temperature set" facility built into this controller incorporates both fast and slow scroll modes to allow for easy temperature setting over large temperature ranges.
- The digital pressure meter of the pyrolyzer implements a solid-state pressure transducer capable of measuring gas pressure from 0 to 100 psi (0–690 kPa) with a resolution of 0.1 psi (1 kPa).

The pyrolysis products obtained depend primarily on the temperature of the pyrolysis furnace and the residence time of the sample in the furnace. With the continuous-mode pyrolyzer, both of these parameters may be adjusted to obtain the desired pyrolysis profile.

The injector allows the sample to be placed at the beginning of the chromatographic column in a narrow zone loaded with the pyrolysates. The pressure of the carrier gas (head column pressure) may be utilized to control the gas flow in the analytical column.

The usual injection port system adapted for Py-GC is, however, the split/splitless one. With this,

1. Either the sample can be transferred from the pyrolyzer into the injection port of the GC and then to the analytical column, which is connected at the bottom of the injection port; or
2. A piece of fused silica capillary connects the pyrolyzer and the analytical column through the injection port.

The detector of a GC is an important part in Py-GC. Among the GC detectors, the most frequently utilized is the FID.

1.3.2 Pyrolysis Mass Spectrometry

Mass spectrometers are used as detectors in conjunction with gas chromatography, offering an exceptionally good capability of compound identification and giving reproducible results. When a pyrolyzer is used at the front edge of the chromatograph, no visible problems related to the GC/MS analysis are added. In this case, the only role of the pyrolyzer is to provide a pyrolysate to the GC. Py-GC/MS is an excellent tool for polymer analysis.

1.3.2.1 Online Pyrolysis Gas Chromatography/Mass Spectrometry (Py-GC/MS)

Several online procedures have been developed. The author's laboratory has recently introduced this technique to enhance the analytical facility for gas-phase pyrolytic reactions, especially for polymeric compounds.

The pyrolysis of polymeric materials is carried out in a pyrolyzer (Figure 1.10). This is a double-shot analysis, where the sample cup free-falls into the micro furnace in a few milliseconds, first for thermal desorption at a given temperature with a certain hold time. The sample can then be pyrolyzed using different programs in the micro furnace from 40 to 800 °C. The pyrolyzer is closely coupled to the GCMS with a completely inert sample path made up of quartz/ultra alloy, which ensures that there is no loss of compounds at active sites or cold spots. Any desired zone of the thermogram can be automatically collected as a heart cut and introduced in the GCMS. The evolved gas analysis allows online MS analysis over a wide range of temperature programming. The peak identification is performed by a quadrupole mass spectrometer directly coupled to the GC system.

Tertiary recycling of polymers; de-polymerization or partial degradation of polymers leading to monomer or other valuable, secondary materials; and the mechanism of the mutual stabilizing effect of the components in polymeric blends and composites have been studied in our laboratory using this technique [30, 31].

1.3.3 FVP with Spectroscopy

For FVP reactions that involve reactive intermediates such as carbenes, nitrenes, and unstable molecules, it is essential to have spectroscopic evidence for characterization and identification of such intermediates. Wentrup has developed a general-purpose FVP apparatus using IR, UV, and ESR spectroscopies; the details of this apparatus are fully described

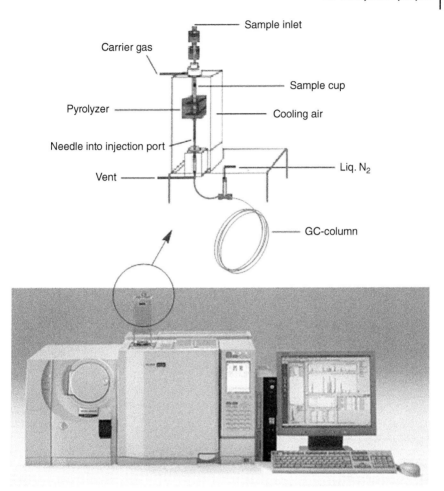

Figure 1.10 Pyrolysis GC/MS technique showing the details of the pyrolyzing unit.

and explained in his recent review [28a]. The pyrolysis vacuum is usually on the order of 10^{-4} nPa. The apparatus allows the FVP reaction to be routinely monitored by both IR and mass spectrometer. Such apparatus can also help in determining different reactive intermediates formed at different temperatures and determining whether these are formed in parallel or sequentially. A representative example is the tautomerization of Meldrum's acid derivatives and its enols: the latter eliminates acetone in FVP to form three ketones sequentially (Scheme 1.6). Each of the ketones was detected by IR spectroscopy at 77 K, and they were isolated as esters by trapping with methanol.

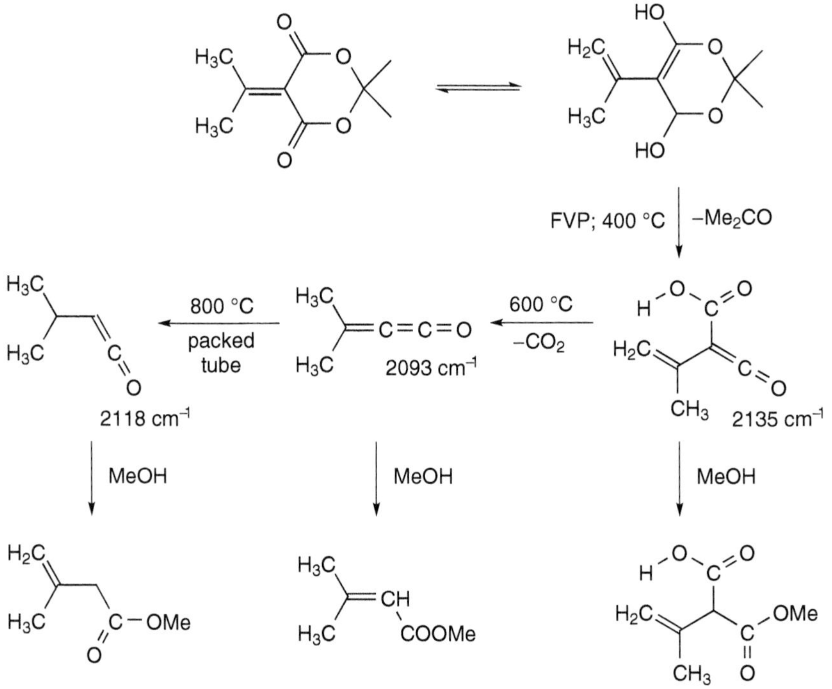

Scheme 1.6

1.3.4 Catalytic Gas-Phase Pyrolysis

Gas-phase pyrolysis reactions are often carried out over a solid support simply in order to increase the contact times and hence lower the temperature needed for a reaction to occur. The support used is often quartz chips, short sections of quartz tubing (e.g. 10 mm length, 4 mm ID), or quartz wool. Whereas the purpose of such supports is to increase the collision numbers, the possibility of surface-catalyzed reactions also increases.

Denis and co-workers [32] have used vacuum gas–solid reactions extensively; the apparatus has been described in detail. They elegantly used the apparatus to carry out two- and three-step catalytic reactions. For example, ketenimine, $CH_2{=}C{=}NH$, was synthesized in a three-step reaction: (i) *N*-chlorination of 1-aminonorbornene over *N*-chlorosuccinimide, (ii) dehydrohalogenation over *t*-BuOK, and finally (iii) FVP at 850 °C (Scheme 1.7).

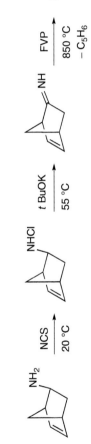

Scheme 1.7

Dehydrohalogenations have also been performed using a column packed with *t*-BuOK on Chromosorb W at 240 °C 10^{-2} hPa, e.g. in the preparation of methylenecyclopropene from 2-chloromethylenecyclopropane [33a, b, 34].

Chapman reported the synthesis and matrix-isolation of benzocyclobutadiene by reaction of *Z*-1,2-diiodobenzocyclobutane over Zn dust at 230 °C [35] (Scheme 1.8).

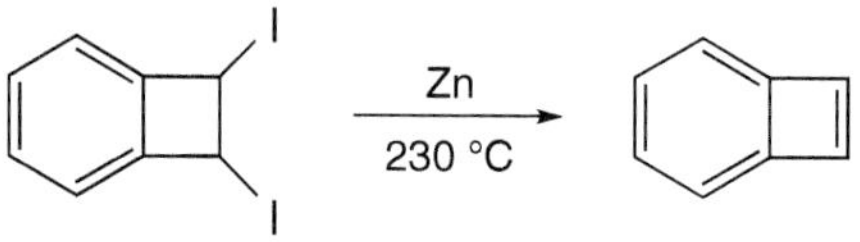

Scheme 1.8

Aitken has described a procedure for subliming magnesium onto glass wool and using this in an FVP apparatus to transform benzyl chlorides to bibenzyls and benzal chlorides to stilbenes [36]. The dehydration of oximes to nitriles was conveniently carried out by FVP over molecular sieves (3 Å, 350 °C). This simple reaction has been applied in the syntheses of pyrrole-, thiophene-, and indole-carbonitriles, for example [37].

Pyrolysis over heterogeneous catalysts has numerous applications in the petrochemical industry [38–40]. Cyclization/dehydrogenation of aliphatic hydrocarbons by pyrolysis over solid supports such as chromium oxide or zeolites is an important industrial route to the aromatic compounds [41], which can be applied advantageously in preparative organic chemistry. For example, it has long been known that carbazole is readily obtained by pyrolysis of diphenylamine over a Pt catalyst at 300 °C [42], whereas a red-hot tube is required for the low-yielding, uncatalyzed reaction [43].

Pyrolysis over zeolites has been used as a means of reducing the reaction temperatures in the FVP reactions of several heterocyclic compounds, such as pyrazoles [44]. While reaction temperatures of the heterogeneous reactions over zeolites were lower than in the conventional FVP reactions, the products obtained were different. The normal course of events for pyrolysis of pyrazoles is elimination of nitrogen and isomerization of the resulting 1,3-biradicals or carbenes to cyclopropenes, allenes, acetylenes, or 1,3-dienes [45]; but additional fragmentation and rearrangement reactions were observed in the zeolite-catalyzed processes, including rearrangement to an imidazole. Further studies using both acidic and basic supports (zeolites, hydrotalcites, and mesoporous materials) have been reported, and the presence of acidic sites was found to be particularly important in determining the reaction outcomes [46]. Recently, a one-pot synthesis of dibenzo[*b,d*]azepin-7-one from 1-phenacylbenzotriazole was achieved using FVP over materials with a scheelite structure of the type ABO_4 (A = Ca^{2+}, Sr^{2+}, Ba^{2+} and B = Mo^{6+}, W^{6+}) and fergusonite $BiVO_4$. These oxides promoted high conversion of the starting material at lower temperatures relative to the uncatalyzed pyrolyses. A yield of up to 87% of the desired azepinone was obtained with $BaWO_4$ at 400 °C [47] (Scheme 1.9).

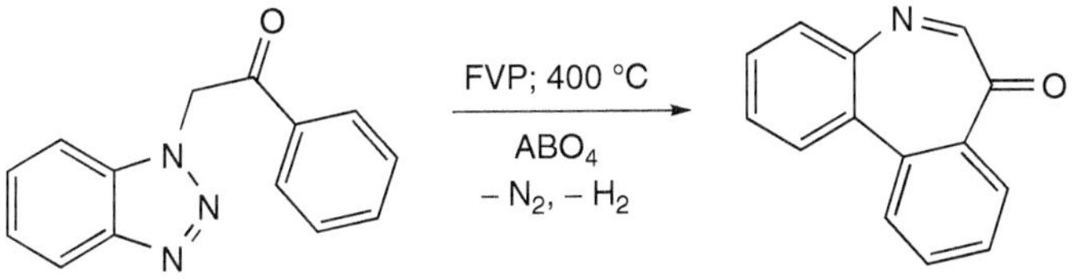

Scheme 1.9

In all these reactions, the outcomes, yields, and required temperatures are highly dependent not only on the composition but also on the amount and method of packing the pyrolysis tube with the solid support and catalyst, the vacuum, and the residence time. All parameters have to be worked out by trial and error.

References

1 Karpf, M. (1986). *Angew. Chem. Int. Ed.* 25: 414–430.

2 Funk, R.L. and Vollhardt, K.P.C. (1980). *Chem. Soc. Rev.* 9: 41–61.

3 Oppolzer, W. and Snieckus, V. (1978). *Angew. Chem. Int. Ed.* 17: 476–496.

4 Depuy, C.H. and King, R.W. (1960). *Chem. Rev.* 60: 431–457.

5 (a) Bigley, D.B. and Wren, C.M.J. (1972). *J. Chem. Soc. Perkin Trans. II* 8: 926–928. (b) Al-Awadi, N.A. and Bigley, D.B.J. (1997). *J. Chem. Soc. Perkin Trans. II*: 497–500.

6 (a) Al-Awadi, N.A., Al-Bashir, R., and El-Dusouqui, O.M. (1989). *J. Chem. Soc. Perkin II*: 579–581. (b) Al-Awadi, N.A., Al-Bashir, R., and El-Dusouqui, O.M. (1990). *Tetrahedron* 46: 2903–2910.

7 Saunders, J.W.H. and Cockerill, A.F. (1973). *Mechanisms of Elimination Reactions*. New York: Wiley.

8 (a) Taylor, R. (1968). *J. Chem. Soc. B* (12): 1397–1401. (b) Taylor, R. and Katritzky, A.R. (1990). *Advances in Heterocyclic Chemistry*, 47. London: Academic Press, Inc., Ch. 5.

9 (a) Taylor, R. (1971). *J. Chem. Soc. B* (2): 255–257. (b) Taylor, R. (1968). *J. Chem. Soc. B* 8: 1397–1401.

10 Smith, G.G. and Bagley, F.D. (1961). Taylor, R. *J. Am. Chem. Soc.* 83: 3647–3653.

11 Al-Awadi, N.A. and Taylor, R. (1986). *J. Chem. Soc. Perkin* II: 1581–1583.

12 Brown, R. (1980). *Pyrolytic Methods in Organic Chemistry*. London: Academic Press, Inc.

13 Schaden, G. (1982). *J. Anal. Appl. Pyrolysis* 4: 83–101.

14 (a) McNab, H. (2004). *Aldrichimica Acta* 37 (1): 19–26. (b) Wentrup, C. (2014). *Aust. J. Chem.* 67: 1150–1165.

15 Schaden, G.J. (1985). *J. Anal. Appl. Pyrolysis* 8: 135–151.

16 Wiersum, U.E. (1984). *Aldrichimica Acta* 17 (2): 31–41.

17 George, B.J., Dib, H.H., Abdallah, M.R. et al. (2006). *Tetrahedron* 62: 1182–1192.

18 (a) McNab, H. (1996). *Contemp. Org. Synth.* 3: 373–396. (b) Koch, R., Blanch, R.J., and Wentrup, C. (2014). *J. Org. Chem.* 79: 6978–6986. (c) Wentrup, C., Fischer, S., Maquestiau, A., and Flammang, R. (1985). *Angew. Chem. Int. Ed. Eng.* 24: 56–57. (d) Cadogan, J.I.G., Hickson, C.L., and McNab, H. (1986). *Tetrahedron* 42: 2135–2165. (e) Rzepa, H.S. and Wentrup, C. (2013). *J. Org. Chem.* 78: 7565–7574. (f) Vu, Y., Chrostowska, A., Huynh, T.K.X. et al. (2013). *Chem. Eur. J.*: 14983–14988. (g) Justyna, K., Lesniak, S., Nazarski, R.B. et al. (2014). *Eur. J. Org. Chem.*: 3020–3027.

19 Klunder, A.J.H. and Zwanenburg, B. (1997). *Gas Phase Reactions in Organic Synthesis* (ed. Y. Vallèe). Amsterdam: Gordon and Breach Science Publishers.

20 Graebe, C. and Liebermann, C. (1968). *Ber. Dtsch. Chem. Ges.* 1: 49–51.

21 Hurd, C.D. (1929). *The pyrolysis of Carnon Compounds*. New York: Chemical Catalogue Co.

22 Duffy, E.F., Foot, J.S., McNab, H., and Milligan, A.A. (2004). *Org. Biomol. Chem.* 2: 2677–2683.

23 Clancy, M., Hawkins, D.G., Heshabi, M.M. et al. (1982). *J. Chem. Res. Synop.* 3: 78.

24 Meth-Cohn, O. (1987). *Acc. Chem. Res.* 20 (1): 18–27.

25 Meth-Cohn, O., Patel, D., and Rhouati, S. (1982). *Tetrahedron Lett.* 23 (48): 5085–5088.

26 Rubin, Y., Lin, S.S., Knobler, C.G. et al. (1991). *J. Am. Chem. Soc.* 113: 6943–6949.

27 Régimbald-Krnel, M. and Wentrup, C. (1998). *J. Org. Chem.* 63: 8417–8423.

28 (a) Wentrup, C. (2017). *Angew. Chem. Int. Ed.* 56: 14808–14835. (b) Wentrup, C., Becker, J., and Winter, H.W. (2015). *Angew. Chem. Int. Ed.* 54: 5702–5704.

29 (a) Kvaskoff, D., Becker, J., and Wentrup, C. (2015). *J. Org. Chem.* 80: 5030–5034. (b) Becker, J., Diehl, M., and Wentrup, C. (2015). *J. Org. Chem.* 80: 7144–7149.

30 Ahmad, D., Al-Awadi, N.A., and Al-Sagheer, F. (2008). *Polym. Degrad. Stab.* 93: 456–465.

31 Ahmad, D., Al-Sagheer, F., and Al-Awadi, N.A. (2010). *J. Anal. Appl. Pyrolysis* 87: 99–107.

32 Denis, J.M. and Gaumont, A.C. (1997, Chapter 4). Vacuum Gas-Solid Reactions (VGSR): applications to the synthesis of unstable species. In: *Gas Phase Reactions in Organic Synthesis* (ed. Y. Vallée), 195–238. Amsterdam: Gordon and Breach.

33 (a) Billups, W.E., Lin, L.J., and Casserly, E.W. (1984). *J. Am. Chem. Soc.* 106: 3698–3699. (b) Billups, W.E. and Lin, L.J. (1986). *Tetrahedron* 42: 1575–1579.

34 Staley, S.W. and Norden, T.D. (1984). *J. Am. Chem. Soc.* 106: 3699–3700.

35 Chapman, O.L., Chang, C.C., and Rosenquist, R.N. (1976). *J. Am. Chem. Soc.* 98: 261–262.

36 Aitken, R.A., Hodgson, P.K.G., Oyewale, A.O., and Morrison, J.J. (1997). *J. Chem. Soc. Chem. Commun.*: 1163–1164.

37 Aitken, R.A. and Boubalouta, Y. (2015). *Adv. Heterocycl. Chem.*: 93–150.

38 Kaup, G. and Matthies, D. (1988). *Mol. Cryst. Liq. Cryst.* 161: 119–143.

39 Albright, L.F., Crynes, B.L., and Corcoran, W.H. (1983). *Pyrolysis: Theory and Industrial Practice.* New York: Academic Press.

40 Donahue, W.S. and Brandt, J.C. (2009). *Pyrolysis: Types, Processes and Industrial Sources and Product.* New York: Nova Science Publishers, Inc.

41 Hansch, C. (1953). *Chem. Rev.* 53: 353–396.

42 Zelinsky, N.D., Titz, I., and Gaverdowskaja, M. (1926). *Ber. Dtsch. Chem. Ges.* 59: 2590–2593.

43 Wentrup, C. and Gaugaz, M. (1971). *Helv. Chim. Acta* 54: 2108–2111.

44 Moyano, E.L. and Yranzo, G.I. (2001). *J. Org. Chem.* 66: 2943–2947.

45 Wentrup, C. and Crow, W.D. (1971). *Tetrahedron* 27: 361–366.

46 Moyano, E., del Arco, M., Rives, V., and Yranzo, G. (2002). *J. Org. Chem.* 67: 8147–8150.

47 Lener, G., Carbonio, R.E., and Moyano, E.L. (2013). *ACS Catal.* 3: 1020–1025.

2

Synthesis and Applications

This chapter explores the wide use of gas-phase pyrolysis reactions as a valuable alternative and environmentally friendly synthetic strategy in which waste is reduced or eliminated. An added advantage is that this technique is effectively free from solvation, hydrogen bonding, and protonation effects common in solution reactions. Gas-phase pyrolysis reactions offer important routes for novel heterocyclization and selective synthesis. Various pyrolytic reactions that involve loss of thermodynamically stable small gaseous fragments – HX, CO, CO_2, N_2 – leading to interesting organic compounds are described.

2.1 Flash Vacuum Pyrolysis in Organic Synthesis

Gas-phase pyrolysis is a special synthetic method that is usually clean, convenient, and efficient, and that has advantages over other synthetic methods for accomplishing the same goals. Flow gas-phase pyrolysis has long been used to provide a very powerful and useful alternative methodology in synthetic organic chemistry [1].

Gas-Phase Pyrolytic Reactions: Synthesis, Mechanisms, and Kinetics,
First Edition. Nouria A. Al-Awadi.
© 2020 John Wiley & Sons, Inc. Published 2020 by John Wiley & Sons, Inc.

Organic chemists have been performing flash vacuum pyrolysis (FVP) reactions with the aim of synthesizing new compounds and/or to study reactive intermediates [2]. Two key parameters are important to be considered – temperature and duration or residence time – which are adjusted to optimize yield, conversion, and avoidance of intractable products [3].

FVP is a technique for intramolecular reactions such as eliminations and cyclizations, and it is ideal for unimolecular reactions [4]. Also, the radical reactions are the principal biomolecular processes usually observed under FVP [5].

FVP has been successfully used to synthesize reactive compounds or novel products that are considered very difficult to obtain by conventional synthetic methodologies [6a, b]. The unstable dihydrothiophene **1** shown in Scheme 2.1 was obtained in >85% yield under FVP conditions [7].

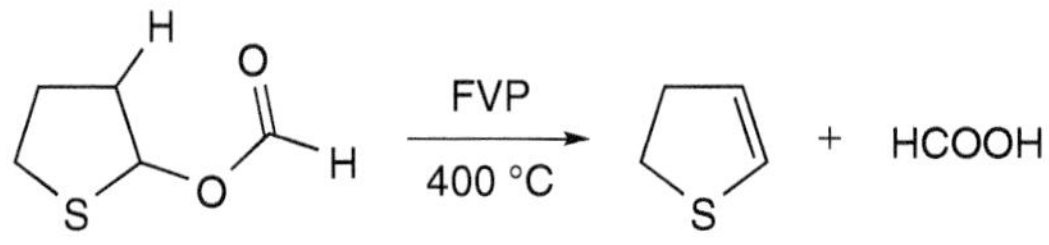

Scheme 2.1

Another example is the preparation of acetylenes **2** by an interesting rearrangement of the vinylidene intermediates **3** obtained by FVP of isoxazolones **4** [8] (Scheme 2.2). It was elegantly shown by Wentrup et al. that FVP permits the synthesis of substituted acetylenes whose preparation by conventional methods is rather difficult. Substituted acetylenes **2** that were prepared by FVP of isoxazolones **4a–h** at 700–800 °C gave a high yield of acetylenes **2a–h** [9], most probably via carbenes **3** (Scheme 2.2). This is considered the best general FVP route to kinetically unstable but thermodynamically stable alkynes [4].

The products from the FVP of isoxazolone **5** (Scheme 2.3) were monitored at different furnace temperatures. The major product was thioketenes **6** at temperature ~500 °C, whereas at higher temperatures of 800–900 °C, complete conversion to the iminopropadienethiones R-N=C=C=C=S **7** took place [10].

Generally, substrates with unsaturation units undergo either concerted or homolytic rearrangement or retro fragmentation; this type of reaction will be considered in Chapter 3 with several examples. Thermal elimination reactions with loss of thermodynamically stable small gaseous fragments – e.g. HX, CO, CO_2, N_2, C_2H_4 – take place

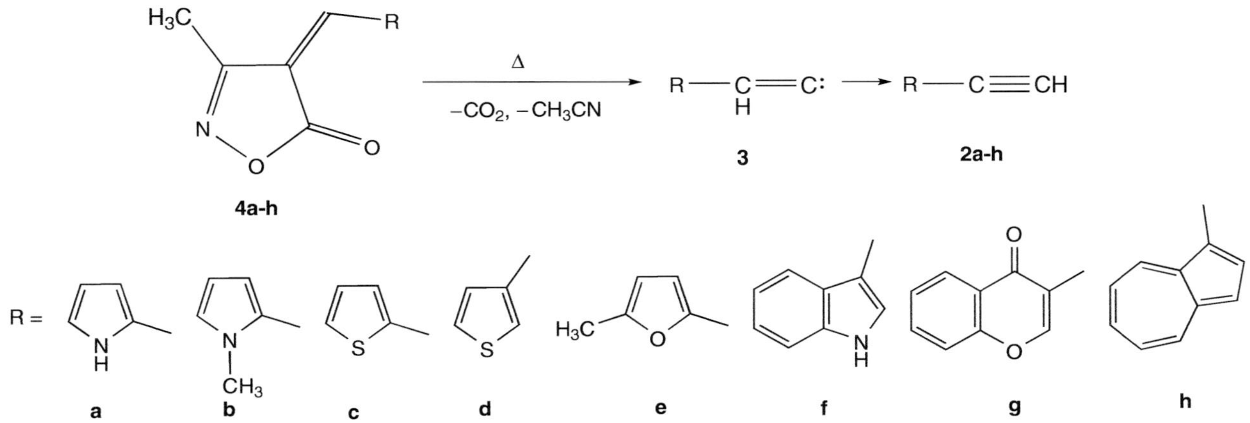

Scheme 2.2

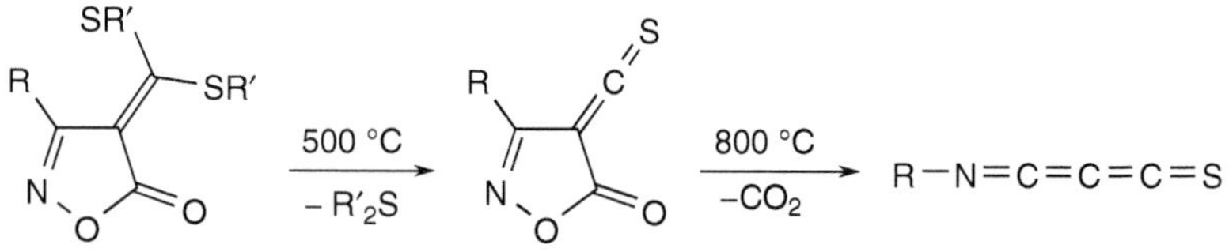

Scheme 2.3

easily under FVP conditions, resulting in various types of organic molecules. This chapter contains different examples of pyrolytic reactions that involve loss of HX, CO, and N_2, leading to interesting organic compounds.

2.2 Elimination of HX

Gas-phase pyrolysis of alkyl halides frequently results in dehydrohalogenation reactions [11] (Scheme 2.4).

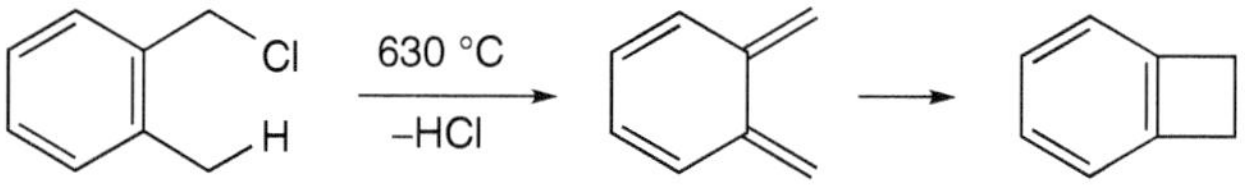

Scheme 2.4

Ortho-alkylated benzoic acid chlorides **8** undergo 1,4-elimination of HCl to generate benzocyclobutenones **9** [12] (Scheme 2.5).

Gas-phase pyrolysis of ortho-methylated aromatic acid chlorides resulted in a wide variety of substituted benzocyclobutenones **10** (Scheme 2.6) with a reasonable yield, which makes them useful intermediates in organic synthesis [13, 14].

2-Propenylbenzaldehydes **11–13** eliminate from acid chlorides **14** as a result of an uncatalyzed 1,4-elimination of HCl followed by a 1,5-hydrogen shift [15] (Scheme 2.7).

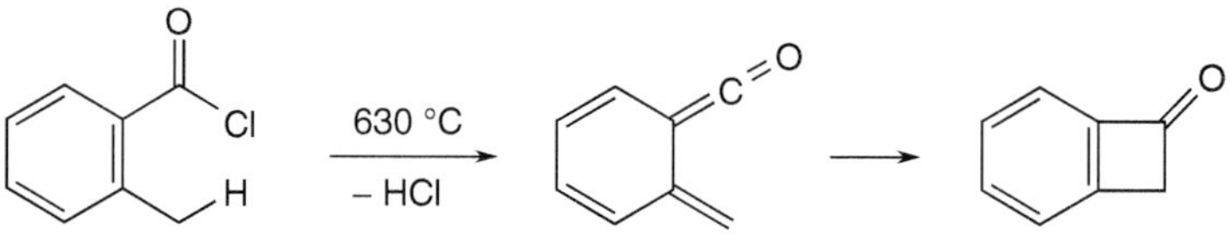

Scheme 2.5

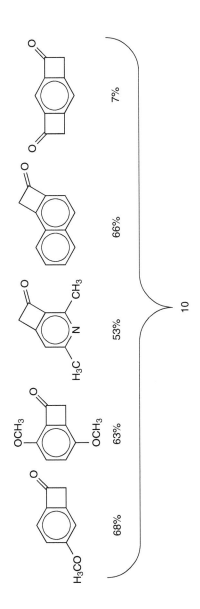

Scheme 2.6

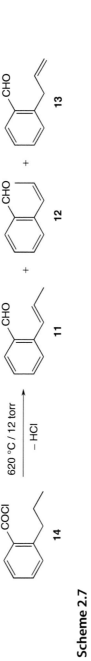

Scheme 2.7

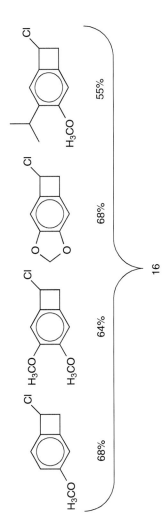

68% 64% 68% 55%

16

Scheme 2.8

Pyrolysis of substituted 2-methylbenzylchlorides **15** has resulted in a wide variety of α-chlorobenzocyclobutenes **16**. Some examples are shown in Scheme 2.8 with the yield obtained for each product. The halogen in **16** can be replaced by substitution with nucleophiles [15]; this provides access to the synthesis of benzocyclobutenes with different substituents in the butene ring (Scheme 2.9).

Some examples of α-chlorobenzocyclobutenes **16** obtained in this way are shown in Scheme 2.9 together with the yield from thermal elimination of HCl. Many of these highly functionalized benzocyclobutenes, especially those carrying a cyano group in the four-membered ring, have found use as building blocks in the convergent synthesis of polycyclic natural products by means of intramolecular cycloaddition [16, 17].

Benzocyclobutenones **17** carrying various substituents on the aromatic ring were carried out on a preparative scale (Table 2.1). The merit of the procedure to synthesize benzocyclobutenones lies in the simplicity of generating the strained bicyclic ring system from inexpensive precursors [18].

2.3   Elimination of CO and CO_2

Wentrup and his group reported on the conversion of phthalide **18a** and benzocyclopentene **18b** to identical products of fulvenallene **19** and ethynylcyclopentadiene **20** through a mechanism involving 2-methylenecyclohexadienylidene as a common intermediate [19] (Scheme 2.10). α-Coumaranone **21**, an isomer of phthalide **18**, behaved differently: it gave, in addition to CO, fulvene **19** and benzene, the trimer of the *ortho*-quinomethane intermediate **22** [19] (Scheme 2.11).

The gas-phase pyrolysis of homophthalic anhydride **23a** at 515–545 °C resulted in 45% pure benzocyclobutenone **9** (Scheme 2.12), while at a slightly higher temperature (570 °C), pure fulveneallene **19** was obtained in 36% yield without any additional products. This pyrolysis reaction may be considered a convenient, one-step process for the preparation of compounds **9** and **19** [20a]. This is in accordance with what was reported on the pyrolysis of 1,2-indandione **23b**, which gave ketone **9** and fulveneallene **19** as major products [20b].

Pyrolysis of 4,4-dimethylhomophthalic anhydride **24a** has been compared with analogous 3,3-dimethyl-1,2-indandione **24b** [20a]: the latter gave o-(2-propenyl)benzaldehyde **25** as the major product [20c]. Pyrolysis of **24a** also gave benzaldehyde **25** in 65% yield according to a mechanism shown in Scheme 2.13.

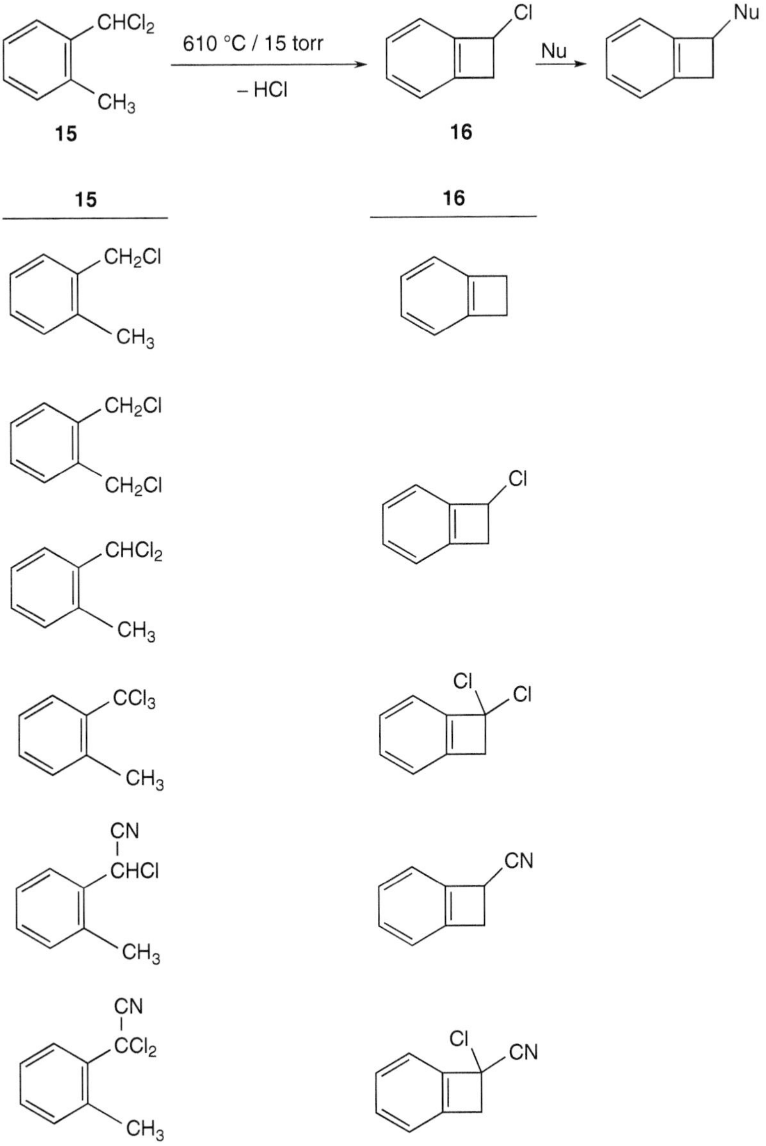

Scheme 2.9

Table 2.1 Synthesis of benzocyclobutenones 17.

	2	3	4	5	°C mm^{-1}
	H	H	H	H	630/13
	H	CH$_3$	H	CH$_3$	560/13
	CH$_3$	CH$_3$	CH$_3$	CH$_3$	550/13
	H	OCH$_2$O	OCH$_2$O	H	570/13

H	OCH$_3$	H	H	570/13
H	H	H	OCH$_3$	590/0.1
OCH$_3$	H	H	OCH$_3$	590/0.1
OCH$_3$	OCH$_3$	OCH$_3$	H	530/13
H	H	CH=CH–CH=CH	CH=CH–CH=CH	550/13

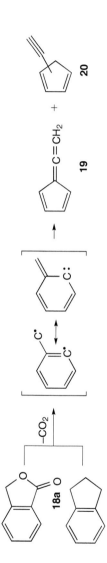

18a

18b

$-CO_2$

19

$C{=}CH_2$

20

Scheme 2.10

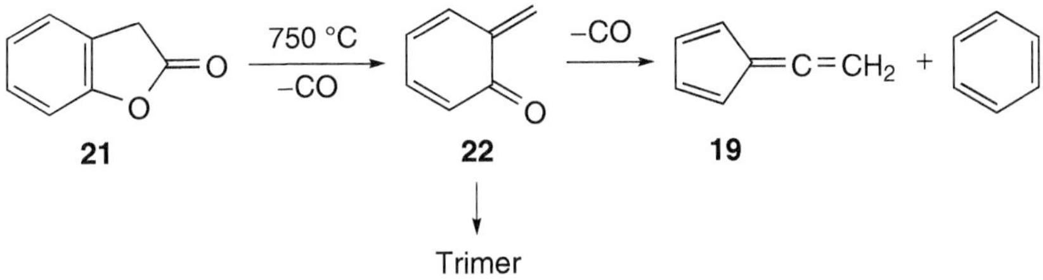

Scheme 2.11

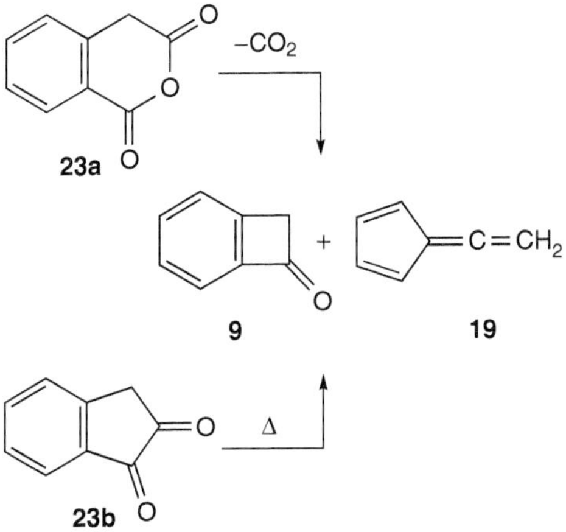

Scheme 2.12

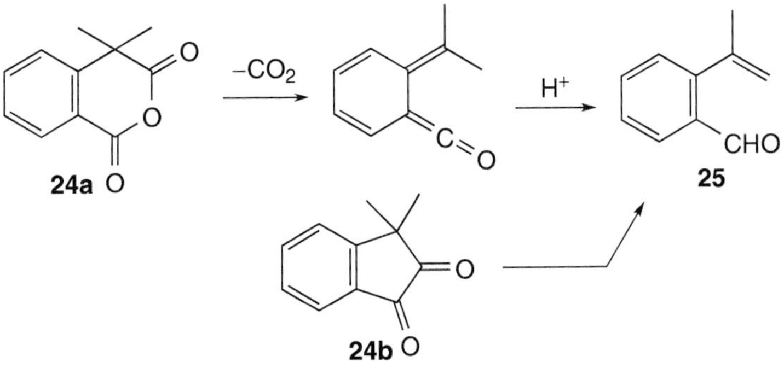

Scheme 2.13

FVP of the cyclic ester 5-benzylidene-2,2-dimethyl-1,3-dioxan-4,6-diones **26** at 430 °C/0.1 mm is a simple and convenient method for synthesizing substituted methyleneketenes **27** [21], which upon warming to room temperature dimerize to 2,4-dimethylenecyclobutan-1,3-dione **28** (Scheme 2.14). On the other hand, FVP of the same cyclic ester **26** at 560 °C/0.1 mm gave phenylacetylene (Scheme 2.14), which is consistent with a mechanistic pathway that involves carbene intermediate; the latter undergoes either hydrogen or phenyl migration to give phenylacetylene [22]. This indicates the ready migration of either hydrogen or phenyl.

Wentrup et al. in their 1994 review have reported on synthetically useful products for the generation of α-oxoketenes that cannot be isolated or even observed under conventional reaction conditions [23]. They are of considerable interest because of their use as building blocks in organic synthesis. FVP provides an ideal alternative process for clean and convenient synthesis of α-oxoketenes. Thermal extrusion of CO from furan-2,3-diones **29a** has proved to be a widely applicable and straightforward process for generation of α-oxoketenes **30a** (Scheme 2.15). Similarly, α-oxoketenes **30b** can be readily obtained by FVP at 500 °C/10^{-3} mbar from the corresponding furan-2,3-diones **29b**. It is noteworthy that identification of these α-oxoketenes **33b** was based on low-temperature IR and high-resolution mass spectrometry. Formation of α-oxoketenes **31** can be easily observed by FVP (500 °C, 10^{-3} mbar) for the corresponding furan-2,3-diones **32** [23] (Scheme 2.16).

Gas-phase pyrolysis of **33a** at 250 °C eliminates CO and gives **34a** (X=O) (Scheme 2.17), which was confirmed from the strong IR absorption at 2140 cm^{-1} [23]. On the other hand, the sulfur analogue 4-benzoyl-5-phenyl-2,3-dihydrothiophen-2,3-dione **33b** at 500 °C resulted in benzoyl (thiobenzoyl)ketene **34b** (X=S) with only weak IR absorption at 2117 cm^{-1}. This was justified by the formation of the thermodynamically more stable thietone **35**. Similarly, extrusion of CO from the pyrrole-2,3-diones **36** under the same conditions gave **34c**, which was confirmed by low-temperature IR; **34c** readily cyclized to quinolones **37** in the gas phase [23] (Scheme 2.17).

α-Oxoketene **38** was generated in quantitative yield by elimination of CO from furandione **39** at 350 °C [23]. **38** dimerizes into more-stable α-oxoketene **40** (Scheme 2.18).

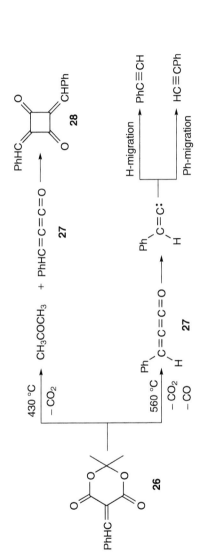

Scheme 2.14

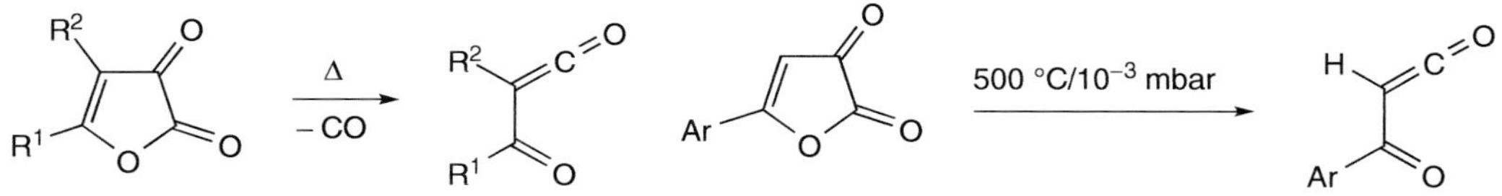

R^1 = Aryl, t-Bu
R^2 = H, Aryl, Arylimino, Thiobenzoyl, Aroyl

Scheme 2.15

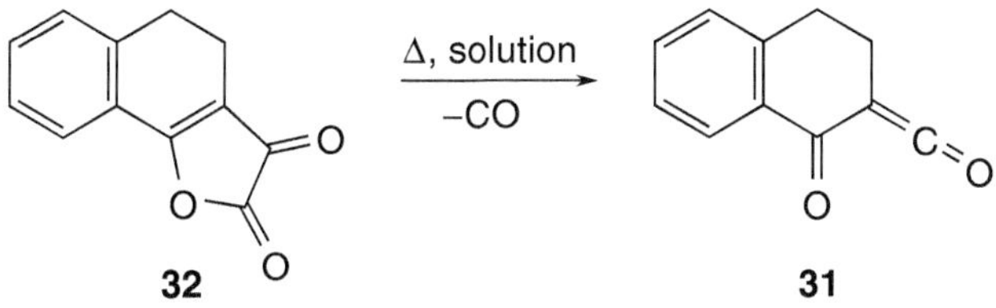

Scheme 2.16

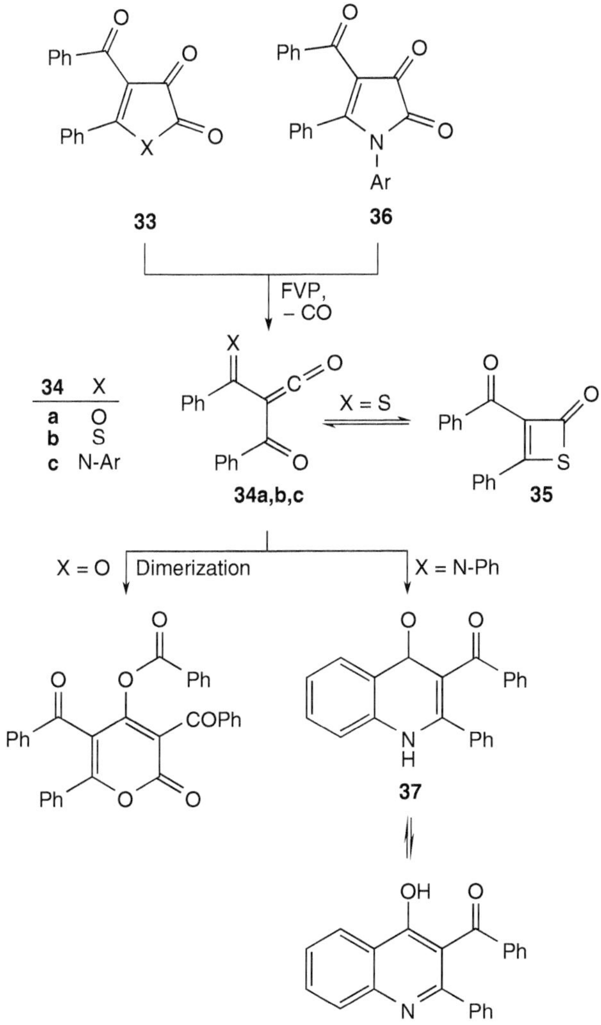

34	X
a	O
b	S
c	N-Ar

Scheme 2.17

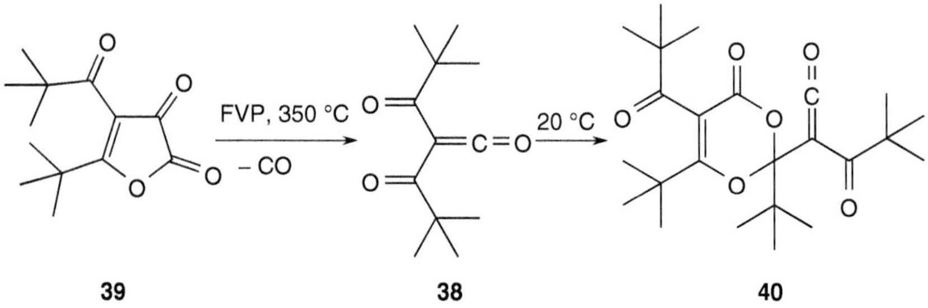

Scheme 2.18

Furandiones are very reactive and decompose at 350 °C under the conditions of FVP. The thiophenediones and pyrrolediones decompose above 500 °C on FVP [24]. In a similar way, the cheletropic extrusion of CO from 4-methyleneoxazolones (azlactones) **41** gives N-acylketenimines **42** in good yields on FVP at 600 °C [25a] (Scheme 2.19).

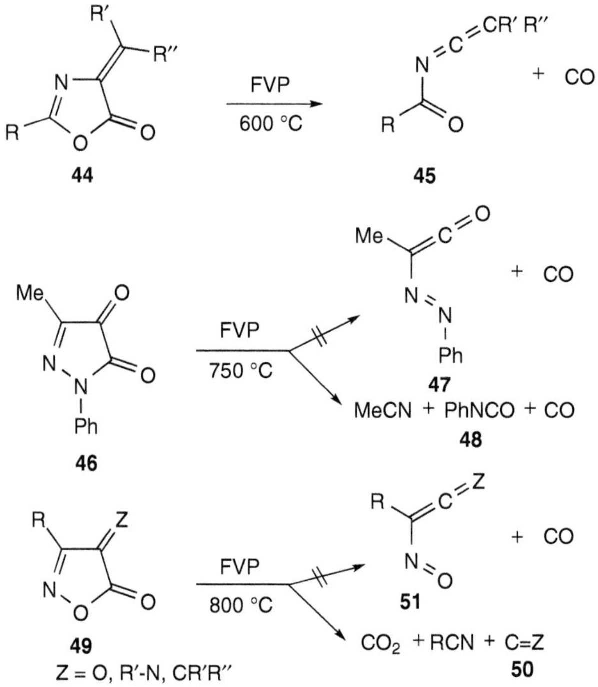

Scheme 2.19

Accordingly, pyrazole-4,5-diones **43** was expected to follow the same pattern to yield azoketenes **44**. Instead, it eliminates CO on FVP at 750 °C, producing phenyl isocyanate **45**, MeCN, and CO [25b, 26] (Scheme 2.19). This behavior was observed with substituted isoxazole-5(4H)-ones **46** that result in an analogous fragmentation to CO$_2$, RCN, and a vinylidene-type species Z=C **47**. Cheletropic extrusion of CO to form nitrosoketene **48** does not take place [25b]. The reactivities of the pyrazolones and isoxazolones are undoubtedly due to the cleavage of the relatively weak N—N and N—O bonds [25b].

Synthesis of aryl- and heteroaryl-acetylenes **49** in high yield was achieved by ring-opening with extrusion of CO$_2$ from the FVP of 4-methylene-5(4H)-isoxazolones **50** [8, 27], leading to vinylnitrenes **51**, azirenes **52**, and vinylidenes **53**, which upon 1,2-shifts provided acetylenes with low activation barriers [28]. The formation of propadieneimine **54** could also have been expected but was not observed [23, 24] (Scheme 2.20).

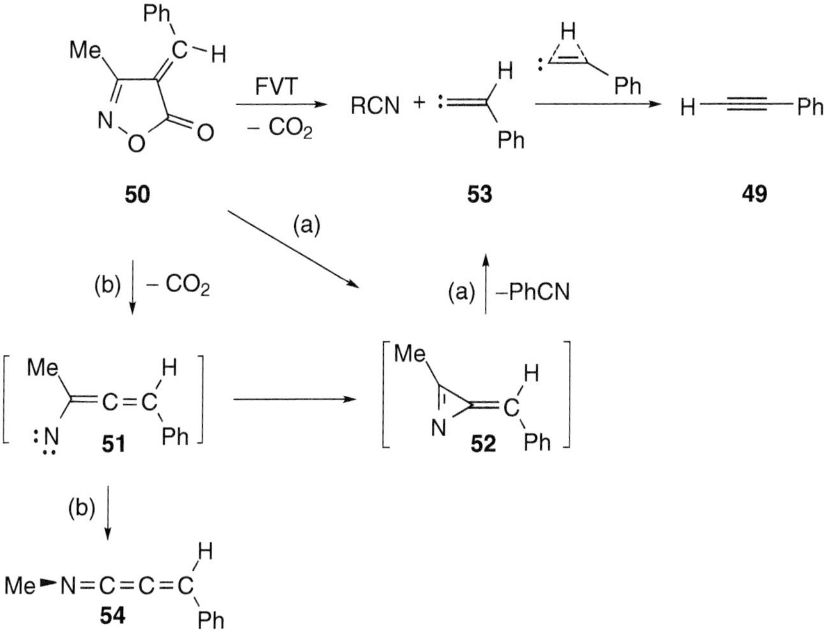

Scheme 2.20

The 3-methyl-4-[bis(methylthio)methylene]isoxazolone **55** eliminates CO$_2$ to produce bis-(methylthio)ethyne **56** on FVP, presumably

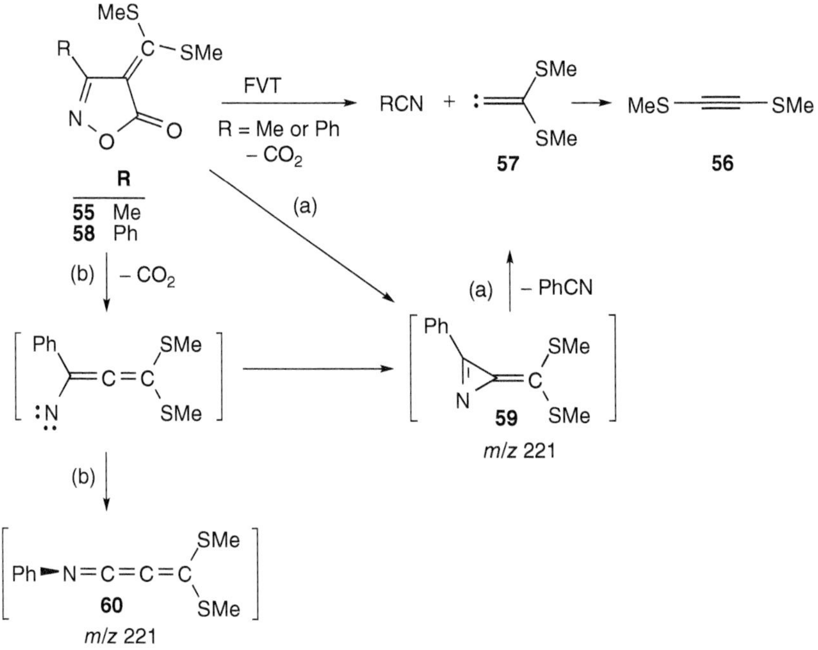

Scheme 2.21

via the vinylidene **57** (Scheme 2.21). The 3-phenyl derivative **58** similarly gave ethyne **56**, but in addition, either azirine **59** or cumulene **60** (m/z 221) was observed from collisional activation mass spectrometry (CAMS) [10].

4-(Arylaminomethylene)isoxazolones **61** gave bisiminopropadienes **62** by FVP [29] (Scheme 2.22) through intermediate ketenimines **63**, which was observed at 500–600 °C by matrix IR and CAMS. At higher temperatures, the ketenimines **63** fragments to generate the bis-isomer **62** with extrusion of CO_2 [30].

FVP of furanones **64**, which involves cheletropic extrusion of CO, produced furan **65** as a main product (Scheme 2.23). Allene **66** was observed by IR absorptions. Cyclization of this allene is likely to lead to intermediate cyclic vinylcarbenes **67**, which will finally give furan **65** [31].

Similarly, an additional acyl substituent (**68**, **69**) leads to furans **70** and **71** [31] (Scheme 2.24).

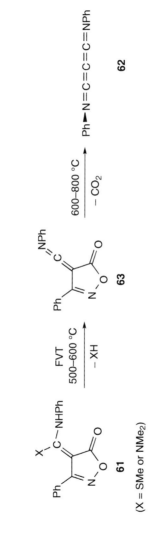

Ph—N=C=C=C=NPh

62

$$\xrightarrow[-CO_2]{600–800\ °C}$$

63

$$\xrightarrow[-XH]{\substack{FVT \\ 500–600\ °C}}$$

61

(X = SMe or NMe₂)

Scheme 2.22

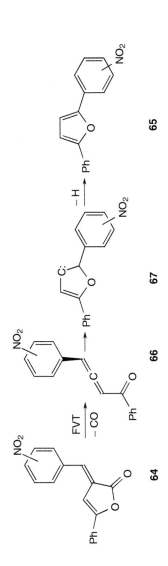

Scheme 2.23

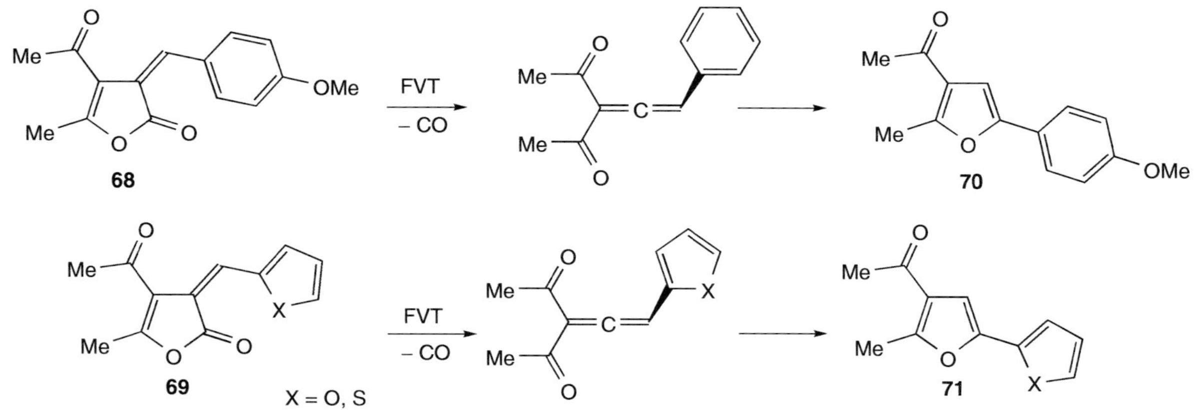

Scheme 2.24

2.4 Pyrolysis of Meldrum's Acid Derivatives

Meldrum acid decomposes in the melt to give many products [32], but on FVP at 430 °C, it fragments cleanly to give acetone, carbon dioxide, and ketene [21] (Scheme 2.25).

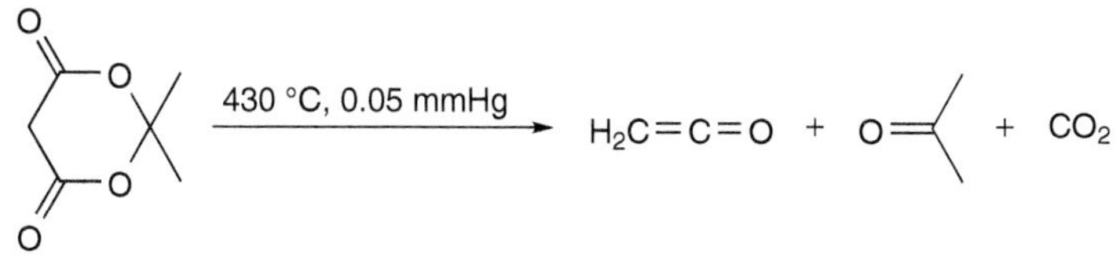

Scheme 2.25

Matrix-isolation techniques have successfully been used to characterize the wide range of structurally diverse ketene derivatives that can be generated by such reactions [33]. Many synthetically useful reactions out of Meldrum's acid pyrolysis are carried out with the 5-methylidene derivatives as a source of propadienone (methyleneketene) intermediates (Scheme 2.26).

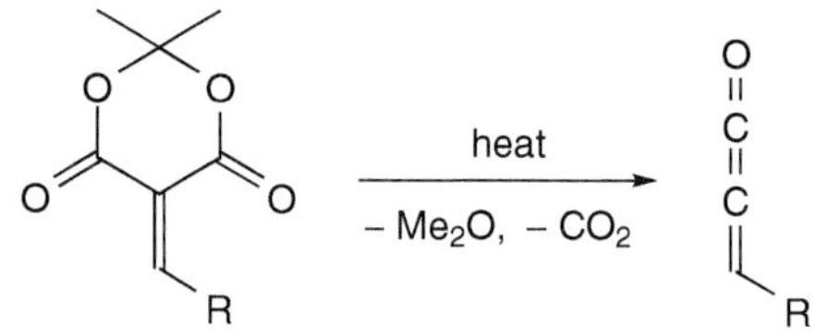

Scheme 2.26

The generation of dimethylketene by pyrolysis of the 5,5-dimethyl derivative of Meldrum's acid was first noted by Ott [34]. FVP of these derivatives over the temperature range of 400–600 °C is a useful general method for the generation of a variety of substituted ketenes [32]. The pyrolytic chemistry of Meldrum's acid and its use in the generation of methyleneketenes have been reviewed by Brown [35]. The synthetic value of methyleneketenes and vinylidenes (methylenecarbenes) was worked out by McNab [36], and the main features of this work have been summarized [37]. The availability of versatile synthetic routes to 5-methylene-Meldrum's acid derivatives (in particular with the R group containing heteroatoms), together with the unique ability of methyleneketenes to undergo internal hydrogen shifts and subsequent cyclization, has led to the preparation of a diverse range of heterocyclic compounds [38], including some unusual systems. Following are some of the representative examples.

FVP of Meldrum's acid derivatives can be used for synthesis of heterocyclic compounds that are difficult to prepare otherwise. A range of alkyl and aryl substituted pyrrol-3-ones have been prepared [38]. The hydrogen transfer in the intermediates formed from FVP of aminomethylene derivatives of Meldrum's acid at 600 °C, 0.01 torr, followed by cyclization provided a general synthetic approach for pyrrol-3-ones with a general yield of 60–80% (Scheme 2.27).

This convenient two-step synthesis is in marked contrast to the multistep sequences of conventional syntheses [39, 40] and provides access to typical 1,2,2-trisubstituted pyrrolones in multigram quantities.

A detailed matrix-isolation study of the mechanism by IR spectroscopy has shown that an intermediate anhydride **86** is formed as the precursor of the methyleneketene **87** on FVP of **88** [41]. The subsequent formation of 1H-pyrrol-3(2H)-ones **89** from methyleneketene **87** (Scheme 2.28) requires a 1,4-hydrogen transfer from a site adjacent to the nitrogen atom to generate a 1,5-dipolar intermediate **90**; this was confirmed by deuterium labeling [42]. An electrocyclic ring closure then produces **91** [43, 44].

FVP of N-arylaminomethylene Meldrum's acid **92** derivatives provides an efficient route from anilines to quinolin-4-ones **93**. Matrix-isolation studies have confirmed that both methyleneketene **94** and iminoketene **95** (characterized by IR spectroscopy) are initially formed in the pyrolysis, but at higher temperatures, electrocyclization and aromatization give rise to quinolone-4-one (Scheme 2.29). The sequence is of considerable value in the synthesis of quinolin-4-ones and related heterocyclic systems [45].

The use of substituted arylamines produces the corresponding quinolinones substituted in the benzene ring [46–48].

FVP of the oxazolidine derivatives **96** at 600 °C, 0.01 hPa represents a special case [49]. Formation of the methyleneketene **97** takes place in the usual way, but the resulting fused pyrrolone is unstable under the given reaction conditions. The 1,3-dipole **98** was believed to be formed by **97** ring opening, and a subsequent hydrogen shift and skeletal

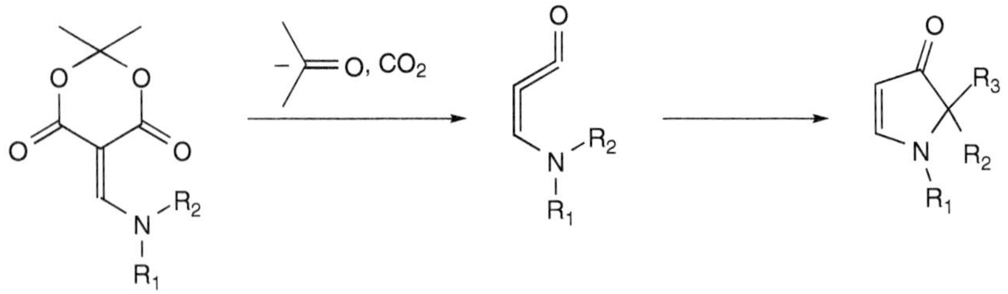

	R1	R2	R1	R2	R3
72	Pr^i	Pr^i	Pr^i	Me	Me
73	Cycloheptyl	Cycloheptyl	Cycloheptyl	$-(CH_2)_6-$	
74	Pr^i	Cyclohexyl	Cycloheptyl	Me	Me
74			Pr^i	$-(CH_2)_5-$	
75	Et	Et	Et	Me	H
76	CH$_2$Ph	CH$_2$Ph	CH$_2$Ph	Ph	H
77	Et	Ph	Ph	Me	H
78	CH$_2$Ph	Ph	Ph	Ph	H
79	Pr^i	CH$_2$Ph	Pr^i	Ph	H
80	Me	Cyclohexyl	Cyclohexyl	H	H
80			Me	$-(CH_2)_5-$	
81	Et	Cyclohexyl	Cycloheptyl	Me	H
81			Me	$-(CH_2)_5-$	
82	Bu^t	Me	Bu^t	H	H
83	Me	Ph	Ph	H	H
84	Me	Me	Me	H	H

85 $-(O)C_6H_4-C_6H_4(O)CH_2-$

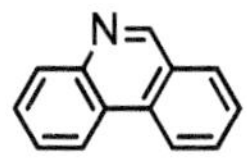

Scheme 2.27

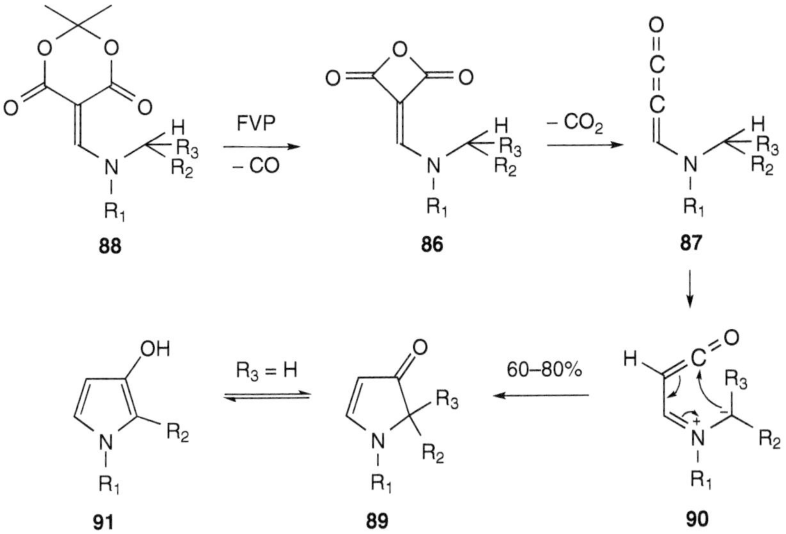

Scheme 2.28

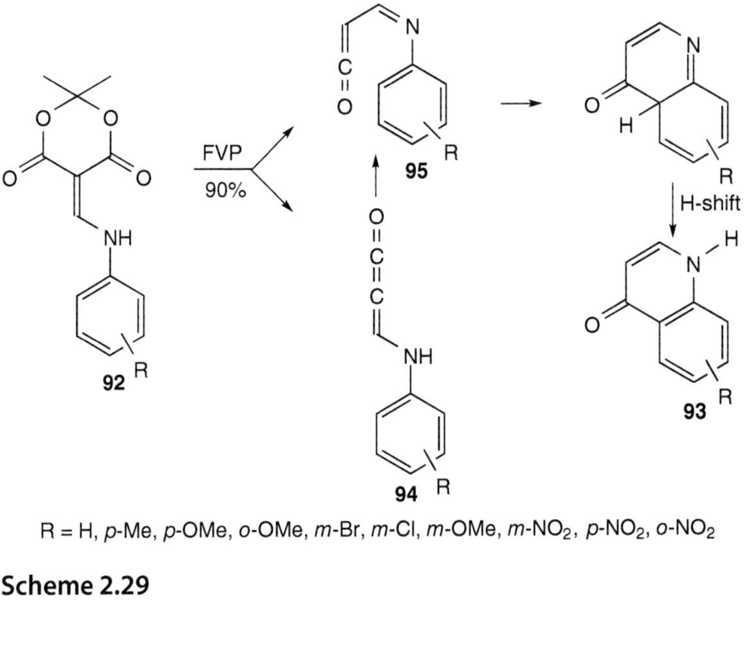

R = H, *p*-Me, *p*-OMe, *o*-OMe, *m*-Br, *m*-Cl, *m*-OMe, *m*-NO$_2$, *p*-NO$_2$, *o*-NO$_2$

Scheme 2.29

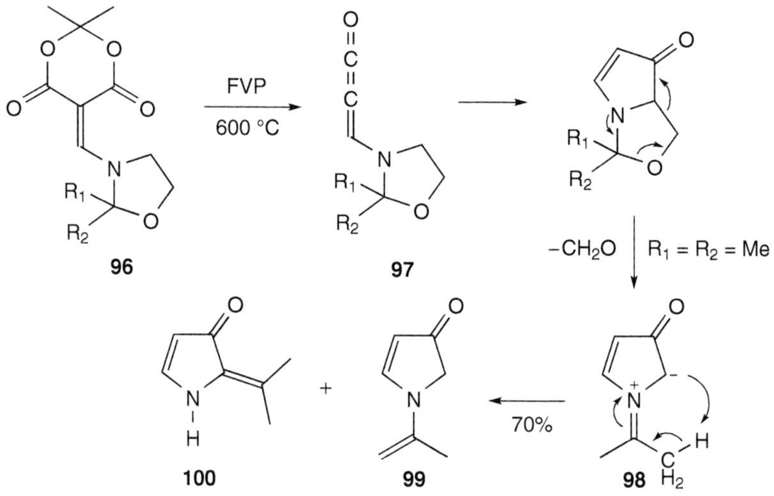
FVP
600 °C
96
97
$-CH_2O$
$R_1 = R_2 = Me$
70%
100
99
98

Scheme 2.30

rearrangement were postulated to account for the products **99** and **100** [49] (Scheme 2.30).

FVP of Meldrum's acid derivatives **101** and **102** at 600 °C (0.01 hPa) produced the hydroxycarbazoles **103** and **104** in 40 and 89% yield, respectively (Scheme 2.31).

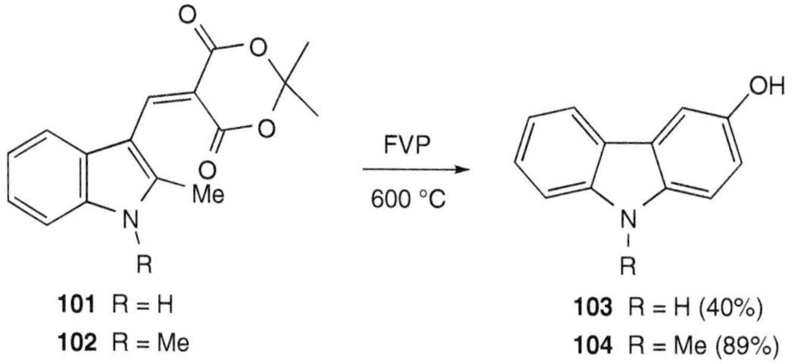
FVP
600 °C
101 R = H
102 R = Me
103 R = H (40%)
104 R = Me (89%)

Scheme 2.31

A 1,5-hydrogen shift is involved in the thermal cyclization of the hydrazine derivatives **105** at 550 °C (0.01 hPa) to yield pyridazine-3-ones **106** [50, 51]. This is considered a general route to 2-aryl pyridazines with an optional substituent at the 4- or 5-position providing 38–83% yield (Scheme 2.32).

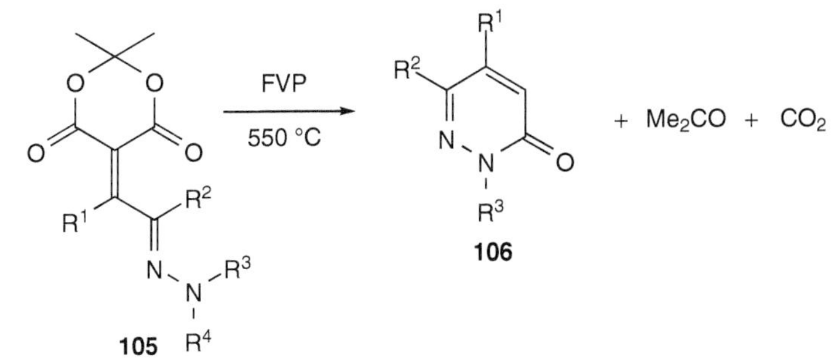

R^1	R^2	R^3	R^4	R^1	R^2	R^3
H	H	H	Ph	H	H	Ph
H	H	H	p-MeC$_6$H$_5$	H	H	p-MeC$_6$H$_5$
H	H	Me	Ph	H	Me	Ph
Me	H	H	Ph	Me	H	Ph

Scheme 2.32

At higher pyrolysis temperatures (750 °C, 0.01 torr), the *tert*-butyl substituents are eliminated in a retro-ene process to provide a route to 2-unsubstituted pyridazin-3-ones **107** in good yield [50] (Scheme 2.33).

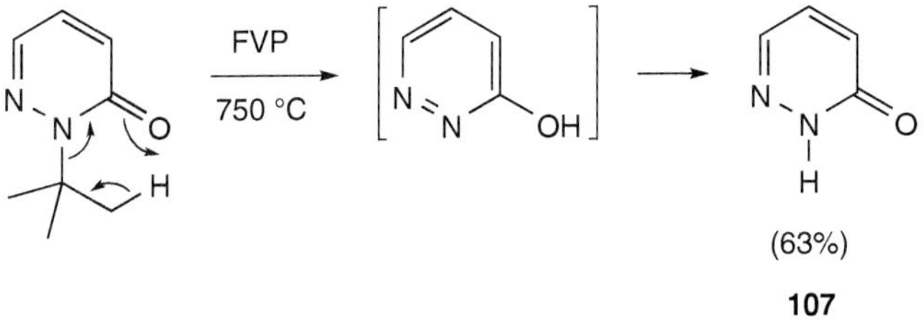

Scheme 2.33

Meldrum acid derivatives **108–110** cyclize to fused α-pyrones **111–113** upon FVP (600 °C, 0.01 hPs) [52] (Scheme 2.34). 1,5-Hydrogen shifts are formally involved in these reactions.

The pyrolysates of FVP reactions of the 5-membered lactams N-pyrrolyl derivatives **114–118** were characterized as pyrrolizinedione derivatives and provide fused pyrrolones **119–123** in 62–80% yields [53]. The optimum temperature for the pyrolysis was 700 °C, whereas the pyrolysis of **117** was complete at 600 °C (Scheme 2.35).

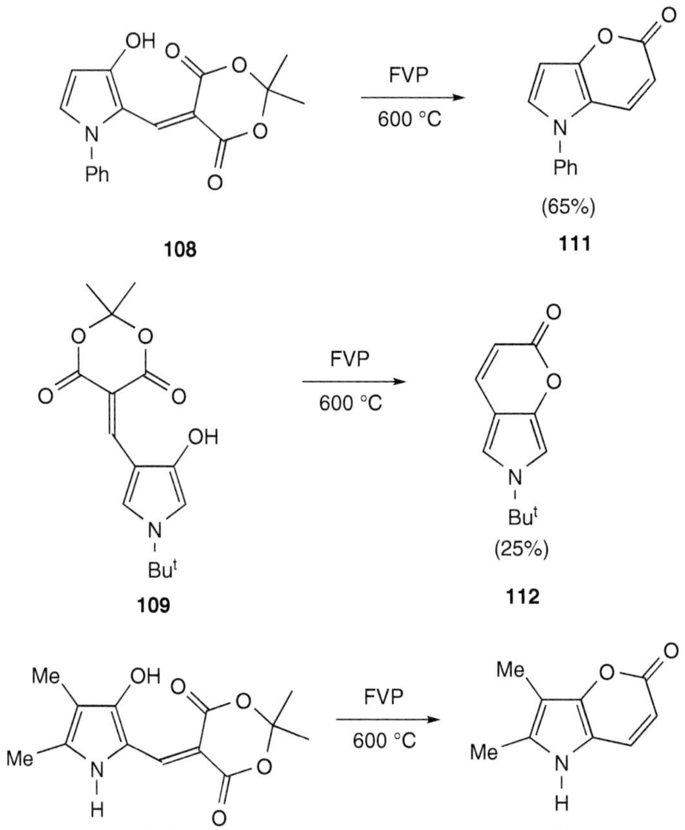

Scheme 2.34

2.5 Elimination of N$_2$

Carbonyl azide **124** eliminates two nitrogen molecules at 400 °C by FVP to produce a relatively stable diazirinone **125** (Scheme 2.36) with a half-life of over one hour in the gas phase at room temperature, which will then decompose to N$_2$ and CO [54].

Azide **126** eliminates N$_2$ by FVP at 350 °C to produce indolone **127–128** [55] (Scheme 2.37).

Extrusion of N$_2$ from α-aminodiazoketones **129** by FVP at 300 °C followed Wolf rearrangement to give the ketene; the latter then undergoes intramolecular addition of NH (Scheme 2.38), resulting in formation of indolenes **130**. Quinoline derivative **131** follows the same approach in FVP at 400 °C to give fused quinoline analogues [56].

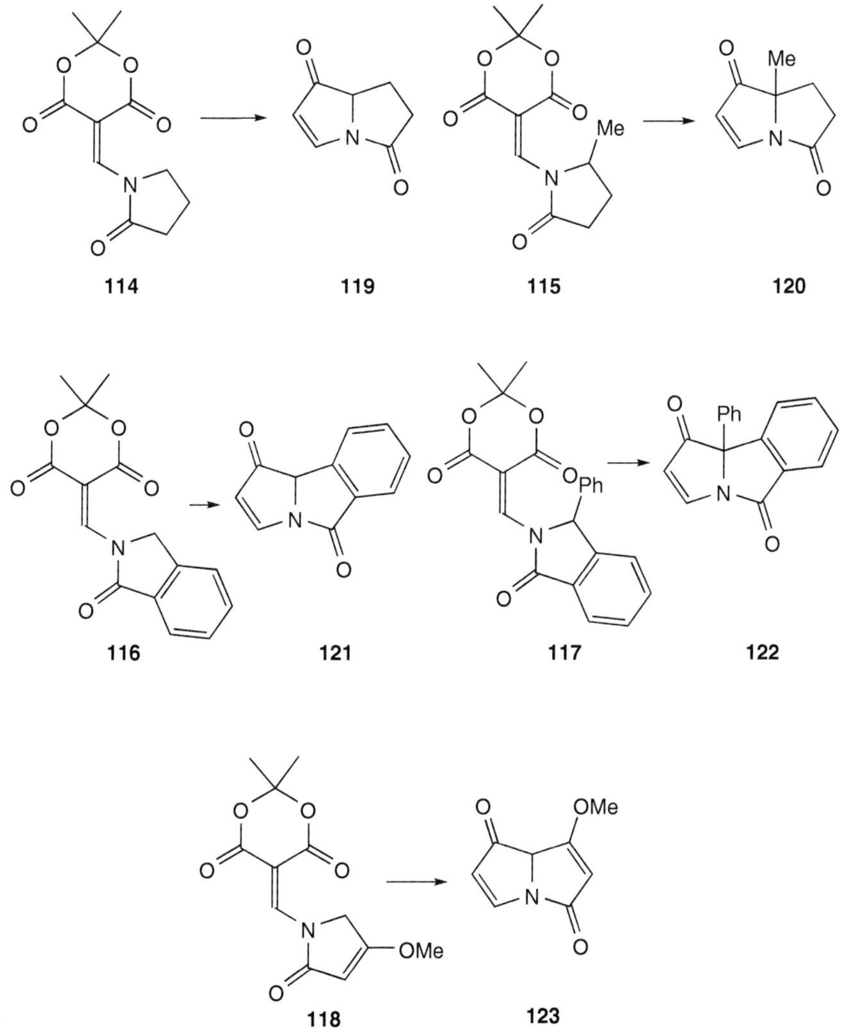

Scheme 2.35

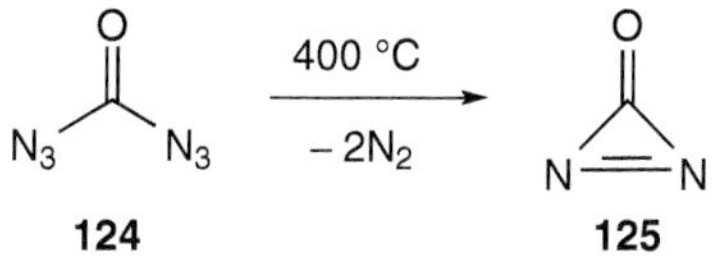

Scheme 2.36

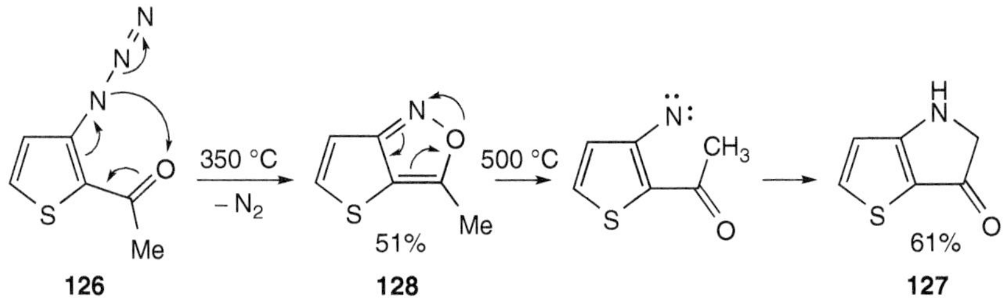

Scheme 2.37

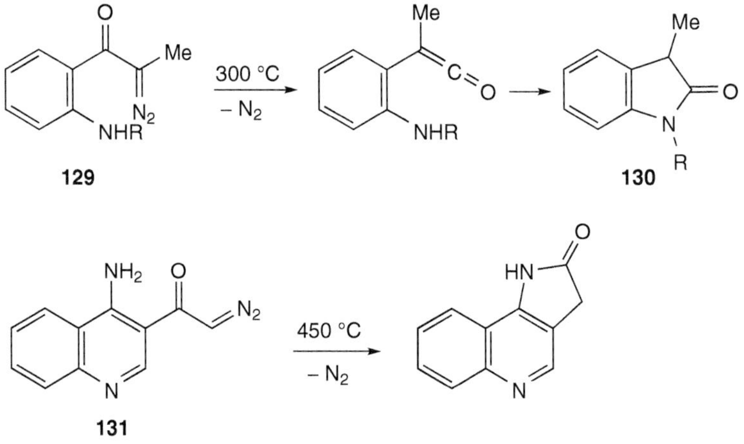

Scheme 2.38

Elimination of N_2 molecules from a number of selected substrates has been utilized effectively in generating heteroindoxyles. Tetrazopyridine **132** eliminates N_2 to give **133** by FVP at 680 °C via rearrangement of the isoxazole **134** [55b] (Scheme 2.39). Thiophene **135** and benzothiophene **136** similarly follow the same approach to produce thiophene- and benzothiophene-fused systems **137** and **138** [55a] (Scheme 2.40).

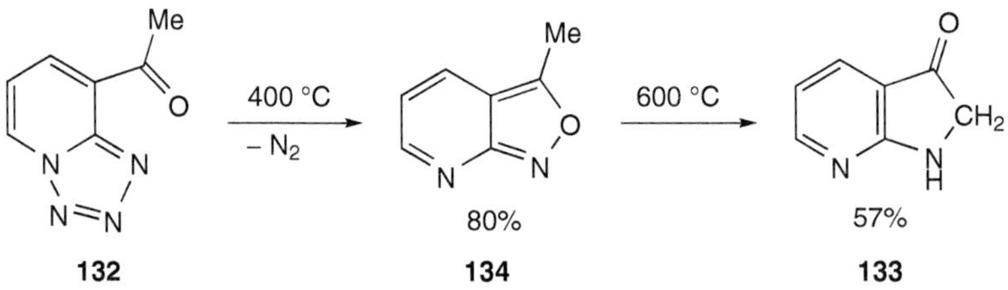

Scheme 2.39

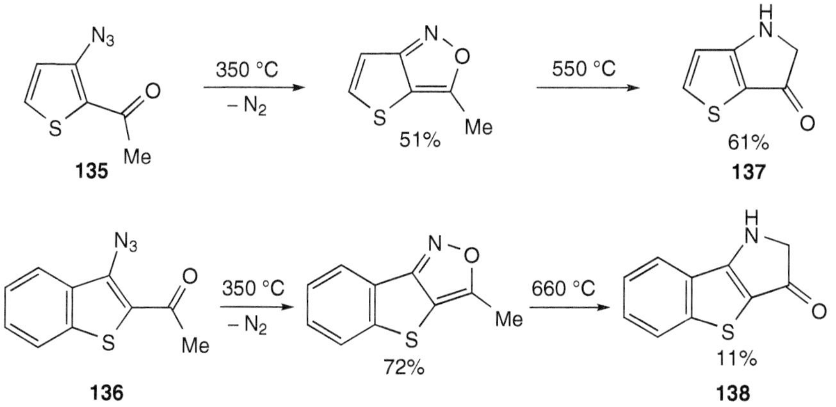

Scheme 2.40

2.5.1 Deazetization Reactions of Allylic Diazenes

Nitrogen extrusion from diazenes is a facile process with potential synthetic applications [57, 58]. Sorensen et al. obtained cyclohexenes by nitrogen extrusion reactions with alkene transposition from the diazenes **139** [59] (Scheme 2.41). This was followed by DFT calculations to study the transition-state character and to elucidate the reaction mechanism of these retro-ene reactions [60]. This reaction was tested for groups other than hydrogen in similar reactions. For the reaction to occur, it is necessary that allyldiazene be in the *cis*-form, whereas the calculations show that the anti-*trans* form is more stable, as it has 4.1 kcal mol^{-1} less energy than the *cis*-form (Scheme 2.42).

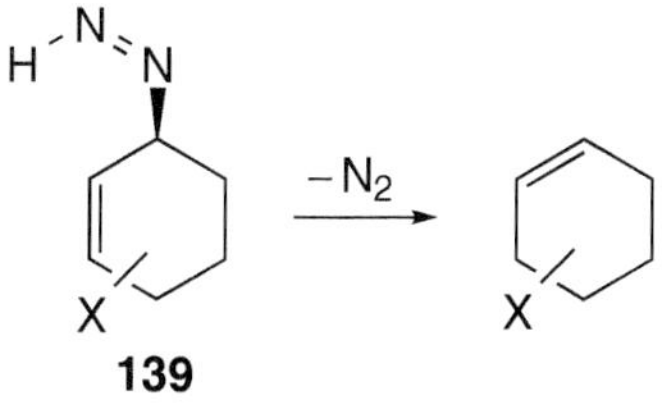

Scheme 2.41

DFT calculation shows that the retro-ene reaction given in Scheme 2.43 proceeds via a six-center cyclic transition state and requires activation energy of 4.5 kcal mol^{-1}, and that the elimination of nitrogen is an exothermic process by 61 kcal mol^{-1}.

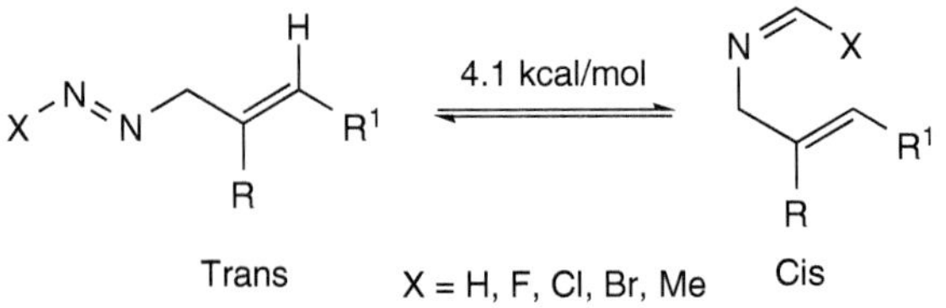

Trans   X = H, F, Cl, Br, Me   Cis

Scheme 2.42

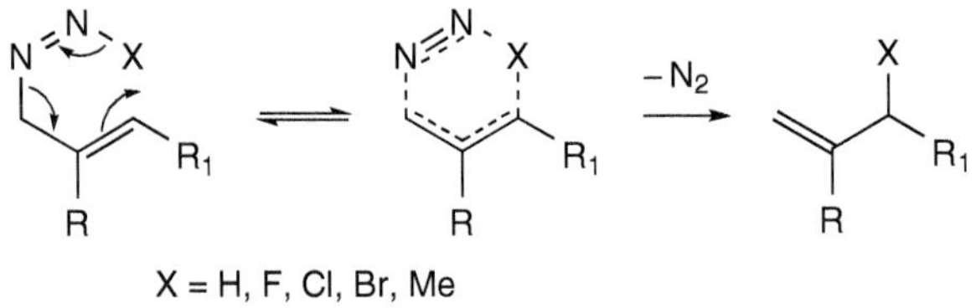

X = H, F, Cl, Br, Me

Scheme 2.43

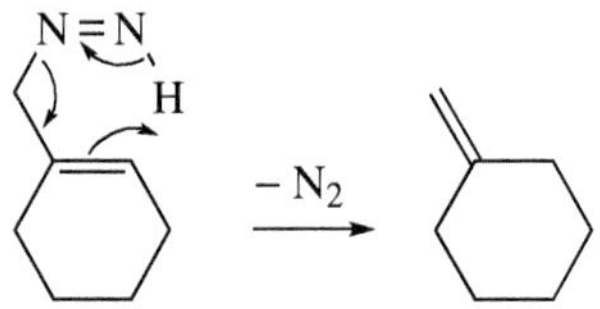

140

Scheme 2.44

The concerted reaction is highly favored. The retro-ene reaction of **140** (Scheme 2.44) has a calculated activation energy of $5.4\,\text{kcal}\,\text{mol}^{-1}$. The *trans* form of **141** is $4.0\,\text{kcal}\,\text{mol}^{-1}$ lower in energy than the *cis* form. The activation barrier for elimination of nitrogen from the *cis* isomer is $5.8\,\text{kcal}\,\text{mol}^{-1}$ (Scheme 2.45). All these reactions are highly exothermic [60].

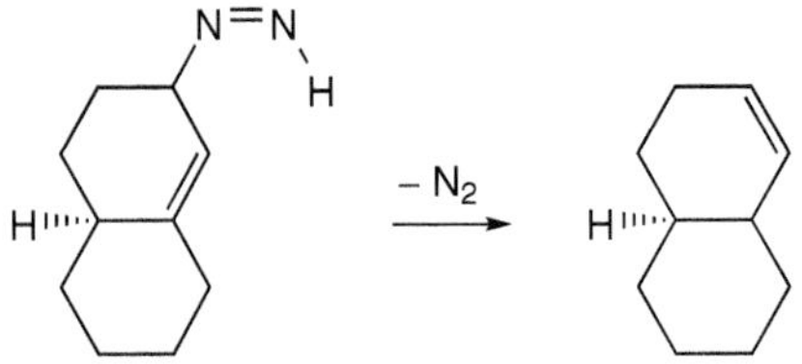

141

Scheme 2.45

2.5.2 Pyrolysis of Benzotriazole Derivatives

Pyrolysis of N=N heterocyclic compounds has been shown to start mainly by N$_2$ loss, leading to reactive diradical intermediates that undergo subsequent transformation leading to many interesting products. These pyrolytic reactions have been reported for five-membered heterocyclic rings including pyrazoles [61], thiadiazoles [62], triazoles [32], and tetrazoles [63], and also six-membered heterocyclic rings including benzotriazoles [64], naphtho[1,8-de][1–3]triazines [65], cinnolines [66], 1,2,3-benzotriazines [67], and 1,2,4-benzotriazines [68].

The well-known Graebe–Ullmann synthesis of carbazol from FVP of 1-phenylbenzotriazole [69] is shown in Scheme 2.46.

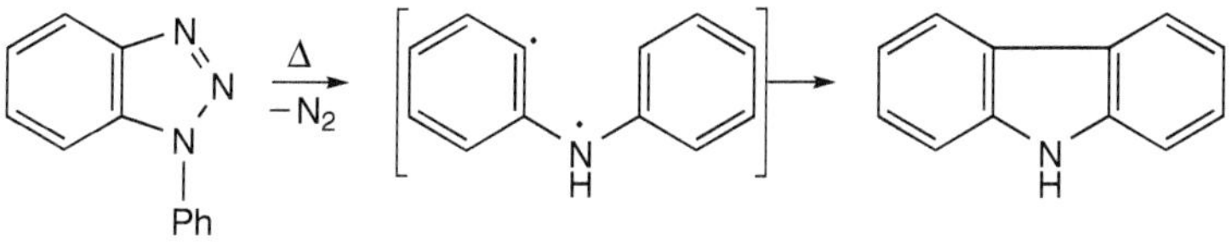

Scheme 2.46

The behavior of benzotriazole and its derivatives under pyrolytic conditions has received considerable attention. α-Carboline **142** derivatives known for their potential anticarcinogenic and antiviral activity [70] were produced from the pyrolysis of 1-(2-pyridin-2-yl)-1H-benzotriazole **143** in the presence of polyphosphoric acid, as shown in Scheme 2.47.

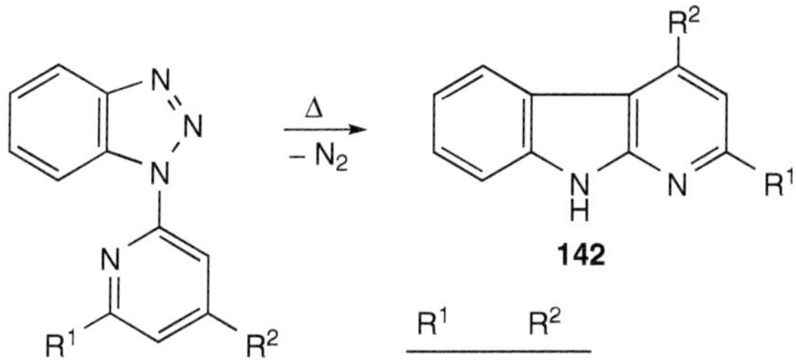

R^1	R^2
H	C$_2$H$_5$
CH$_3$	C$_6$H$_5$
H	NH$_2$
H	COOH
H	COOEt
H	CN

Scheme 2.47

Pyrolysis of 1-(pyrimidin-2-yl)1*H*-benzotriazole **144** in the presence of aqueous polyphosphoric acid at 150 °C gave benzo[4,5]imadazo[1, 2-a] pyrimidine **145** [71], as shown in Scheme 2.48.

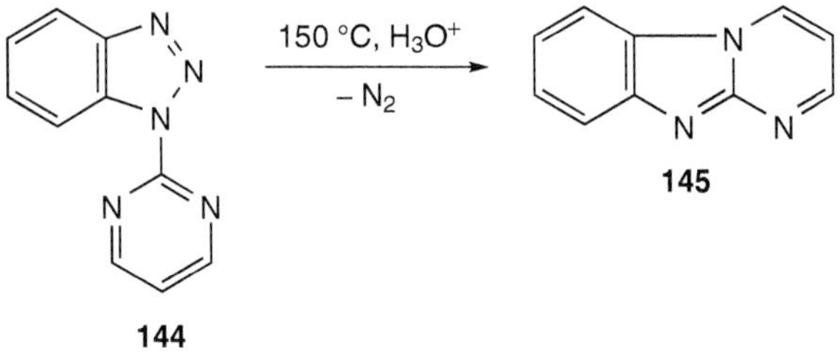

Scheme 2.48

Under the same conditions, 6-benzotriazol-1-ylpyridazin-3-ol **146** gave benzo[4,5]imadazo[1,2-b]pyridazin-2-ol **147** [71] (Scheme 2.49).

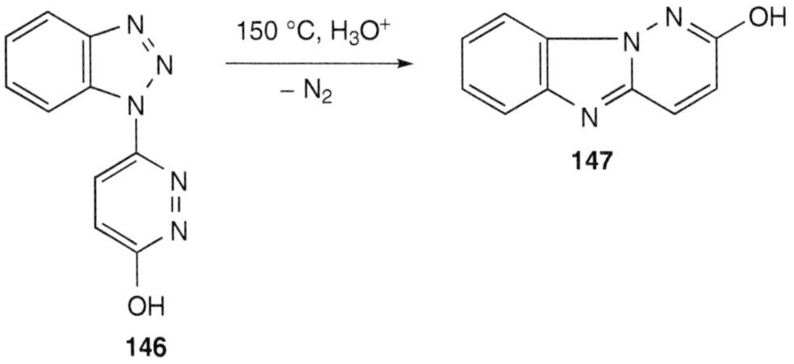

Scheme 2.49

FVP of 1-(benzotriazol-1-yl)isoquinoline **148** produces benzo[4,5] imidazo[2,1-a]isoquinoline **149** in high yield [72] (Scheme 2.50).

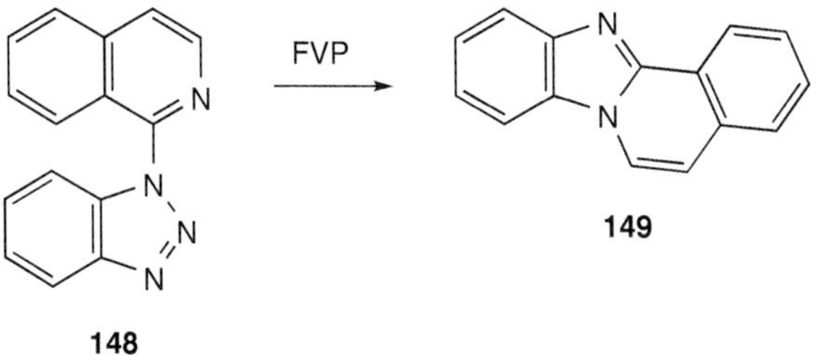

Scheme 2.50

FVP of 1-(benzothiazol-2-yl)-1*H*-benzotriazole **150** at 750 °C gives benzimidazo[2,1-b]benzothiazole **151** in 46% yield [73] (Scheme 2.51).

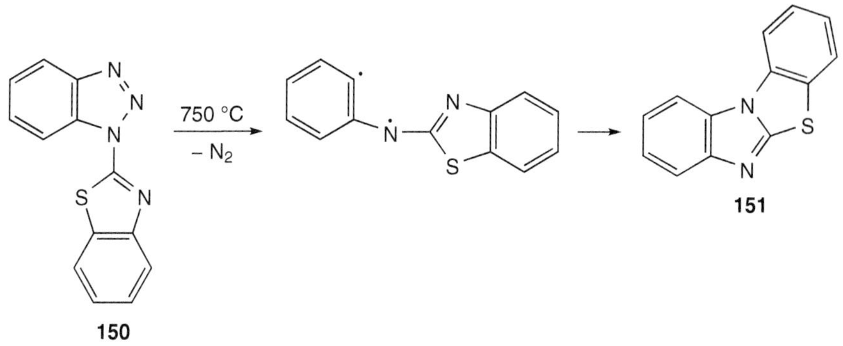

150

151

Scheme 2.51

FVP of 2-(benzotriazole-1-yl)benzonitrile **152** at 700 °C proceeds via hydrogen transfer to give 2-phenyliminomethylbenzonitrile **153** [74] (Scheme 2.52).

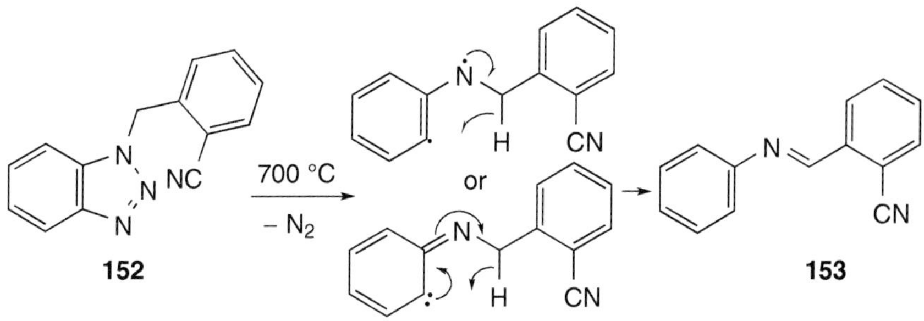

152

153

Scheme 2.52

Wentrup and his group studied the FVP of 1-(benzotriazol-1-yl)ethanone **154** at various temperatures and reported the major product at lower temperature to be the *N*-cyclopenta-2,4-dienylidenemethyleneacetamide **155**, with increasing amounts of 2-methylbenzooxazole **156** at higher temperatures [75]; the maximum yield of the latter was only 24% [76] (Scheme 2.53).

FVP of benzotriazol-1-yl-(3-chloromethylphenyl)methanone **157** at 520 °C yielded acridine **158** (65%). The triplet diradical intermediate formed by the loss of nitrogen appears to prefer hydrogen atom transfer to the amidyl radical center rather than to the aryl radical center, which

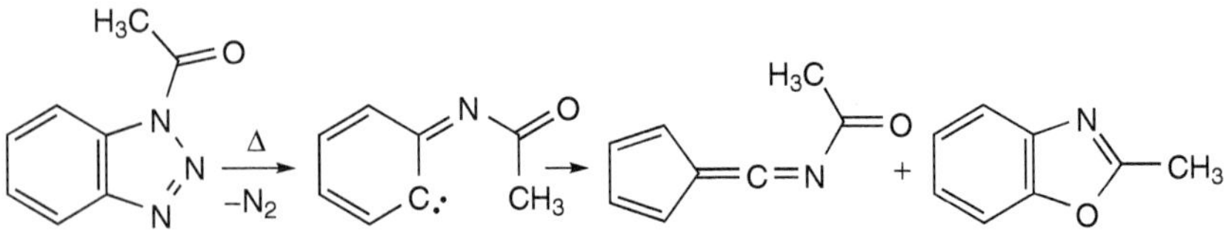

Scheme 2.53

would have given the 3-chloro-2-phenyl-2,3-dihydroisoindol-1-one **159** but was absent. These transformations are shown in [77, 78] Scheme 2.54.

It has been reported that 1-vinylbenzotriazole **160** is converted into indole **161** under FVP conditions at 700 °C [79] (Scheme 2.55).

Based on pyrolysis-mass spectrometry, Katritzky [80] reported that 1-vinylbenzotriazole **160** undergoes FVP to N-phenyl ketenimine and indole in parallel (Scheme 2.56) and not through consecutive reactions.

Wentrup et al., contrary to this report, reported that [81] phenylketen-imine does not rearrange to indole to any significant extent and that the thermal formation of ketenimines and indoles from 1-vinylbenzotriazoles are in fact competing and not consecutive reactions (Scheme 2.57).

At high temperature, N-phenylketenimines undergo homolytic frag-mentation to Ph· and CH_2CN, which recombine to phenylacetonitrile, abstract hydrogen, to form benzene and acetonitrile, and dimerize to succinonitrile and biphenyl [82].

The first kinetic investigation carried out on gas-phase pyroly-sis of the benzotriazoles was in 2004 (Table 2.2) [83], where the gas-phase pyrolytic reactions of 1-(benzotriazol-1-yl)propan-2-one **162**, 3-(benzotriazol-1-yl)-2-one **163** and their phenyl analogues **164** and **165** were investigated (Scheme 2.58). This study was followed by extensive kinetic studies [84a, b] on benzotriazole derivatives in gas phase.

2.5.3 Pyrolysis of Triazine Derivatives

Flash vacuum pyrolysis of 1,2,3-benzotriazenes **166** (Scheme 2.59) was shown to give direct and easy access to many interesting compounds, some of which are otherwise difficult to obtain. The primary step in the pyrolysis of **166** involves mainly N_2 elimination, resulting in

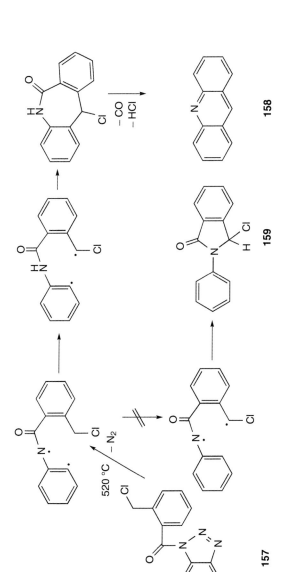

Scheme 2.54

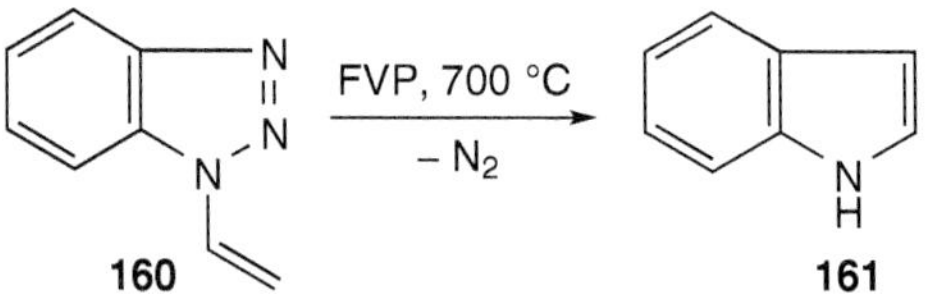

Scheme 2.55

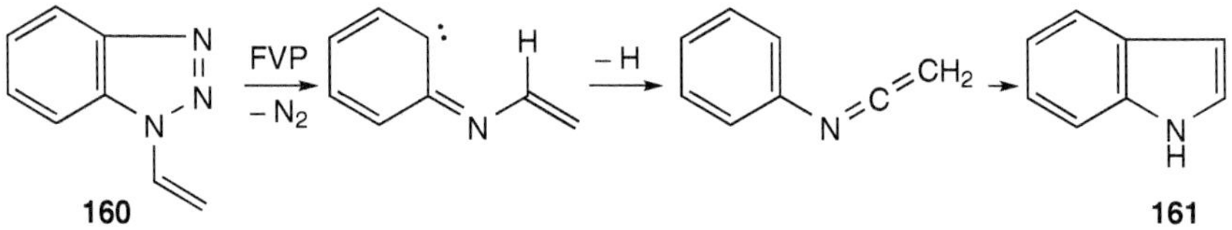

Scheme 2.56

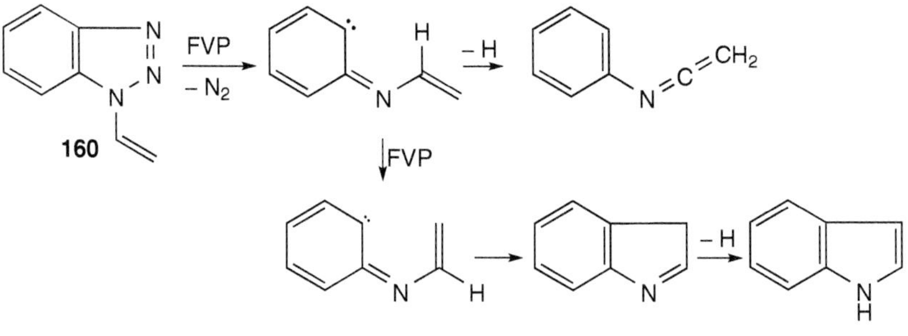

Scheme 2.57

Table 2.2 Rate coefficients ($k\,s^{-1}$) and Arrhenius parameters of α-benzotriazolyl ketones (**162–165**).

Substrate	Log $A\,s^{-1}$	$E_a/k\,J\,mol^{-1}$	$k\,s^{-1}$
162	10.03 ± 0.11	156.7 ± 1.34	4.591×10^{-7}
163	9.934 ± 0.31	158.0 ± 3.64	2.674×10^{-7}
164	12.68 ± 0.47	169.9 ± 5.02	8.592×10^{-6}
165	9.452 ± 0.12	144.4 ± 1.38	2.322×10^{-6}

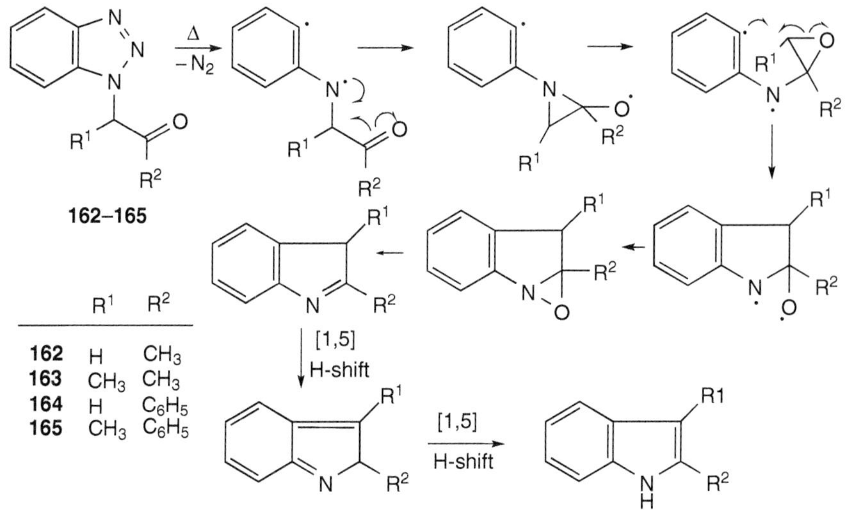
Δ
−N2
[1,5]
H-shift
[1,5]
H-shift
162–165
R1 R2
162 H CH3
163 CH3 CH3
164 H C6H5
165 CH3 C6H5

Scheme 2.58

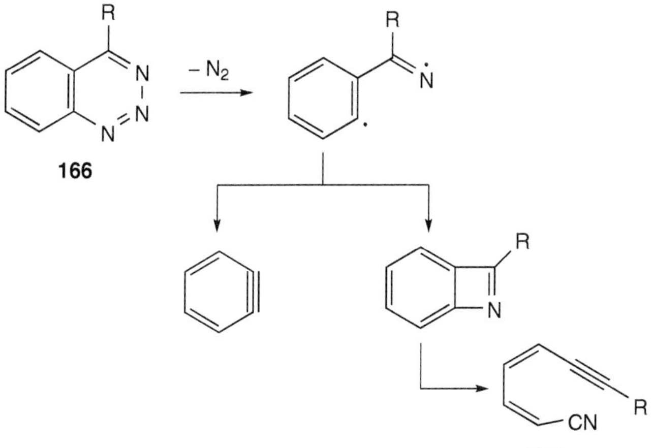
– N2
166

167

Scheme 2.59

diradical intermediates that subsequently combine intramolecularly into a condensed cyclobutenes that will further collapse to the stable cyano-acetylene **167** [67a, 68a].

Pyrolysis of 1-arylnaphtho[1,8-*de*][1,2,3]triazines **168** has been shown to give the corresponding 7*H*-benz[*kl*]acridine derivatives **169** through biradical intermediate **170** [85], with elimination of N_2

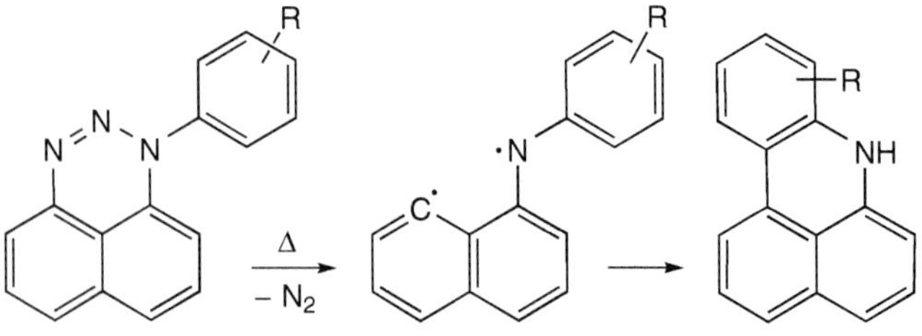

Scheme 2.60

followed by intramolecular cyclization with the appropriate substituent (Scheme 2.60).

Accordingly, pyrolysis of the appropriately substituted naphtho[1,8-*de*][1,2,3]triazine **171a–f** derivatives has opened an important synthetic access to many condensed heterocyclic systems condensed on the *peri* positions of the naphthalene ring [86]. FVP of **171a–f** at 500 °C and 0.02 torr gave the corresponding pure 2-substituted naphtho[1,8-de]-1,3-oxazines **172a–f** in 48–65% yields [65] (Scheme 2.61).

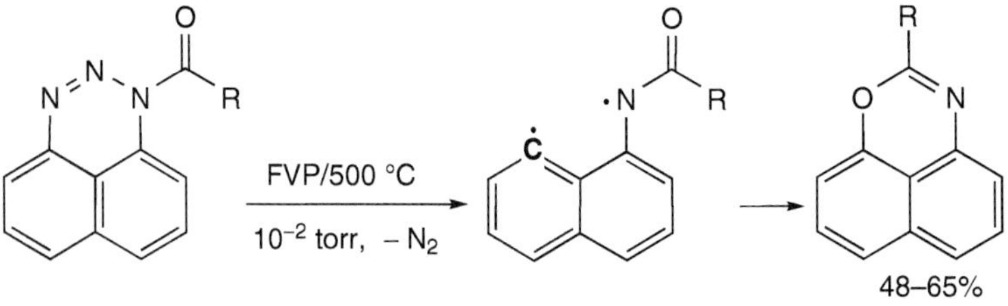

a : CH_3 ; b: C_6H_5 ; c: p-ClC_6H_4 ; d: p-$CH_3C_6H_4$; e: p-$CH_3OC_6H_4$; f: o-CH_2=$CHCH_2C_6H_4$

Scheme 2.61

References

1 Klunder, A.J.H. and Zwanenburg, B. (1997). *Gas-Phase Reaction in Organic Synthesis* (ed. Y. Vallée). Amsterdam: Gordon and Breach Science Publishers.

2 Wentrup, C. (2013). *Aust. J. Chem.* 66: 852–863.

3 McNab, H. (2004). *Aldrichimica Acta* 37: 19–26.

4 McNab, H. (2006). *Contemp. Org. Synth.* 3: 373–396.

5 Hedaya, E. (1969). *Acc. Chem. Res.* 2: 367–373.

6 (a) Moldoveanu, S.C. (1998). *Analytical Pyrolysis of Natural Organic Polymers* (Vol. 20). Amsterdam: Elsevier Science. (b) Wiersum, U.E. (1982). *Recueil des Travaux Chimiques des Pays-Bas* 101 (10): 317–332.

7 Sauer, N.N., Angelici, R.J., Huang, Y.C.J., and Trahanovsky, W.S. (1986). *J. Org. Chem.* 51: 113–114.

8 Wentrup, C. and Winter, H.W. (1978). *Angew. Chem. Int. Ed. Engl.* 17: 609–610.

9 Wentrup, C., Wiedenstritt, M., and Winter, H.W. (2015). *Aust. J. Chem.* 68: 1233–1236.

10 Kvaskoff, D. and Wentrup, C. (2010). *Aust. J. Chem.* 63: 1694–1702.

11 (a) Maccoll, A. (1969). *Chem. Rev.* 69: 33. (b) Smith, G.G. and Kelly, F.W. (1971). *Progress in Physical Organic Chemistry*, vol. 8, 75–233. New York: Wiley-Interscience.

12 Schiess, P. and Heitzmann, M. (1977) (Angew. Chem., 89, 485). *Angew. Chem. Int. Ed. Engl.* 16: 469–470.

13 Schiess, P. (1987). *Thermochim. Acta* 112: 31–46.

14 Schiess, P., Eberle, M., Huys-Francotte, M., and Wirz, J. (1984). *Tetrahedron Lett.* 25: 2201–2204.

15 Schiess, P., Rutschmann, S., and Toan, V.V. (1982). *Tetrahedron Lett.* 23: 3669–3672.

16 Oppolzer, W. (1978). *Synthesis* (11): 793–802.

17 Kametani, T. and Nemoto, H. (1981). *Tetrahedron* 37: 3–16.

18 (a) Schiess, P., Barve, P.V., Dussy, F.E., and Pfiffner, A. (1998). *Organic SynthesesColl.* 9: 28. (b) (1995) 72, 116–116.

19 Wentrup, C. and Müller, P. (1973). *Tetrahedron Lett.* 31: 2915–2918.

20 (a) Spangler, R.J., Beckmann, B.G., and Kim, J.H. (1977). *J. Org. Chem.* 42 (18): 2989–2996. (b) Hedaya, E. and Kent, M.E. (1970). *J. Am. Chem. Soc.* 92: 2149–2151. (c) Brown, R.F.C. and Butcher, M. (1969). *Aust. J. Chem.* 22: 1457–1469.

21 Brown, R.F.C., Eastwood, F.W., and Harrington, K.J. (1974). *Australian J. Chem.* 27: 2373–2384.

22 Brown, R.F.C., Eastwood, F.W., Harrington, K.J., and McMullen, G.L. (1974). *Aust. J. Chem.* 27: 2393–2402.

23 Wentrup, C., Heilmayer, W., and Kollenz, G. (1994). *Synthesis* (12): 1219–1248.

24 Wentrup, C., Finnerty, J.J., and Koch, R. (2010). *Curr. Org. Chem.* 14: 1586–1599.

25 (a) Berstermann, H.M., Harder, R., Winter, H.W., and Wentrup, C. (1980). *Angew. Chem. Int. Ed. Engl.* 19: 564–566. (b) Rzepa, H.S. and Wentrup, C. (2013). *J. Org. Chem.* 78 (15): 7565–7574.

26 Kappe, C.O., Kollenz, G., and Wentrup, C. (1993). *Acta Chem. Scand.* 47: 940–942.

27 Wentrup, C., Briehl, H., Lorenzak, P. et al. (1988). *J. Am. Chem. Soc.* 110: 1337–1343.

28 (a) Schulenburg, W., von der Graf, Hopf, H., and Walsh, R. (1999). *Angew. Chem. Int. Ed.* 38: 1128–1130. (b) Ebrahimi, A., Deyhimi, F., and Roohi, H.J. (2001). *J. Mol. Struct. (Theochem)* 246: 207–215.

29 Wolf, R., Stadtmuller, S., Wong, M.W. et al. (1996). *Chem. Eur. J.* 2: 1318–1329.

30 Downing, J., Murray-Rust, P., Tonge, A.P. et al. (2008). *J. Chem. Inf. Model.* 48: 1571–1581.

31 Koch, R., Berstermann, H.M., and Wentrup, C. (2014). *J. Org. Chem.* 19: 65–71.

32 Brown, R.F.C. (1980). *Pyrolytic Methods in Organic Chemistry* (ed. H. Wassermann), 141–148. New York: Academic.

33 Wentrup, C., Gross, G., Berstermann, H.M., and Lorencak, P.J. (1985). *Org. Chem.* 50: 2877–2881.

34 Ott, E. (1913). *Justus Leibigs Ann. Chem.* 401: 159–177.

35 Brown, R.F.G. and Eastwood, F.W. (1980). *The Chemistry of Ketenes and Allenes*, 757–778. New York: Wiley.

36 McNab, H. (2004). *Aldrichimica Acta* 37 (1): 19–26.

37 McNab, H. and Monahan, L. (1988). *J. Chem. Soc. Perkin Trans.* 1: 863–867.

38 Gaber, A.M. and McNab, H. (2001). *Synthesis* (14): 2059–2074.

39 Momose, T., Tanaka, T., and Yokota, T. (1977). *Heterocycles* 6: 1827–1833.

40 Momose, T., Tanaka, T., Yokota, T. et al. (1979). *Chem. Pharm. Bull.* 27: 1448.

41 Lorencak, P., Pommelet, J.C., Chuche, J., and Wentrup, C. (1986). *J. Chem. Soc. Chem. Commun.* (5): 369–370.

42 (a) Hickson, C.L., Keith, E.M., Matin, J.C. et al. (1986). *J. Chem. Soc. Perkin Trans.* 1: 1465–1469. (b) Kappe, C.O., Wong, M.W., and Wentrup, C. (1993). *Tetrahedron Lett.* 34: 6623–6626.

43 McNab, H. and Mohahan, L.C. (1987). *J. Chem. Soc. Chem. Commun.* (3): 138–139.

44 McNab, H. and Mohahan, L.C. (1988). *J. Chem. Soc. Perkin Trans.* 1: 869–873.

45 Al-Awadi, N.A., Abdelhamid, I.A., Al-Etaibi, A.M., and Elnagdi, M.H. (2007). *Synlett* 14: 2205–2208.

46 Hermecz, I., Keresztur, G., and Vasvari-Debreezy, L. (1992). *Adv. Heterocycl. Chem.* 54: 295–335.

47 Delfourne, E., Roubin, C., and Bastide, J. (2000). *J. Org. Chem.* 65: 5476–5479.

48 Walz, A.J. and Sundberg, R.J. (2000). *J. Org. Chem.* 65: 8001–8010.

49 Hunter, G.A. and McNab, H. (1993). *J. Chem. Soc. Chem. Commun.* (9): 794–795.

50 McNab, H. and Stobie, I. (1982). *J. Chem. Soc. Perkin Trans.* 1: 1845–1853.

51 McNab, H. (1983). *J. Chem. Soc. Perkin Trans.* 1: 1203–1207.

52 Derbyshire, P.A., Hunter, G.A., McNab, H., and Monahan, L.C. (1993). *J. Chem. Soc. Perkin Trans.* 1: 2017–2025.

53 McNab, H., Morrow, M., Parons, S. et al. (2009). *J. Org. Biomol. Chem.* 7: 4936–4942.

54 Zeng, X., Beckers, H., Willner, H., and Stanton, J.F. (2011). *Angew. Chem. Int. Ed.* 50: 1720–1723.

55 (a) Gaywood, A.P. and McNab, H. (2009). *J. Org. Chem.* 74: 4278–4282. (b) Gaywood, A.P. and McNab, H. (2010). *Org. Biol. Chem.* 8: 5166–5173.

56 Aitken, R.A. and Boubalouta, Y. (2015). *Advances in Heterocyclic Chemistry*, vol. 115 (eds. E.F.V. Scriven and C.A. Ramsden), 93–150. Amsterdam: Elsevier (incl. Pergamon).

57 (a) Sato, T., Homma, I., and Nakamura, S. (1969). *Tetrahedron Lett.* 10 (11): 871–874. (b) Corey, E.J., Cane, D.E., and Libit, L. (1971). *J. Am. Chem. Soc.* 93: 7016–7021.

58 Myers, A.G., Finney, N.S., and Kuo, E.Y. (1989). *Tetrahedron Lett.* 30: 5747–5750.

59 Sammis, G.M., Flamme, E.M., Xie, H. et al. (2005). *J. Am. Chem. Soc.* 127: 8612–8613.

60 Jabbari, A., Sorensen, E.J., and Honk, K.N. (2006). *Org. Lett.* 8 (14): 3105–3107.

61 Moyano, E.L., Yranzo, G.I., and Elguero, J. (1998). *J. Org. Chem.* 63: 8188–8191.

62 Seybold, G. (1977). *Angew. Chem. Int. Ed. Engl.* 16: 365–373.

63 Banciu, M.D., Popescu, A., Simion, A. et al. (1999). *J. Anal. Appl. Pyrolysis* 48: 129–146.

64 Dib, H.H., Al-Awadi, N.A., Ibrahim, Y.A., and El-Dusouqui, O.M.E. (2003). *Tetrahedron* 59: 9455–9464. and references cited therein.

65 Al-Awadi, H., Ibrahim, M.R., Dib, H.H. et al. (2005). *Tetrahedron* 61: 10507–10513.

66 (a) Ibrahim, Y.A., Al-Awadi, N.A., and Kaul, K. (2001). *Tetrahedron* 57: 7377–7381. (b) MacBride, J.A.H. (1972). *J. Chem. Soc. Chem. Commun.* (22): 1219–1220. (c) MacBride, J.A.H. (1974). *J. Chem. Soc. Chem. Commun.* (9): 359–360. (d) Chambers, R.D., Clark, D.T., Holmes, T.F. et al. (1974). *J. Chem. Soc. Perkin Trans.* (1): 114–125. (e) Barton, J.W. and Walker, R.B. (1975). *Tetrahedron Lett.*: 569–572. (f) Barton, J.W. and Walker, R.B. (1978). *Tetrahedron Lett.* 19 (1): 1005–1008. (g) Kanoktanoporn, S. and MacBride, J.A.H. (1977). *Tetrahedron Lett.*: 1817–1818. (h) Kilcoyne, J.P., MacBride, J.A.H., Muir, M., and Write, P.M. (1990). *J. Chem. Res. Synop.*: 6–7. Kilcoyne, J.P., MacBride, J.A.H., Muir, M. et al. (1990). *J. Chem. Res.*, Miniprint, 215. (i) Adams, D.B., Kilcoyne, J.P., MacBride, J.A.H., and Muir, M. (1990). *J. Chem. Res. Synop.*: 172–173. Adams, D.B., Kilcoyne, J.P., MacBride, J.A.H. et al. (1990). *J. Chem. Res.*, Miniprint, 1301. (j) Shepherd, M.K. (1994). *J. Chem. Soc. Perkin Trans.* 1: 1055–1059. (k) Wilcox, C.F. Jr., Lassila, K.R., and Kang, S. (1988). *J. Org. Chem.* 53: 4333–4339. (l) Coates, W., Pyridazines, G., Derivatives, T.C. et al. (eds.) (1996). *Comprehensive Heterocyclic Chemistry II*, vol. 6, 87. London: Pergamon.

67 (a) Adgar, B.M., Keating, M., Rees, C.W., and Storr, R.C. (1975). *J. Chem. Soc. Perkin Trans.* 1: 41–45. (b) Werstiuk, N.H., Roy, C.D., and Ma, J. (1995). *Can. J. Chem.* 73: 146–149.

68 (a) Adgar, B.M., Rees, C.W., and Storr, R.C. (1975). *J. Chem. Soc. Perkin Trans.* 1: 45–52. (b) Riedl, Z., Hajos, G., Pelaez, W.J. et al. (2003). *Tetrahedron* 59: 851–856.

69 Graebe, C. and Ullmann, F. (1896). *Justus Liebigs Ann. Chem.* 291: 16–17.

70 Nantka-Nemirski, P. and Kalinowski, J. (1975). *Uniw. Adama Mickiewicza Poznaniu, Wydz.Mat., Fiz. Chem. Ser. Chem.* 18: 259–271.

71 Hurbert, A. and Reimlinger, H. (1970). *Chem. Ber.* 103: 2828–2835.

72 Prager, R.H., Baradarani, M.M., and Khalafy, J. (2000). *J. Heterocyclic Chem.* 37: 631–637.

73 Lin, D.C.K. and DeJongh, D.C. (1974). *J. Org. Chem.* 39 (12): 1780–1781.

74 Khalafy, J. and Prager, R. (1998). *Aust. J. Chem.* 51: 925–929.

75 Maquestiau, V., Beugnies, D., Flaming, R. et al. (1990). *Org. Mass Spectrom.* 25: 197–203.

76 Janowski, W.K., Prager, R.H., and Smith, J.A. (2000). *J. Chem. Soc. Perkin Trans.* 1: 3212–3216.

77 Wiersum, U.E. (1982). *Recl. J. Roy. Nether. Chem. Soc.* 101 (11): 365–381.

78 Baradarani, M.M., Khalafy, J., and Prager, R.H. (1999). *Aust. J. Chem.* 52: 773–780.

79 Lawrence, R. and Waight, E.S. (1970). *Org. Mass Spectrom.* 3: 367–377.

80 Moquestiau, A., Beugnis, D., Flammang, R. et al. (1988). *J. Chem. Soc. Perkin Trans.* 2: 1071–1075.

81 Wentrup, C., Freiermutha, B., and Aylward, N. (2017). *J. Anal. Appl. Pyrolysis* 128: 187–195.

82 Begue, D., Dargelos, A., Berstermann, H.M. et al. (2014). *J. Org. Chem.* 79: 1247–1253.

83 Dib, H.H., Al-Awadi, N.A., Ibrahim, Y.A., and El-Dusouqui, O.M.E. (2004). *J. Phys. Org. Chem.* 17: 267–272.

84 (a) Al-Awadi, N.A., George, B.J., Dib, H.H. et al. (2005). *Tetrahedron* 61: 8257–8263. (b) Al-Awadi, H., Ibrahim, M.R., Al-Awadi, N.A., and Ibrahim, Y.A. (2008). *J. Heterocyclic Chem.* 45: 723–727.

85 Waldmann, H. and Back, S. (1940). *Justus Leibigs Ann. Chem.* 545: 52–58.

86 Flowerday, P. and Perkins, M.J. (1970). *J. Chem. Soc. C* (2): 298–303.

3

Reaction Mechanism

The main feature of gas-phase pyrolysis reactions is fragmentation of molecules. The breakage of one bond leads to a pair of free radicals. The breakage of two or more bonds usually involves concerted reactions taking place via cyclic transition states. The majority of these reactions are either four-membered or six-membered, and numerous thermal fragmentation reactions can be described as symmetry-allowed pericyclic reactions, such as the retro-ene reaction and the Cope rearrangement. An extensive discussion on reaction intermediates – radicals, diradicals, benzynes, carbenes, and nitrenes – is presented.

3.1 Retro-ene Reactions

The ene reaction occurs with the addition of an electron-deficient double bond to an olefin containing an allylic hydrogen atom. According to the Woodward–Hoffmann rules of pericyclic reactions, the ene

Gas-Phase Pyrolytic Reactions: Synthesis, Mechanisms, and Kinetics,
First Edition. Nouria A. Al-Awadi.
© 2020 John Wiley & Sons, Inc. Published 2020 by John Wiley & Sons, Inc.

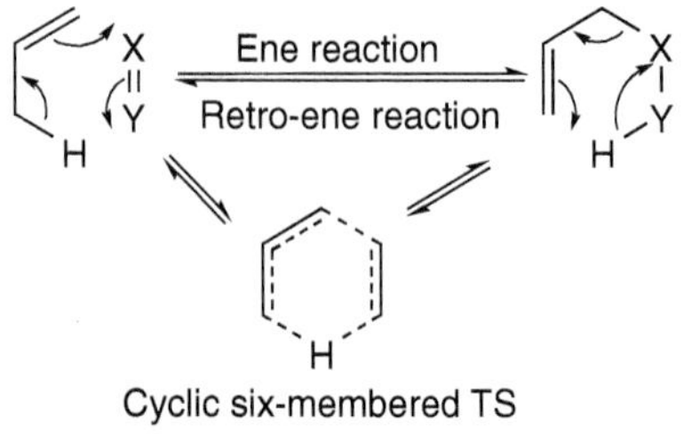

Cyclic six-membered TS

Scheme 3.1

reaction is a concerted $[\pi^2 s + \pi^2 s + \sigma^2 s]$ thermal-symmetry-allowed process involving a six-membered transition state [2]. The retro-ene reaction proceeds through a six-membered transition state involving a 1,5-hydrogen shift [3] (Scheme 3.1). As indicated by experimental data [4] and theoretical calculations [5], the mechanism and synthetic utility of retro-ene reactions have been extensively studied [6]. Thermal retro-ene reactions are a method for accessing unsaturated molecules and parallel the retro-Diels–Alder reaction. However, when a concerted reaction is geometrically impossible, it may occur through a stepwise biradical pathway [4e, 7].

3.1.1 Acetylenic Compounds

Participation of acetylenic bonds in intramolecular reactions was tested in the gas-phase pyrolysis of β-hydroxyacetylenes **1** [8] (Scheme 3.2). Allenes and carbonyl compounds resulting from a 1,5-hydrogen transfer via six-membered cyclic transition state were the only products identified from this pyrolytic reaction. Activation parameters E_a and $\Delta S^{\#}$ lend support to a cyclic transition state. The acetylenic compound was susceptible to pyrolysis seven times faster than the olefinic analogue.

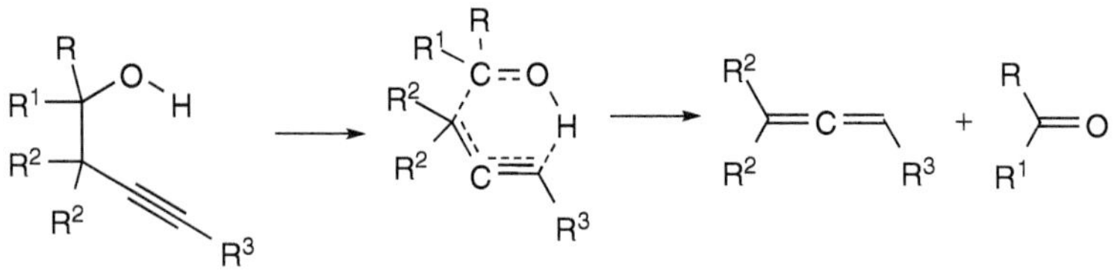

Scheme 3.2

The effects of primary, secondary, and tertiary alkyl substituents, phenyl and vinyl substituents, were investigated by kinetic and product analysis from their pyrolytic reactions (Scheme 3.3), and the results are

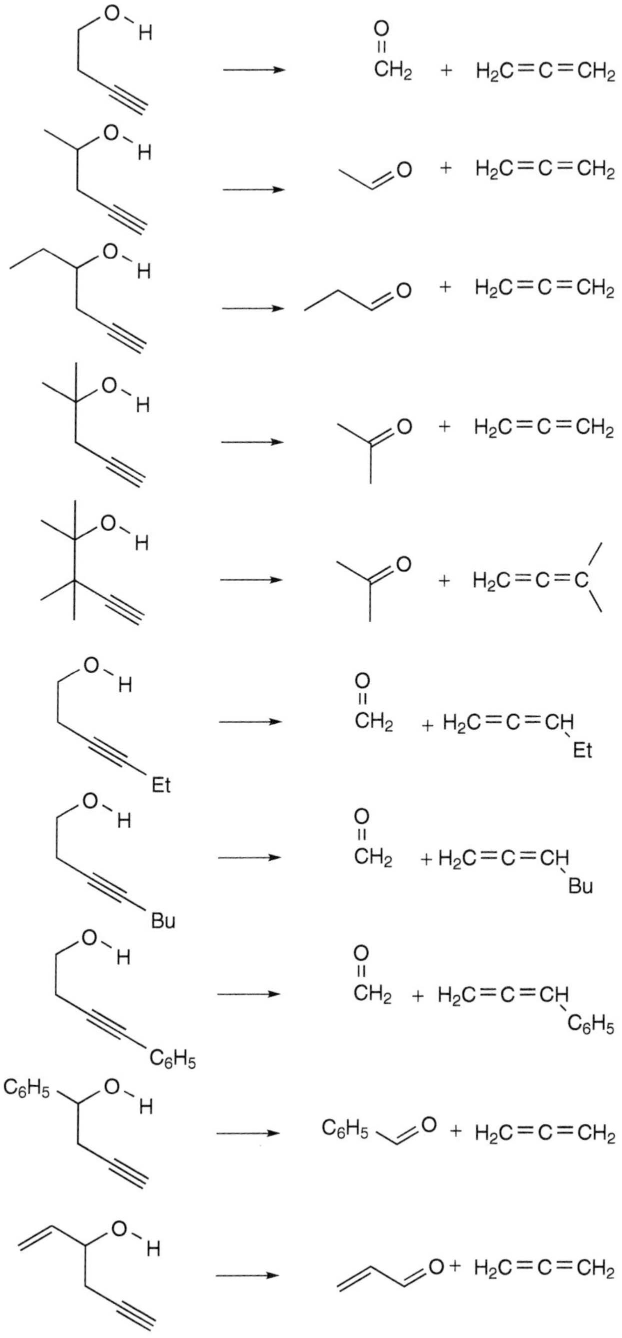

Scheme 3.3

in accordance with a planar transition state with participation of the triple bond.

Claisen rearrangement of phenyl propargyl ether **2** (Scheme 3.4) in gas phase gave indanone **4** as its sole product [9]. It is noteworthy that the same ether **2** converts in solution to benzopyran **3** [10]; as **6** aromatizes rapidly to give **3** through conventional sigmatropic and electrocyclization processes, while a concerted intramolecular pathway is forbidden in gas-phase [11], an alternative pathway via **7** to indanone **4** was followed.

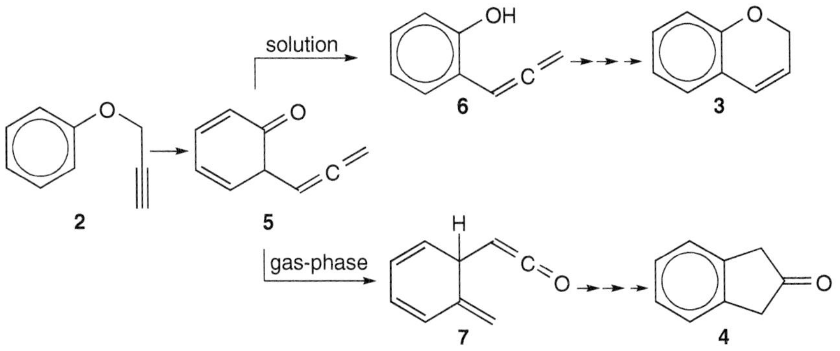

Scheme 3.4

3.1.2 Acyl Group Participation

An interesting intramolecular redox reaction via a six-membered transition state took place in gas-phase pyrolysis of acid chloride **8** to form 2-propenylbenzaldehydes **9–11** [12] (Scheme 3.5).

On the other hand, gas-phase pyrolysis of acid chlorides **12** and **13** would lead to highly strained arenocyclobutenones, and these would give further decarbonylated products [13] (Scheme 3.6).

3.1.3 Cyanates and Isocyanates

Retro-ene reactions of cyanate **14a**, alkylcyanate **14b**, N-dimethylaminocyanate **14c**, N-dimethylaminothiocyanate **14d**, and methylsulfenylthiocyanate **14e** were investigated computationally together with a detailed product analysis [14]. The results are in favor of a six-centered transition state (Scheme 3.7), although in some cases an alternative four-centered 1,2-elimination occurs (Scheme 3.8).

Isomeric isocyanate and isothiocyanates behave in a similar way. Table 3.1 reports the activation barriers for six-membered retro-ene type reactions and the alternative four-centered 1,2-elimination of cyanates, thiocyanates, isocyanates, and isothiocyanates [14].

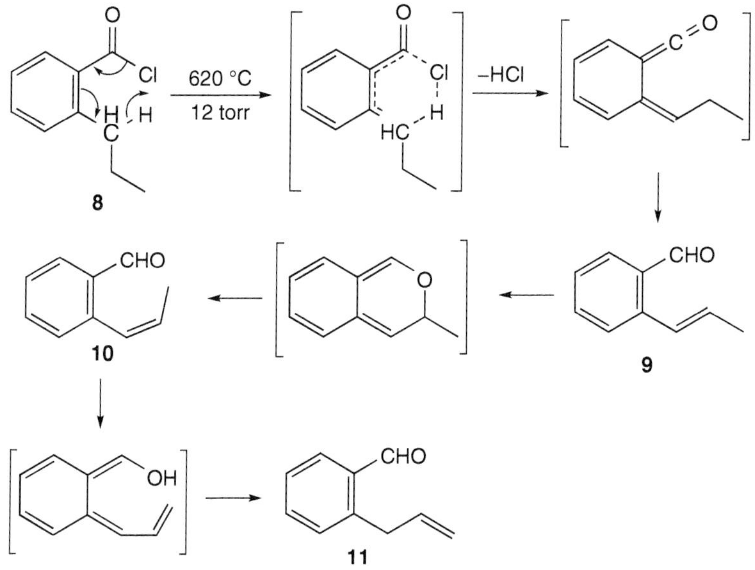

Scheme 3.5

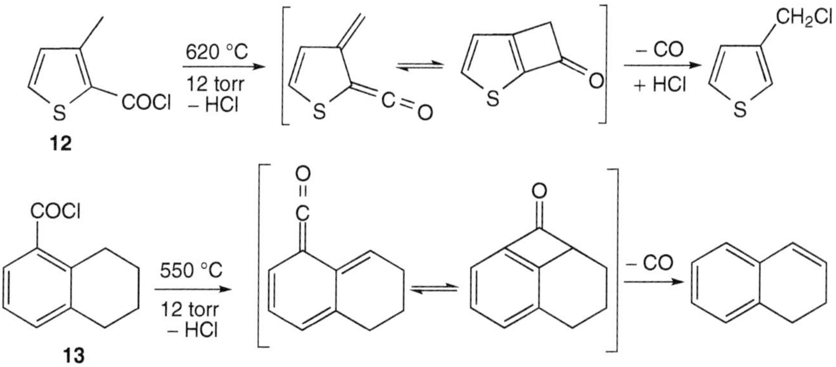

Scheme 3.6

3.1.4 Esters

Esters containing non-vinylic β-hydrogen atoms on the alkyl group undergo gas-phase elimination to an alkene and carboxylic acid. This is one of the oldest known pyrolytic reactions [15], and it has been intensively studied and summarized in a number of general reviews on pyrolytic reactions [1, 16]. The mechanism of the reaction can be represented as shown in Scheme 3.9.

For esters D and F=O, G=alkyl A, B, and E are carbon atoms. Changing D to S will result in thioesters, changing A to C=O results

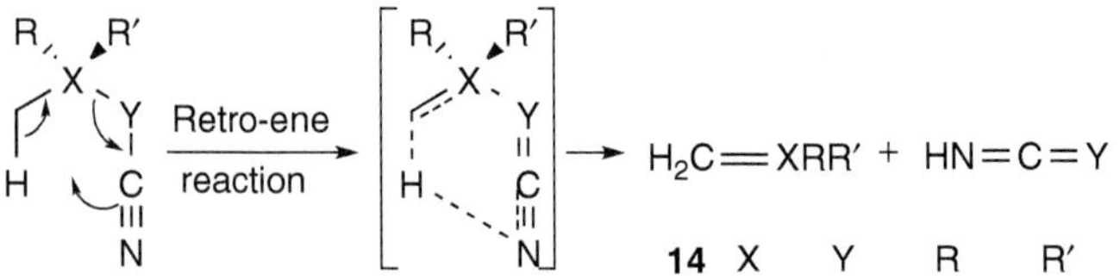

14	X	Y	R	R′
a	C	O	H	H
b	C	O	Me	Me
c	N	O	Me	Me
d	N	S	-	Me
e	S	S	-	-

Scheme 3.7

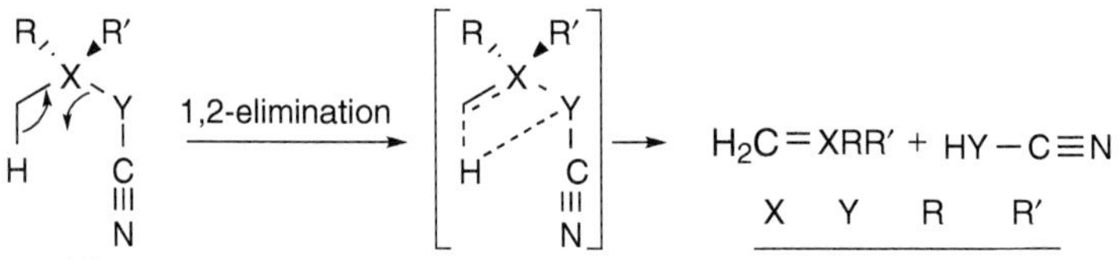

X	Y	R	R′
C	S	H	H
C	S	Me	Me
N	O	Me	-
O	O	-	-
S	O	-	-

Scheme 3.8

in anhydride, changing D from O to NH results in amides, interchanging D and F in amide (D=O, F=NH) results in imino ethers, and interchanging D and F for ketones (D=CH_2, F=O) to (D=O, F=CH_2) results in vinyl ethers. Such functional group interconversion by changing/interchanging atoms forming the six-membered ring and the nature of the groups attached at various points clearly enrich the synthetic potential of gas-phase pyrolytic reactions.

The factors affecting the rate in the gas-phase six-centered elimination reactions have been quantified. Considerable progress has been made in understanding the mechanisms of such processes by considering the relative rates of thermal elimination of analogous compounds. The notable features are: (i) the reaction is aided by electron supply at A so that the order of reactivity of alkyl esters is $3° > 2° > 1°$; (ii) the reaction is aided by electron withdrawing group (G), where the reactivity order is formate > acetate > propanoate (G=H, CH_3, C_2H_5), and carbonate > carbamate > acetate (G=OR, RNH, CH_3); and (iii) reactivity is aided

Table 3.1 Activation barriers (kcal mol^{-1}) for six-centered retro-ene reactions and four-centered 1,2-elimination reactions.

Reactant	Six-centered retro-ene reaction	Four-centered 1,2-elimination
$CH_3—CR_2—OCN$	$H_2C=CR_2 + HNCO$	$H_2C=CR_2 + HOCN$
Alkyl cyanate		
R=H	26	45
R=Me	20	30
$(CH_3)_2N—OCN$	$H_2C=NMe + HNCO$	$H_2C=NMe + HOCN$
N-Dimethylamino	9	24
cyanate		
$(CH_3)_2N—SCN$	$H_2C=NMe + HNCS$	$H_2C=CMe_2 + HSCN$
N-Dimethylamino	33	51
thiocyanate		
$CH_2S—SCN$	$H_2C=S + HNCS$	$H_2C=S + HSCN$
Methylsulfenyl	31	55
thiocyanate		
$CH_3—CR_2—NCO$	$H_2C=CR_2 + HOCN$	$H_2C=CR_2 + HNCO$
Alkyl isocyanate		
R=H	56	64
R=Me	47	50
$CH_3—CR_2—NCS$	$H_2C=CR_2 + HSCN$	$H_2C=CR_2 + HNCS$
Alkyl isothiocyanate		
R=H	45	62
R=Me	36	45
$(CH_3)_2N—NCO$	$H_2C=NMe + HOCN$	$H_2C=NMe + HNCO$
N-Dimethylamino	42	51
isocyanate		
$(CH_3)_2N—NCS$	$H_2C=NMe + HSCN$	$H_2C=NMe + HNCS$
N-Dimethylamino	32	46
isothiocyanate		
$CH_3O—NCO$	$H_2C=O + HOCN$	$H_2C=O + HNCO$
Methoxy isocyanate	39	54
$CH_3O—NCS$	$H_2C=O + HSCN$	$H_2C=O + HNCS$
Methoxy isothiocanate	27	48
$CH_3S—NCO$	$H_2C=S + HOCN$	$H_2C=S + HNCO$
Methylsulfenyl	49	57
isocyanate		

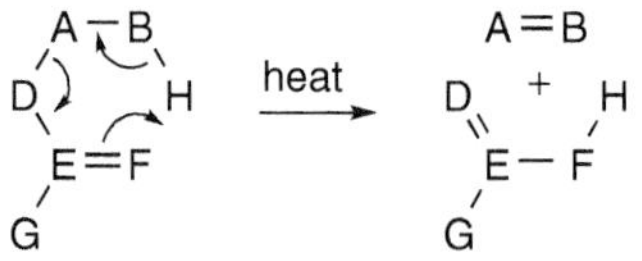

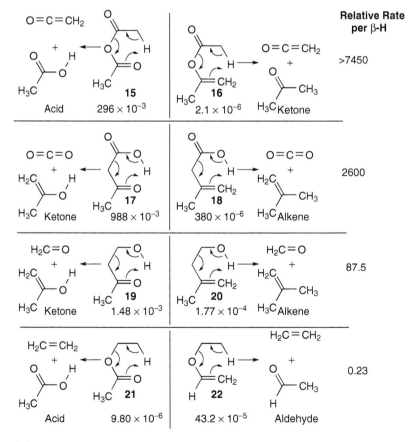

Scheme 3.9

by the increased electronegativity of D. The reactivity order is therefore acetate > thioacetate > amide.

Acetic anhydride **15** undergoes gas-phase elimination very readily [17] to give ketene and acetic acid (Scheme 3.10), whereas vinyl acetate **16** was reported [18] to pyrolyze at a much higher temperature to give ketene and acetone. This large difference in rate parallels that between

Scheme 3.10

β-ketoacids **17** [19] and β-γ-alkenoic acids **18** [20a, b] which results from changing (C=O) of **17** to (C=CH$_2$) in **18**, both eliminating CO$_2$, with the production of ketone from **17** and alkene from **18**. This is also in parallel to the difference between β-hydroxyketones **19** [21] and β-hydroxy alkenes **20** (R. Taylor, unpublished data), both eliminating formaldehyde with acetone from **19** and alkene from **20**, but in contrast with the closely similar reactivities of acetates **21** [22] and vinyl ethers **22** [23].

The protophilic attack by C=O in acetic anhydride **15** proceeds at a much higher rate than C=CH$_2$ of the vinyl acetate **16**; this is expected because of the greater nucleophilic character of C=O. The reactivity is explained [24a] by the fact that in **21** and **22**, the Cα—O bond is polar, and breaking this bond is the principal rate-determining step of the reaction. On the other hand, breaking the β—C—H bond is relatively unimportant, and the difference in basicity of (C=O) of **21** and (C=CH$_2$) of **22** does not significantly affect the rate.

From the rate of acids **17** and **18**, hydroxyketones **19** and hydroxyalkenes **20**, it is evident that the relatively high polarity of the O—H bond makes its breakage the rate-determining step that will enhance the rate of nucleophilic attack of the C=O.

Extensive work on the gas-phase pyrolysis of primary, secondary, and tertiary β-hydroxy alkenes, β-hydroxy esters, and β-hydroxy ketones was carried out by Taylor and his group [24b] to explain the dependence of the transition state structure for six-center eliminations upon the nature of the compound.

It is noteworthy that the low reactivity of the vinyl acetate **16** was compared with its aromatic analogue phenyl acetate, where the C=C is delocalized into the aromatic ring. Heating phenyl acetate **23** at 327 °C showed no significant elimination. 2-Pyridyl acetate **24** with more basic attacking group showed low reactivity [25] (Scheme 3.11).

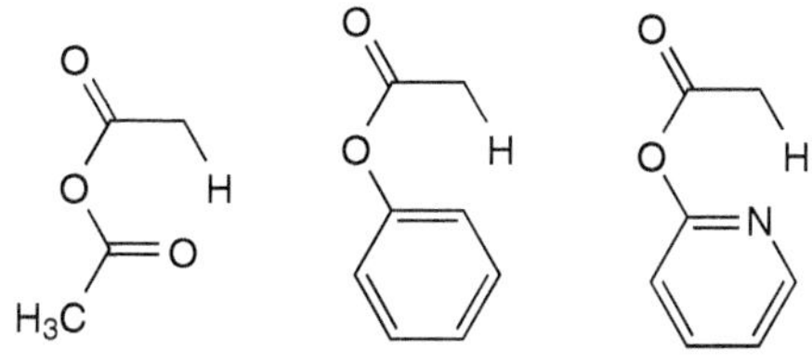

| 16 | 23 | 24 |

Scheme 3.11

The C=O in esters **21** was replaced by the 1,2-aromatic π-bond of pyridine **25**, and the C=C bond of hydroxy-alkenes **20** was replaced by C=N, giving compounds such as 2-(2-hydroxy ethyl) pyridine **26**, an isomer of 2-ethoxy pyridine, whereby groups B and D were interchanged (Scheme 3.12), and their kinetic behavior was investigated.

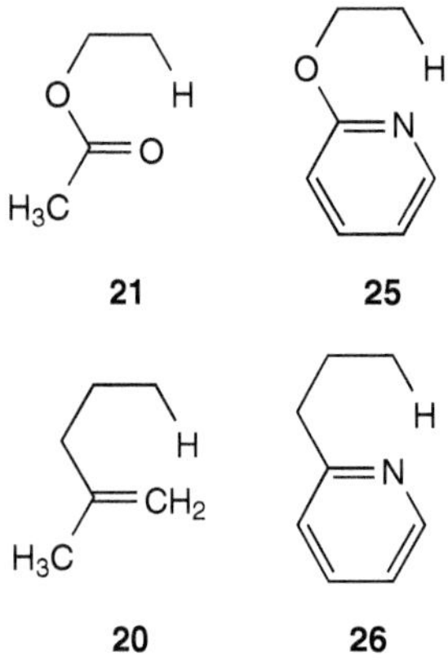

Scheme 3.12

Interchanging B and D produced the rate change shown in Scheme 3.13. The transition state of the reaction involves partially breaking E—D and B—H bonds [26]. Moving the electronegative oxygen from D to B can substantially compensate for the loss of E—D polarity by an increase in the B—H polarity.

When oxygen is at D, breaking the E—D bond is the rate-determining step; and when it is at B, breaking the B—H bond and hence nucleophilic attack upon it is most important. The net result of the interchange of B and D will then largely depend on the nucleophilicity of the attacking group F.

3.1.5 Amides

3.1.5.1 Acetamides and Thioacetamides

Investigation of the gas-phase pyrolysis reactions of amide and diamide systems has shown that they involve a six-membered transition state, as described in Scheme 3.14.

The effect of changing X from O (ester) to NH (amide) has been assessed for the simple t-butyl system [27]. The first-order rate coefficients at 327 °C of 3.25×10^{-2} s^{-1} for t-butylacetate, and 4.73×10^{-7} s^{-1}

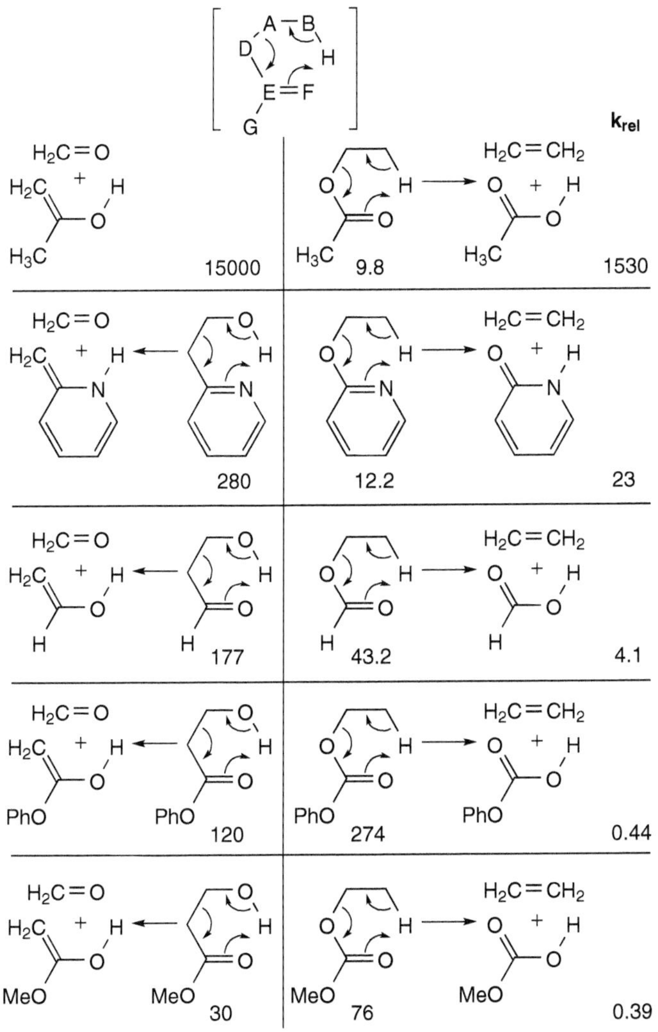

Scheme 3.13

for N-*t*-butylacetamide, yield a relative rate of 68 710 in favor of *t*-butylacetate. This relatively large rate is considered to be a result of the special importance of Cα—O and Cα—NH bond-breaking in the rate-determining step of the elimination pathway [28].

Ketene and thioacetamide were detected as pyrolysis products from N-acetylthioacetamide **27** and thioacetamide from N-*t*-butylthioacetamide **28**.

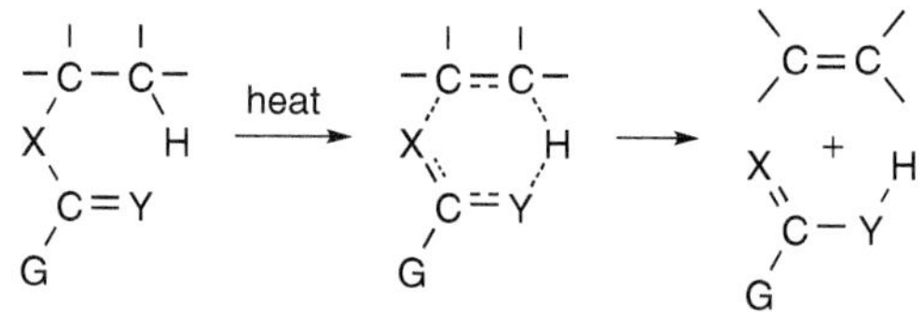

	X	Y	G
Esters	O	O	Alkyl/Phenyl
Acetamide	NH	O	CH_3
Benzamide	NH	O	Ph
Thioacetamide	NH	S	CH_3

Scheme 3.14

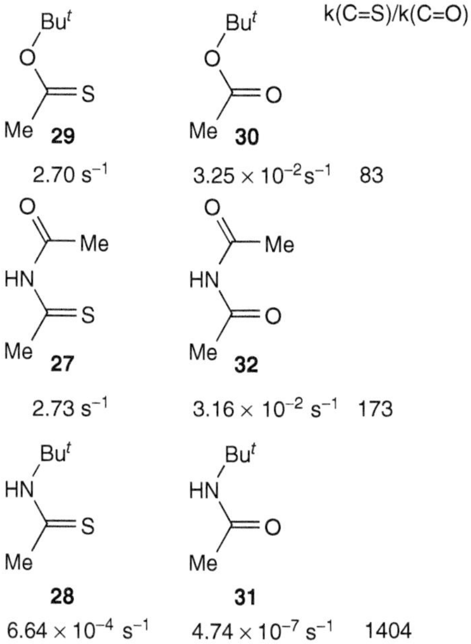

Scheme 3.15

The kinetic data presented in Scheme 3.15 show that all the sulfur-containing compounds are more reactive than their oxygen counterparts; this is justified by the relative thermodynamic stability and the π-bond energy differences of the C=S and C=O bonds [28, 29a].

t-Butyl thioacetate **29** is only 83 times more reactive than *t*-butyl acetate **30**; N-*t*-butylthioacetamide **28** is 1404 times more reactive than N-*t*-butylacetamide **31**.

It is noteworthy that the relative reactivity of 86.4 between **27** and **32** must be doubled to allow for the fact that in the diamide **32**, there are six β-hydrogen atoms available for the elimination compared with only three in amide **27**. This rate difference might have been even larger were it not for the fact that the unshared pair of electrons on the nitrogen atom is delocalized onto two carbonyl oxygen atoms in **32**, whereas in **27**, the electrons will be delocalized preferentially onto the carbonyl oxygen rather than the thione sulfur, and consequently the reactivity of **27** moiety is not enhanced to the expected level.

Scheme 3.16 shows that each of the O-containing compounds is more reactive than its NH-containing analogue due to the greater ease of breaking the more polar C_α—X bond.

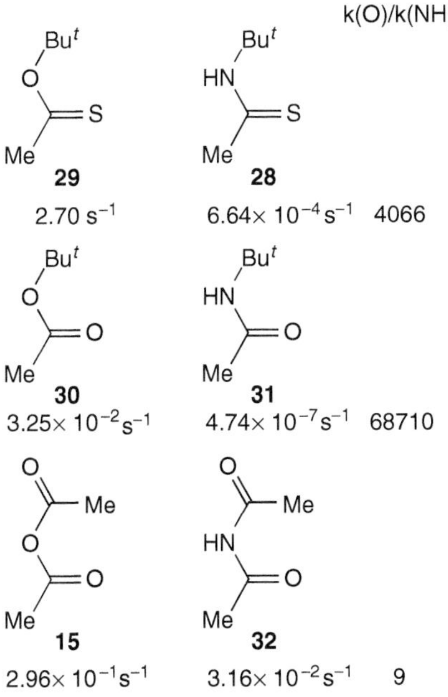

$$k(O)/k(NH)$$

29	**28**	
$2.70\,s^{-1}$	$6.64\times 10^{-4}\,s^{-1}$	4066
30	**31**	
$3.25\times 10^{-2}\,s^{-1}$	$4.74\times 10^{-7}\,s^{-1}$	68710
15	**32**	
$2.96\times 10^{-1}\,s^{-1}$	$3.16\times 10^{-2}\,s^{-1}$	9

Scheme 3.16

The relative rate ratio of 68710 between *t*-butyl acetate **30** and N-*t*-butylacetamide **31** accounts for the reactivity difference between the C—O and C—N bonds. This reactivity difference is reduced to

4066 between the corresponding S-containing compounds **28** and **29**, because of the greater reactivity of the C=S moiety.

The large drop in the reactivity difference between acetic anhydride **15** and diacetamide **32** is because of the resonance between the lone pair on X and the α-carbonyl group, which will increase the C_α—X bond order, thus rendering C—X bond-breaking more difficult (Scheme 3.17). The C=Y moiety attack on the β-hydrogen, therefore, is the more important contributing factor in the rate. A small rate difference between **31** and **14** is, therefore, observed.

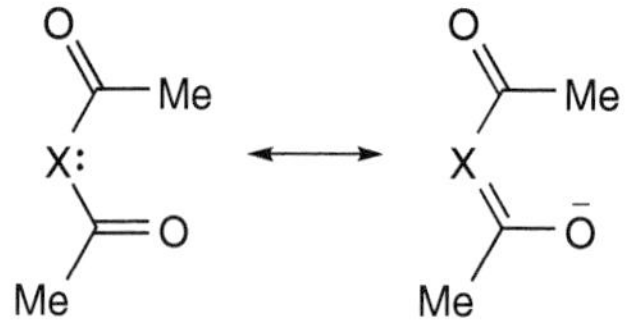

Scheme 3.17

3.1.5.2 Benzamides
The kinetic consequences of changing G from a methyl to a phenyl group are recorded in Scheme 3.18 [29b].

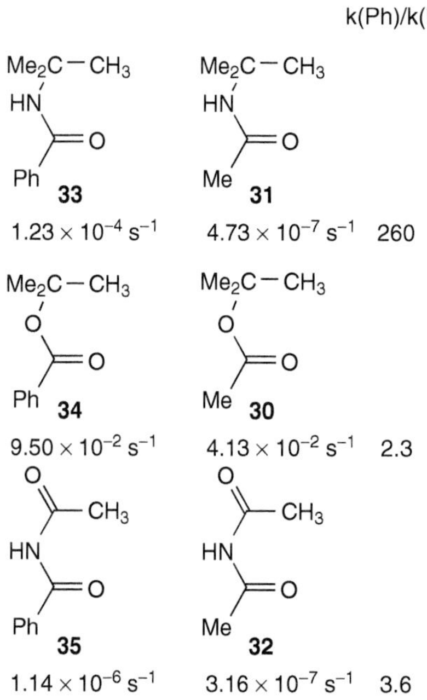

Scheme 3.18

Each of the phenyl-containing compounds is more reactive than its methyl-containing analogue due to the greater electron-withdrawing ability of the phenyl group, which helps in breaking the C_α—X bond.

A greater reactivity difference is observed between *t*-butylbenzamide **33** and *t*-butylacetamide **31**. This relative rate ratio of 260 decreases with increasing C_α—X bond polarity; this ratio is reduced to 2.3 for the corresponding oxygen-containing compounds, **30** and **34**. This effect could be attributed to the electron-withdrawing ability of the phenyl group, which will help in breaking the less polar C_α—NH bond in benzamide in comparison with the C_α—O bond in benzoate.

The relative rate ratio between **35** and **32** is 3.6. This rate difference might have been even larger were it not for same reasons that explain the difference in reactivity between **27** and **32** in Scheme 3.15.

3.1.5.3 N-Substituted Amides

The reactivity of N-acetyl acetamide **32** is highly affected by the electronic nature of the substituent on the nitrogen atom, which will affect the resonance between the electron lone pair on N and the α-carbonyl group.

The kinetic data in Table 3.2 show that each of the N-aryl substituted diacetamides is less reactive than diacetamide itself. We attribute this observation to the resonance-stabilizing effect between the aromatic 6-π electrons and the unshared pair of electrons on the nitrogen. The 4-nitrophenyl substituent in **36** is more effective in this regard and has resulted in a larger rate reduction. A 4-methoxy substituent in **37** is expected to increase the rate, but this has not been observed [29c] (Table 3.2).

Moreover, the jump in the reactivity resulting from introducing a methylene group between the nitrogen and the aromatic ring in

Table 3.2 Rate data at 327 °C for pyrolysis of $GN(COMe)_2$.

G	$10^4\ k\ (s)^{-1}$
H (**32**)	532.9
Phenyl (**35**)	4.3
4-Methoxyphenyl (**37**)	1.7
4-Nitrophenyl (**36**)	0.7
Benzyl (**38**)	314.1

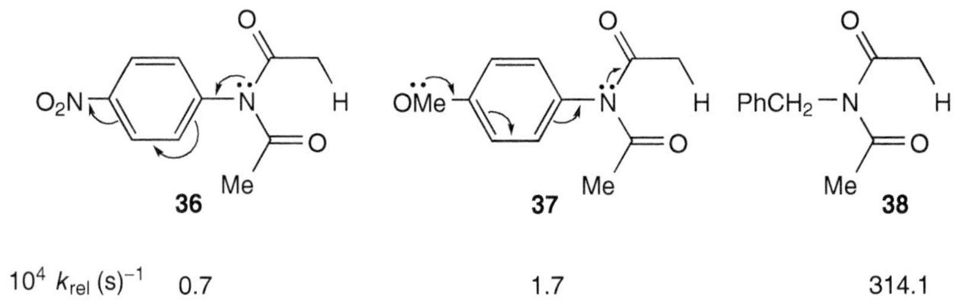

	36	37	38
$10^4\ k_{rel}\ (s)^{-1}$	0.7	1.7	314.1

Scheme 3.19

N-benzyldiacetamide **38** provides further support that the rate-suppression effect of the aryl groups in the gas-phase elimination reactions of the diacetamides is a result of resonance stabilization of the N-substituted diacetamide relative to the unsubstituted diacetamide (Scheme 3.19).

3.2 Reactive Intermediates

3.2.1 Radicals

Azo compounds R—N=N—R′ and peroxides R—O—O—R′, particularly acyl peroxides R—COO—O—R′, are well known as free radical initiators, and gas-phase pyrolysis can be used for the generation of the corresponding radicals R and R′ for spectroscopic and kinetic investigation. A "radical gun" pyrolysis apparatus for the generation of free radicals very close to the ion source of a mass spectrometer was described by Lossing [30]. Although azobenzene **39** is a very stable compound, it has been used as a source of phenyl radicals by flash vacuum pyrolysis (FVP). The radicals were isolated in an Ar matrix and characterized by IR spectroscopy [31] (Scheme 3.20). Phenyl radicals have also been generated and characterized in the pyrolysis of iodobenzene [32].

Sulfones and oxalates **40** have been used to generate free radicals under FVP conditions, especially allyl and benzyl radicals [33] (Scheme 3.21).

Dimerization of benzyl-type radicals to bibenzyls is a common observation, and Vögtle has described an efficient gram-scale synthesis of bibenzyl by pyrolysis of dibenzyl sulfone **41** [34]. When two benzylic moieties are joined in a ring **42**, the sulfone pyrolysis becomes a versatile synthesis of cyclophanes [35, 36] (Scheme 3.22).

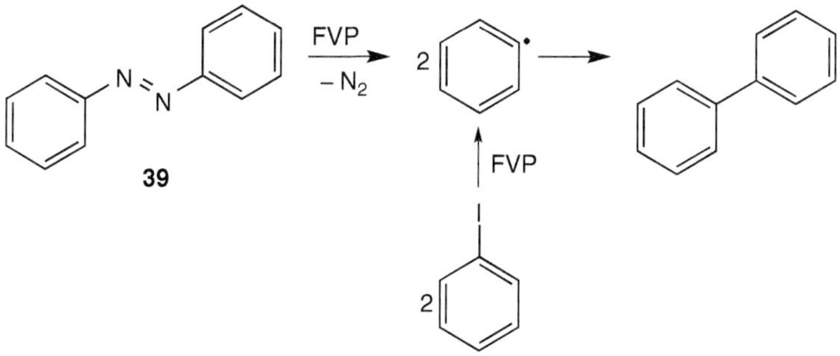

Scheme 3.20

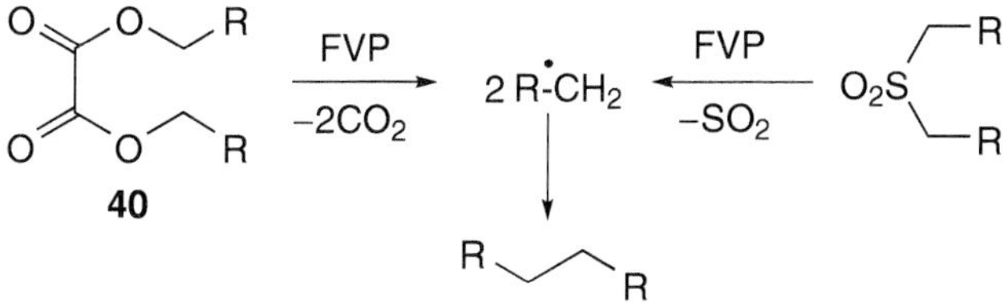

Scheme 3.21

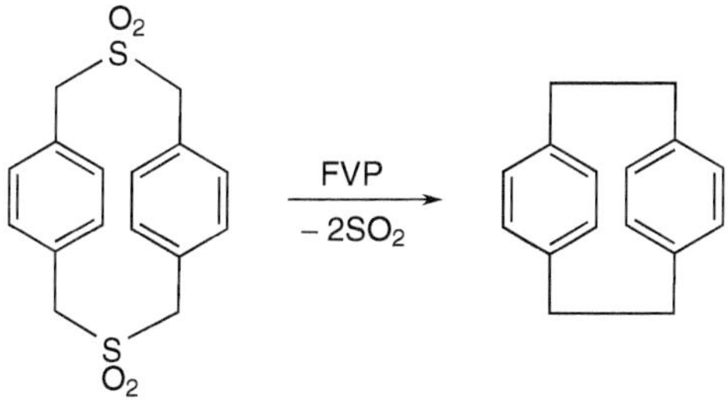

Scheme 3.22

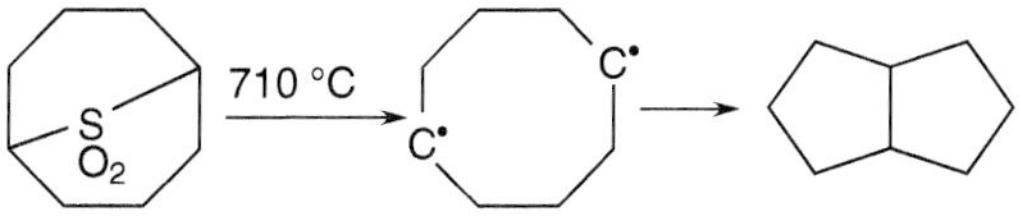

Many other *o*-, *m*-, and *p*-cyclophanes were prepared in this way [34, 35]. Thermal extrusion of SO_2 was also utilized in the synthesis of bicyclo[3.3.0]octane (Scheme 3.22) [37]. A wide variety of spirocyclizations of the neophyl rearrangement type (Scheme 3.23) under FVP conditions have been described. Two examples of spirocyclization [32] are shown in Scheme 3.24.

The pyrolysis of *N*-(*tert*-butyl)imines **43** proceeds by homolysis of a C—C bond in the *tert*-butyl group. The resulting 2-aza-allyl radicals **44** undergo rearrangements and cyclizations to yield heterocyclic compounds: for example, imidazo[1,2-c]pyridines **45** [38, 39], as indicated in Scheme 3.25.

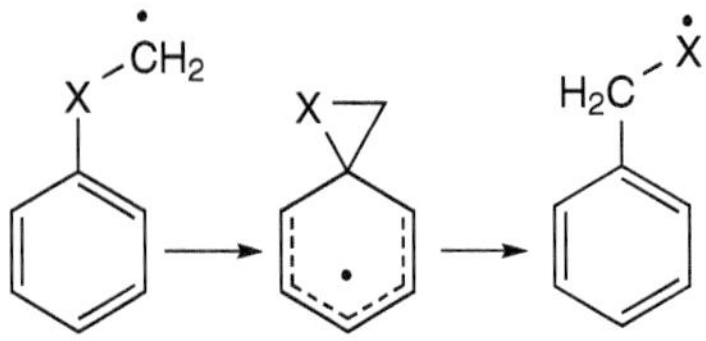

Scheme 3.23

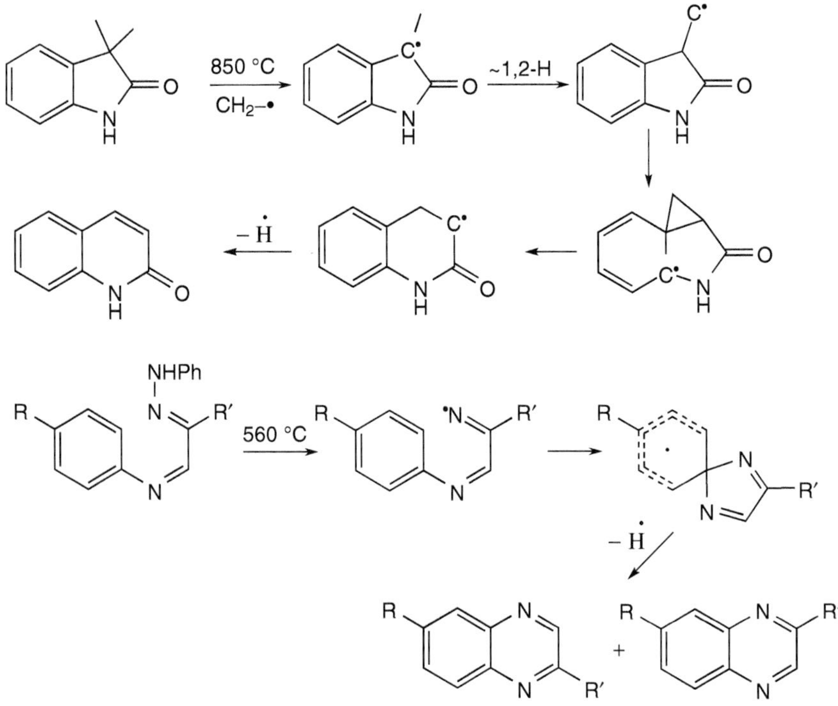

Scheme 3.24

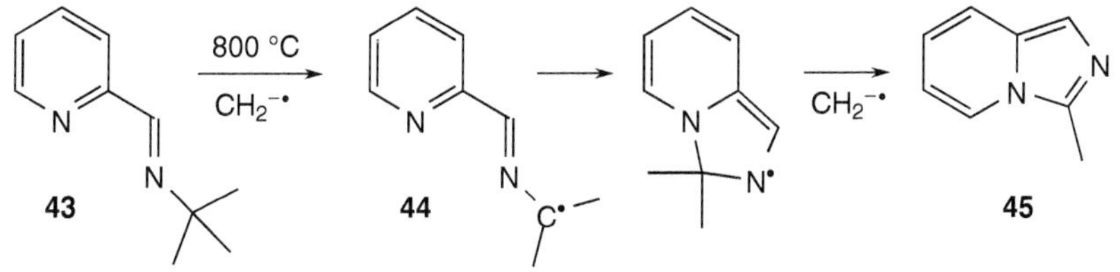

Scheme 3.25

3.2.2 Diradicals

The chemistry of diradicals has been studied in recent years with great attention devoted to the direct characterization of triplet and singlet diradicals, especially using laser flash photolysis and matrix-isolation methods coupled with electron spin resonance (ESR) and UV-Vis spectroscopies [40]. Aliphatic azo compounds undergo easy fragmentation to free radicals and molecular nitrogen with an activation free energy of about $40 \, \text{kcal mol}^{-1}$. Similarly, cyclic diazo compounds can eliminate nitrogen with the formation of a diradical. While mechanistically more important work has been performed for the reactions carried out photochemically, this process can also be performed thermally. Thus, cyclopentane-1,3-diyls **46** are formed on gas-phase thermolysis of 2,3-diazabicyclo[2.2.1]hept-2-enes **47** (Scheme 3.26), whereby the denitrogenation can be either concerted or stepwise via diazenyl diradicals **48** [40, 41] (Scheme 3.26). Depending on the substituents, these 1,3-diradicals may have either triplet or singlet ground states. Electronegative substituents such as F, OH,

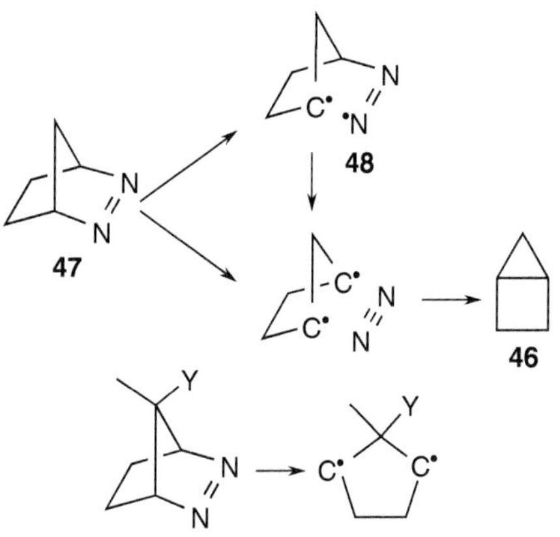

Scheme 3.26

and OR stabilize the singlet states, but so do the electropositive silyl substituents by hyperconjugation [40]. Since the singlet–triplet energy differences are usually small, intersystem crossing is rapid under pyrolysis conditions; therefore, regardless of the ground-state spin multiplicity, the 1,3-diradicals cyclize to bicyclo[2.1.0]pentanes (housanes) (Scheme 3.26). The parent housane is obtained preparatively in 90–93% yield by the pyrolysis of 2,3-diazabicyclo[2.2.1]hept-2-ene at 180–195 °C [42].

This reaction has been the subject of intense investigation [43, 44]. The reaction of the unsubstituted compound has an activation barrier of ca. 37 kcal mol^{-1} for the formation of the cyclopentane-1,3-diyl **46** (Scheme 3.26), which has a triplet ground state. The experimental and computational studies on this and several other bicyclic azo compounds have been described in many works [40–44].

Methylenecyclobutane [45] **49** and spiropentane [46] undergo thermal ring opening to a 1,3-diradical, which can rearrange over a small barrier to an allylic diradical prior to fragmentation to allene and ethene [47] (Scheme 3.27). This process has an experimental activation barrier of ~63 kcal mol^{-1} in a static system at 430–470 °C in the case of methylenecyclobutane. The ring opening/reclosure of spiropentane was revealed by deuterium labeling and has an activation energy of 51.5 kcal mol^{-1} as determined in the static system in the temperature range 304–336 °C; and for the rearrangement and decomposition to allene and ethene, it is ca. 4 kcal mol^{-1} higher [48].

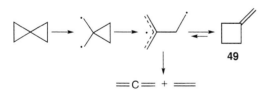

Scheme 3.27

Benzene-1,3-diyl and benzene-1,4-diyl will be dealt with in the following section on benzynes.

3.2.3 Benzynes

Several detailed reviews of benzyne chemistry are available [49], and much of this work will not be repeated here. In particular, a vast amount of research on the photochemical generation of benzyne falls outside the scope of this book. The emphasis will be on methods for the generation of benzynes in pyrolysis reactions.

3.2.3.1 *o*-Benzynes

This compound is usually described as simply benzyne, and this practice will also be used here. Benzynes are enjoying ever-increasing importance in preparative organic chemistry [50]. Benzyne is now recognized to be a ground state singlet molecule, and the currently accepted singlet-triplet splitting is ca. $38\,kcal\,mol^{-1}$ [49d]. *o*-Benzyne has been characterized by infrared spectroscopy in noble gas matrices [51], as well as gas-phase microwave [52], photoelectron and UV spectroscopies [53], and mass spectrometry [54]. The spectroscopic and computational [55, 56] evidence indicates that singlet ground-state *o*-benzyne is best described as a resonance hybrid of the aryne **50** and the cumulene **51**, being closer to the triple-bond structure **50** (Scheme 3.28).

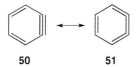

50   **51**

Scheme 3.28

The first evidence for the generation of benzyne in FVP reactions was the isolation of biphenylene **52** from the pyrolysis of iodophenylmercuric iodide **53** and bis(2-iodophenyl)mercury **54** [57–59]. Diphenyliodonium carboxylate **55** underwent FVP at 325 °C, yielding biphenylene as a major product [60] (Scheme 3.29). Catalytic GPPy has been used in the pyrolysis of *o*-diiodobenzene **56** over Zn at 500–550 °C ($2\,hPa\,N_2$ as carrier gas), which produced biphenylene (21%) and triphenylene **57** (10%) [61]. Copyrolysis with anthracene produced the Diels–Alder trapping products triptycene **58** (8%) and biphenylene **52** (10%).

The formation of benzyne **50** by FVP of several carbonyl compounds was reported during 1965–1966: the gas-phase pyrolysis of indanetrione **59** (FVP over quartz tubing at 600 °C/0.2 hPa) yielded up to 40% of **60** [62], the solution-phase pyrolysis of phthalic anhydride **61** in benzene produced 1% biphenylene and ~1% triphenylene [63], and flow pyrolysis of phthalic anhydride (N_2 carrier gas, 50 hPa, ~800 °C) yielded 10–15% biphenylene [61, 64] (Scheme 3.30). The detailed mechanism of

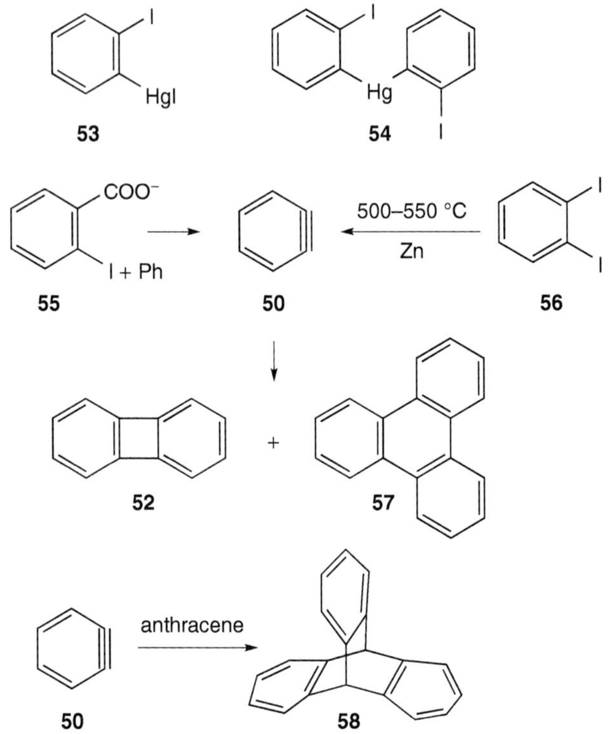

53   **54**

55   **50** $\xleftarrow[\text{Zn}]{\text{500–550 °C}}$ **56**

52   +   **57**

50 $\xrightarrow{\text{anthracene}}$ **58**

Scheme 3.29

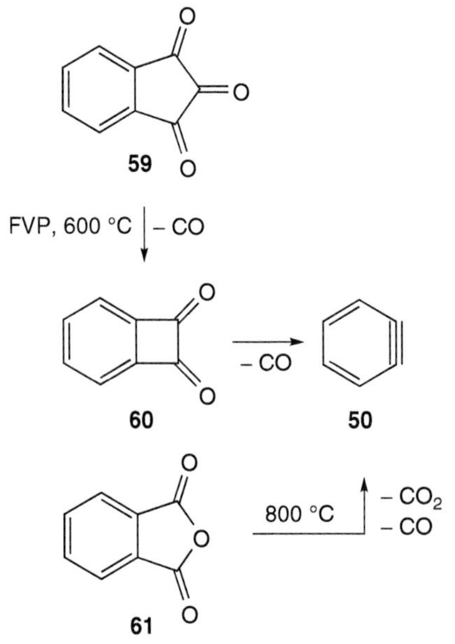

59

FVP, 600 °C | – CO

60 $\xrightarrow{\text{– CO}}$ **50**

61 $\xrightarrow[\text{800 °C}]{}$ $\begin{array}{l}\text{– CO}_2\\\text{– CO}\end{array}$

Scheme 3.30

the thermal formation of benzyne from phthalic anhydride is still being investigated [55, 65, 66].

3.2.3.2 *m*- and *p*-Benzynes

The *m*- and *p*-benzynes **62** and **63** possess diradical properties [67, 68], and for this reason multireference computational methods are required for a proper description [69, 70]. There is computational evidence for significant 1,3-bonding in *m*-benzyne [71] (Scheme 3.31).

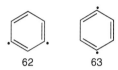

  62   63

Scheme 3.31

The Bergman-cyclization [72] of 3-hexene-1,5-diynes (e.g. **64**) to *p*-benzynes (e.g. **63**) (Scheme 3.32) takes place on heating in a static system at 200 °C for 5 min. The *p*-benzyne cannot be isolated under these conditions, as it lies ca. 14 kcal mol^{-1} higher in energy, but its formation as a transient intermediate was demonstrated by complete scrambling of deuterium between the sp^2 and sp. carbon atoms (**65, 66, 67**). The relatively low activation enthalpy of ca. 28 kcal mol^{-1} means that many such reactions can be performed in solution.

The reaction is biologically important because several natural products with powerful antitumor and/or antibiotic properties such as the calicheamicins, esperamicins, lidamycin, and dynemicins (e.g. **68**, Scheme 3.32) contain such enediyne moieties, and their biological activity can be ascribed to the formation of transient *p*-benzynes. Hydrogen abstraction by a *p*-benzyne **69** generates a more reactive phenyl radical **70**, which abstracts a second H atom from DNA, resulting in **71** [73, 74].

3.2.4 Carbenes

Gas-phase pyrolysis of diazomethane **72** (CH_2N_2) yields methylene, CH_2, and nitrogen [75] (Scheme 3.33). Methylene has a triplet ground state, but the singlet-triplet splitting is small, of the order of 8 kcal mol^{-1}. Addition and insertion reactions of methylene and many other carbenes have been reviewed [76]. Most diazo compounds decompose to carbenes and nitrogen with activation energies in the range 30–40 kcal mol^{-1}.

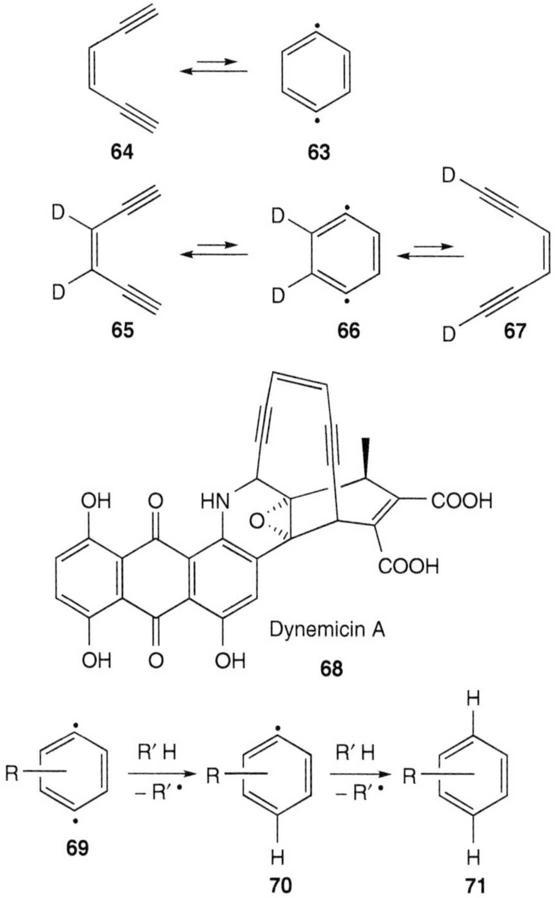

64   **63**

65   **66**   **67**

Dynemicin A
68

69   **70**   **71**

Scheme 3.32

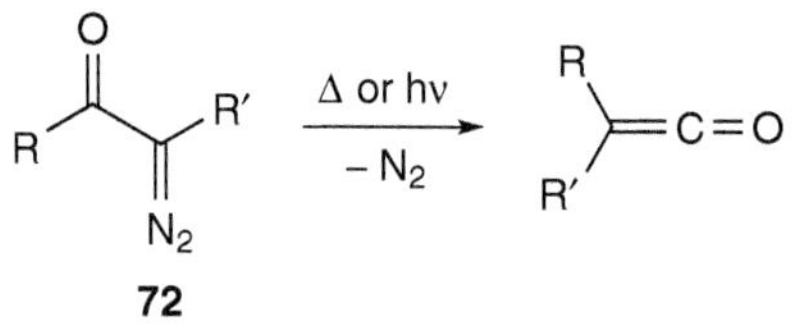

72

Scheme 3.33

The Wolff rearrangement of α-diazocarbonyl compounds to ketenes (Scheme 3.33) is often carried out in solution, either thermally or photochemically. The FVP method offers advantages [77], although under these conditions the α–oxocarbenes usually undergo preferred C—H insertion [78], for example, as shown in Scheme 3.34 [79].

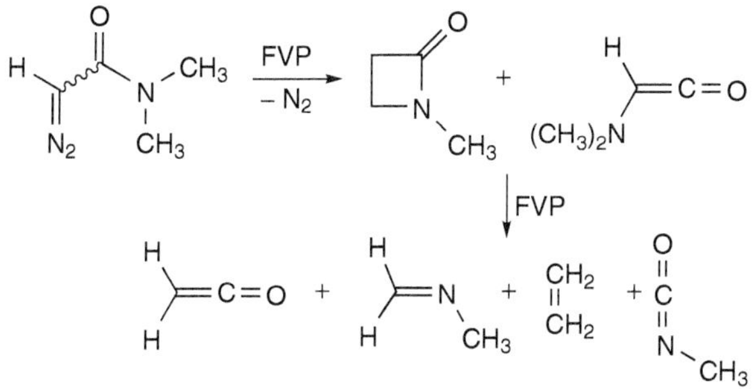

Scheme 3.34

Both phenylcarbene **73a** and phenylnitrene **73b** undergo ring contraction under FVP conditions: the carbene to fulvenallene **74a**, and the nitrene to cyanocyclopentadiene **74b** [80] (Scheme 3.35). The nitrene will be dealt with in the next section.

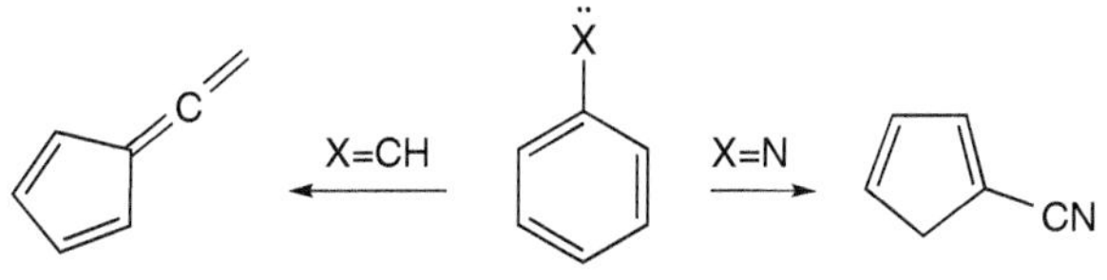

74a   **73**   **74b**

a: X =CH
b: X = N

Scheme 3.35

Thus, fulvenallene **74a** is generated by FVP of phenyldiazomethane **75**. It is an unstable compound, which is conveniently trapped with dimethylamine to yield the stable 6-dimethylamino-6-methylfulvene **76** [80, 81] (Scheme 3.36).

^{13}C-Labeling studies have revealed that the ring contraction is not straightforward; it takes place via ring expansion–ring contraction [80], as illustrated in Scheme 3.37.

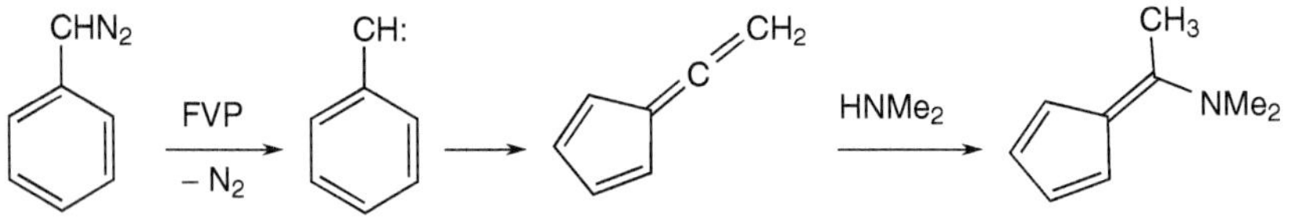

Scheme 3.36

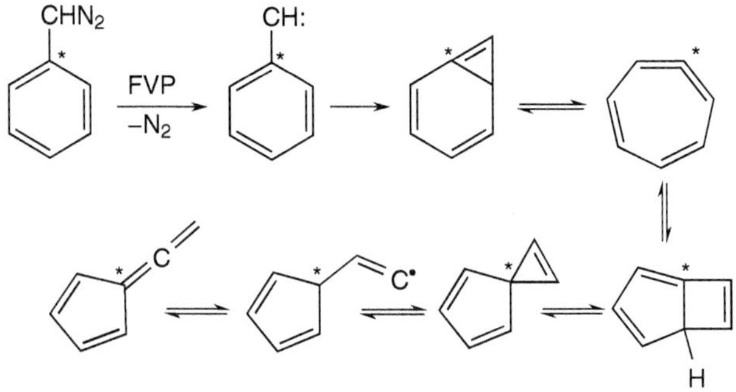

Scheme 3.37

The ring contraction has an overall calculated activation barrier of $35\,\text{kcal}\,\text{mol}^{-1}$ [80]. The non-observed direct ring contraction of **73a** to 5-ethynylcyclopentadiene has a much higher calculated activation barrier ($54\,\text{kcal}\,\text{mol}^{-1}$ relative to phenylcarbene) [80].

For preparative purposes, fulvenallene can be obtained more conveniently and in good yield by FVP of benzocyclopropene **77** at 800 °C (Scheme 3.38). This reaction is assumed to proceed by ring opening to the carbene or biradical **78** followed by a Wolff-type ring contraction to **74** [80, 82]. The same is achieved by FVP of phthalide **79** [82, 83] (Scheme 3.38).

Reversibility of the phenylcarbene ring expansion has the consequence that the tolylcarbenes **80** interconvert under FVP conditions, with the ultimate formation of styrene **81** and benzocyclobutene **82** [84–86] (Scheme 3.39). The rearrangement mechanism has been supported by carbon-labeling studies.

The carbene or diradical **83** formed by FVP of benzofulvenyl-8-diazomethane **84** cyclizes to spirocyclopropene **85**, which is isolable from the reaction at 400 °C. At higher temperatures, either **86** or **87** is formed [80] (Scheme 3.40).

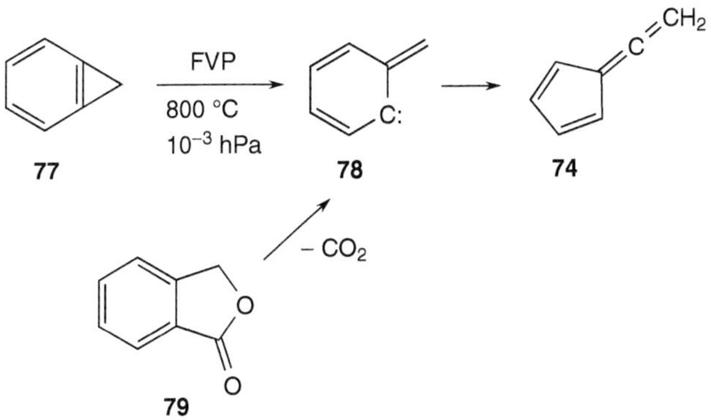

Scheme 3.38

The 1- and 2-naphthylcarbenes **88** and **89** interconvert via the ring-expanded species **90**, and both undergo final cyclization to cyclobutanaphthalene **91** on FVP (Scheme 3.41). Indeed, this is the easiest way to prepare **91** [87].

While the naphthyldiazomethanes can be used for this purpose, their low volatility makes the falling-solid flash vacuum pyrolysis (FS-FVP) method a convenient alternative [88]. Either naphthaldehyde tosylhydrazone salts or 5-naphthyltetrazoles **92** and **93** can be employed in this way (Scheme 3.42).

The pyrolysis of 5-aryltetrazoles such as **92, 93** is a convenient general means of generating arylcarbenes in the gas phase (Scheme 3.42). At higher FVP temperatures, **88** and **89** enter the same energy surface as **83** (Scheme 3.40) with the formation of **86** and **87**. The gas-phase rearrangements of the phenanthrylcarbenes have also been described [89].

The 1- and 2-azulenylcarbenes have been generated in the same way, e.g. 1-azulenylcarbene **90** by FVP of **91** or **92**, as shown in Scheme 3.43. Interestingly, compounds **86, 87**, and **91** are formed by rearrangement of the carbene [90].

Vinylidenes are carbenes of the type $R_2C=C:$ **93**. These are of particular importance because of their interconversion with acetylenes, **94**, under high-temperature FVP conditions [91] (Scheme 3.44). The vinylidenes are often generated by FVP of isoxazolones **95** [88, 92] or Meldrum's acid derivatives **96**, in the latter case via propadienones **97** [91–93].

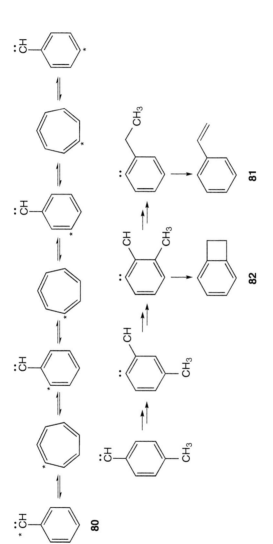

Scheme 3.39

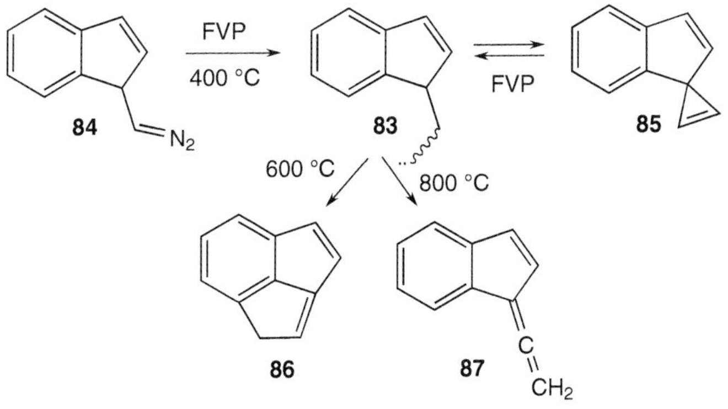

Scheme 3.40

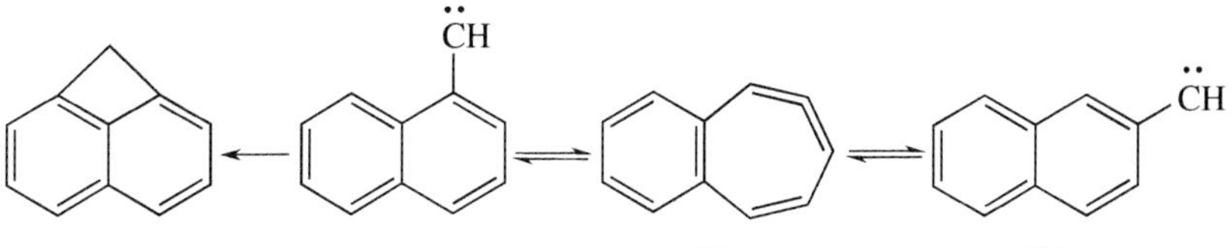

Scheme 3.41

Vinylidene intermediates **98a–98b** have been implicated in azulene-naphthalene rearrangement under low-pressure FVP conditions [94]. **99–102** are formed on FVP of the isoxazolone **103**, and **100–102** are also formed on FVP of azulene **99**. The mechanism was supported by ^{13}C-labeling [95] (Scheme 3.45).

Vinylidenes also feature prominently as intermediates in mechanisms of formation of polycyclic aromatic hydrocarbons (PAHs). For example, pyrolysis of diacetylene **104** at 1000–1100 °C produced corannulene **105** as the only isolated product [95] (Scheme 3.46).

Neilen and Wiersum reported the pyrolytic formation of cyclopent [*h,i*]acephenanthrylene **106** from triphenylene **107** at 1000 °C/0.05 hPa at a flow rate of 0.5 g h^{-1} with a yield of ca. 35% [96a] (Scheme 3.47). FVP of the acetylene **108** also produced **106** [96b]. Compounds **105** and **106** can be considered fragments of C$_{60}$ (Scheme 3.47).

The acetylene–vinylidene interconversion is considered a general mode of formation of cyclopenta-fused PAHs from ethynyl-PAHs [97].

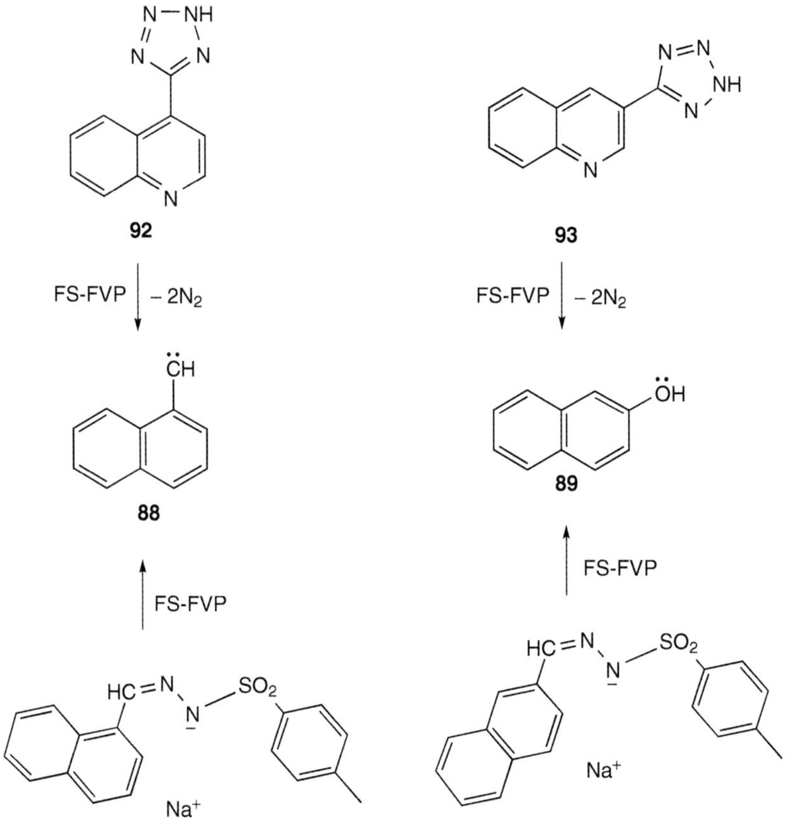

Scheme 3.42

3.2.5 Nitrenes

Methylnitrene, CH_3N, **109** is generated by pyrolysis of methyl azide, CH_3N_3, **110**, and the singlet state of the nitrene has been observed by photoelectron spectroscopy [98]. The nitrene has a triplet ground state, and the singlet triplet splittings for nitrenes are considerably higher than for carbenes, ca. 31 kcal mol^{-1} for CH_3N and of the order of 15–18 kcal mol^{-1} for arylnitrenes. The activation energy for methyl azide decomposition is ca. 38 kcal mol^{-1} [99], and generally for azides the barrier is 35–40 kcal mol^{-1} [100]. The detailed mechanism of pyrolysis of alkyl azides has been investigated computationally [101]. Methanimine $CH_2=NH$ **111** is the primary product of rearrangement of singlet methylnitrene, but due to chemical activation it readily decomposes to HCN/HNC and H_2 [102] (Scheme 3.48).

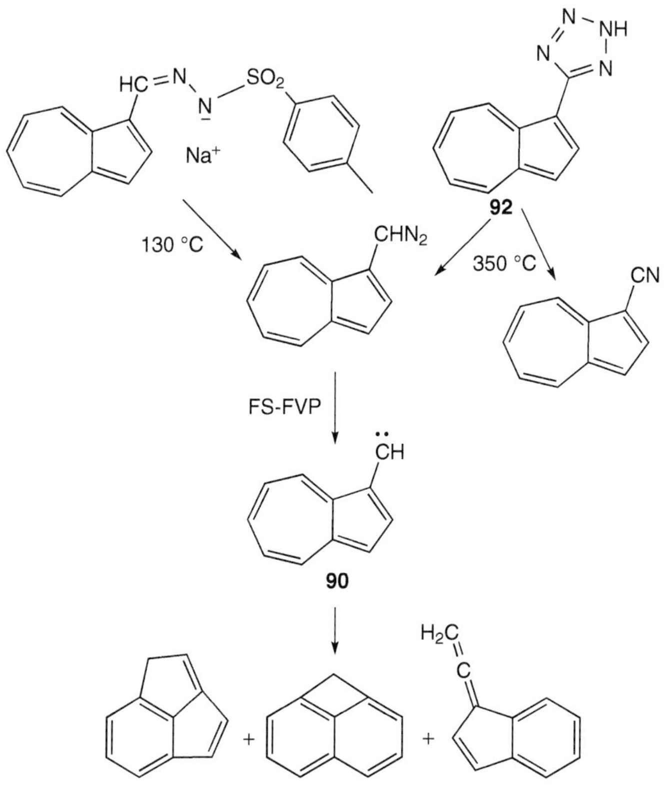

Scheme 3.43

Imine formation is also observed for other alkyl azides, including benzyl azide [103], but C—H insertion and H abstraction also take place [104].

The rearrangements of aryl- and heteroarylnitrenes and the interconversion of arylnitrenes and heteroarylcarbenes under FVP conditions have been studied in detail, and reviews from the Wentrup group are available [105, 106]. FVP of aryl azides generally leads to the formation of arylnitrenes, which can be isolated in low-temperature matrices and observed directly by ESR, UV-vis, and IR spectroscopies [107, 108]. Phenylnitrene **112** undergoes two major types of intramolecular, thermal reactions in the open-shell singlet state: (i) ring expansion to 1-azacyclohepta-1,2,4,6-tetraene **113** [109], which leads to an interconversion with 2-pyridylcarbene **114**; and (ii) ring contraction to

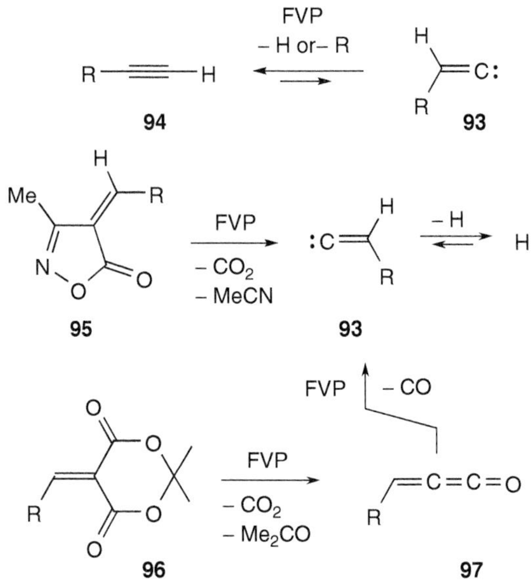

Scheme 3.44

cyanocyclopentadiene **115** [80]. Because the ground state is the triplet, inevitably some degree of triplet reactivity is also observed (route iii), leading to the formation of azobenzene **116** by dimerization and aniline **117** by H-abstraction [80, 105, 106] (Scheme 3.49).

Because of the thermal interconversion of phenylnitrene and 2-pyridylcarbene, a proper description of the chemistry requires consideration of both species. The formation of 2-pyridylcarbene by pyrolysis of 1,2,3-triazolo[1,5-*a*]pyridine **118A** and 2-(5-tetrazolyl)pyridine **119** is described in Scheme 3.50. Both routes are via 2-diazomethylpyridine **118B** [110].

[13]C-Labeling studies demonstrated that the cyanocyclopentadiene **115** formed by high-temperature FVP of 1-^{13}C-phenyl azide was labeled exclusively on the CN carbon, and the aniline **117** also formed was labeled exclusively at C1 [80]. Therefore, phenylnitrene undergoes direct ring contraction (Scheme 3.51).

Cyanocyclopentadiene **115** is also obtained by pyrolysis of either **118** or **119**. Here, ^{13}C-labeling using 2-pyridyl-(^{13}C-carbene) revealed ^{13}C labeling on all carbon atoms, including the CN group, in the product cyanocyclopentadiene **115**. The conclusion from these and related experiments is that there are two routes of ring contraction in 2-pyridylcarbene **114**. The major route leads to phenylnitrene **120** via

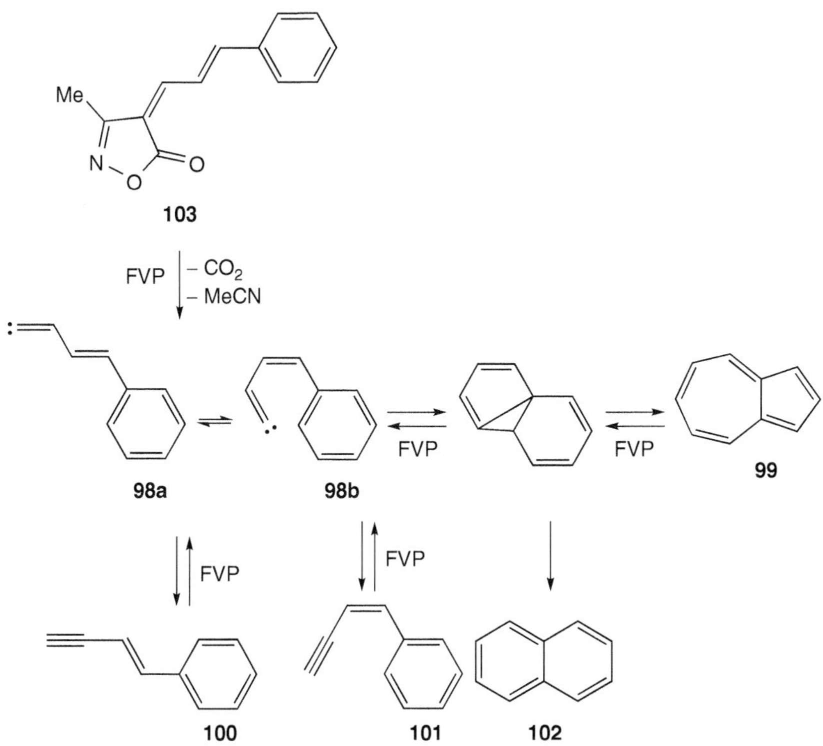

Scheme 3.45

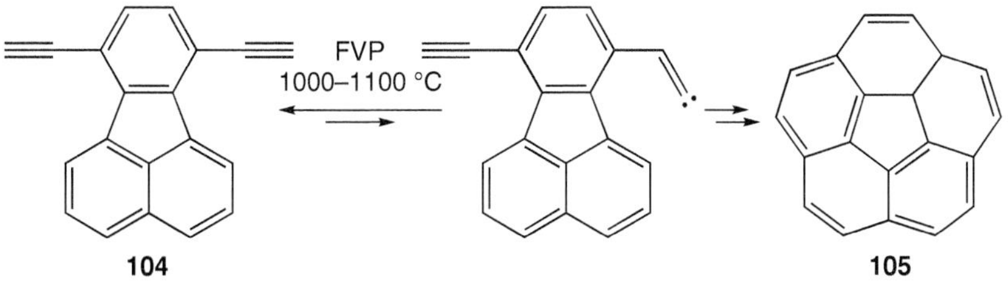

Scheme 3.46

the azacycloheptatetraene **113**; in a second route, carbene **114** first undergoes ring expansion to 4-aza-cyclohepta-1,2,4,6-tetraene **116**, which then rearranges to cyanocyclopentadiene **115** by a mechanism analogous to that of phenylcarbene shown in (Scheme 3.51).

The ring contraction of phenylnitrene to cyanocyclopentadiene is mechanistically important, but for preparative purposes FVP of either **118A** or **119** is more convenient and gives higher yields.

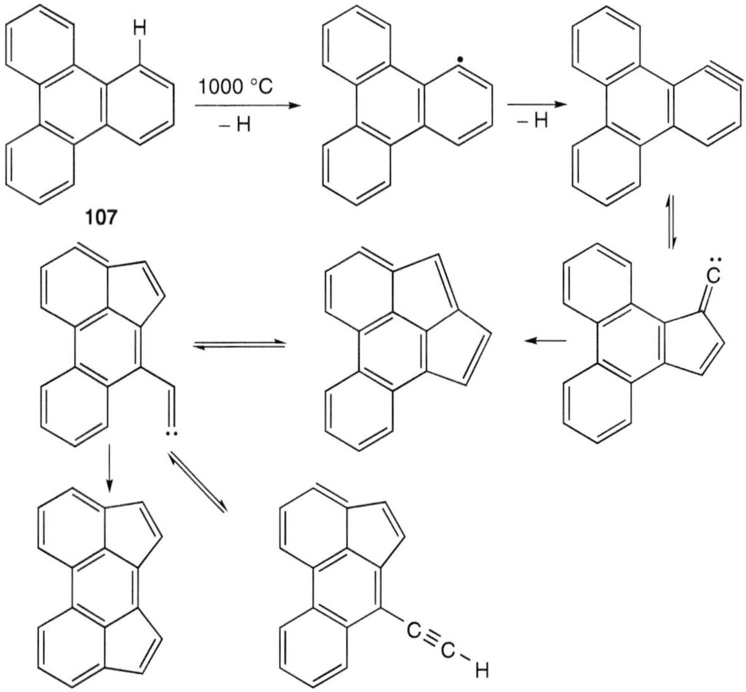
H
1000 °C
– H
– H
C
C≡C–H

107

106

108

Scheme 3.47

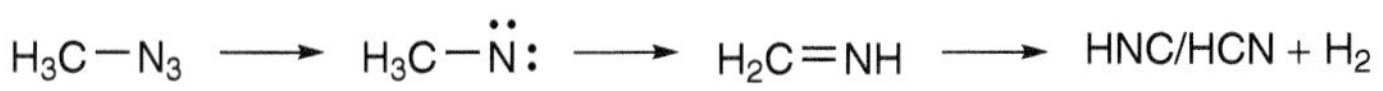
H3C–N3
H3C–N:
H2C=NH
HNC/HCN + H2

110 109 111

Scheme 3.48

Benzotriazole **121**, an isomer of phenyl azide, is an even better precursor, giving virtually a quantitative yield of cyanocyclopentadiene **115** in a Wolff-type ring contraction of the iminocarbene **122** (Scheme 3.52). This reaction allows the preparation of substituted cyanocyclopentadienes in high yields [111]. Isatins **123** are alternative precursors of cyanocyclopentadienes; here, loss of 2 CO generates the same intermediate iminocarbene **122** [85].

As a consequence of the carbene–nitrene interconversion (Scheme 3.49), the 3-aryl-1,2,3-triazolo[1,5-*a*]pyridines **124A** rearrange to carbazoles **125** in near-quantitative yields (Scheme 3.53). The same carbazoles are obtained from the biphenyl azides **126** (via nitrene **127**)

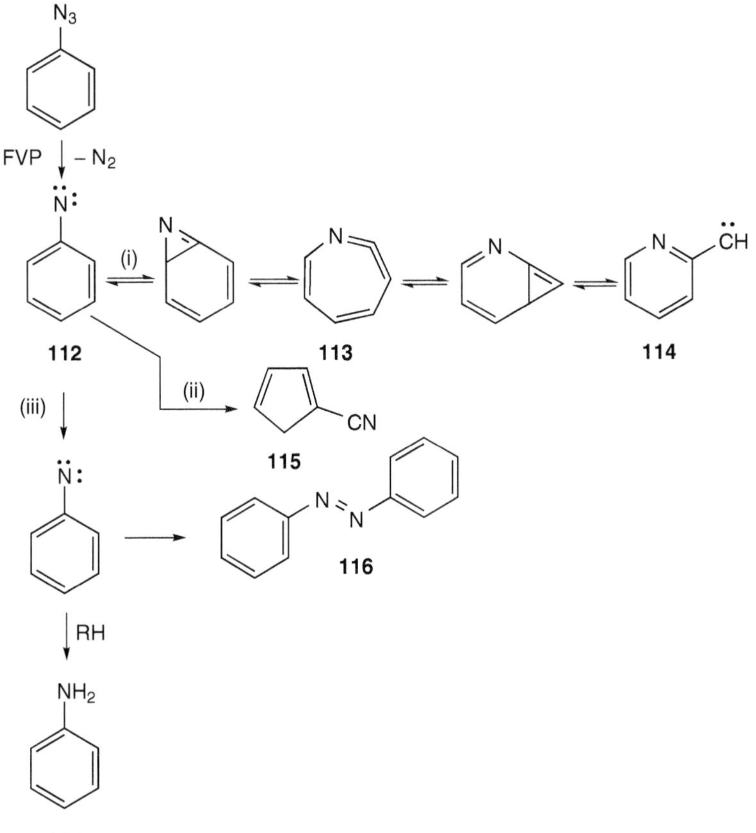

Scheme 3.49

[112] and from 1-arylbenzotriazoles **128**, in the latter case via the iminocarbene (Graebe–Ullmann reaction [113]) (Scheme 3.53).

There is a very big difference between arylcarbenes and arylnitrenes, which is clearly seen in the behaviors of the tolyl derivatives: the tolyl-carbenes interconvert (Scheme 3.39), whereas the tolylnitrenes do not. However, *o*-tolylnitrene undergoes a number of unique rearrangements [114] (Scheme 3.54). FVP at temperatures up to 500 °C leads primarily to the *o*-quinonoid imine **129**. This compound can be interconverted photochemically with benzazetine **130**, but thermally at 500 °C it was not feasible. Higher pyrolysis temperatures lead to the formation of 2-vinylpyridines **131** as a manifestation of the nitrene–carbene rearrangement **132–134**. Nitrenes are thermodynamically more stable

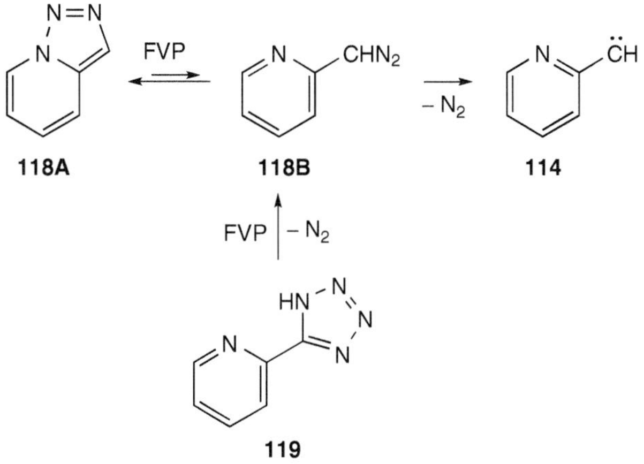

Scheme 3.50

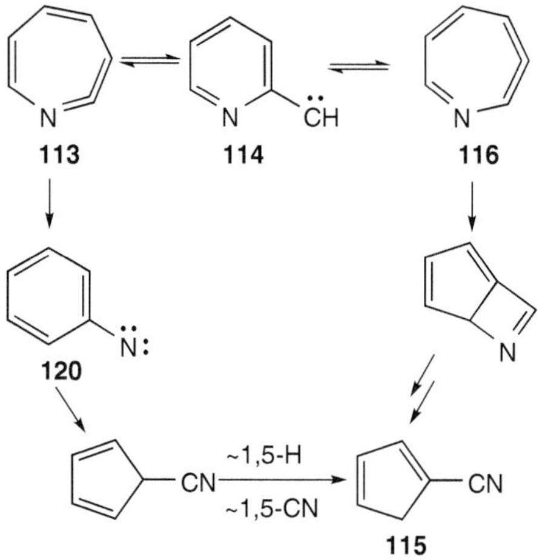

Scheme 3.51

than the isomeric carbenes [115]. Therefore, the thermal equilibrium between **132** and **134** lies on the side of the nitrene, and the yield of vinylpyridine **131** obtained is only ca. 8% [115b]. At high temperatures, a thermal equilibrium between **129** and **130** sets in, and this leads to the formation of benzylidenimine **135** in competition with **131**. The same energy surface can also be entered by pyrolysis of the diazo compound **136** at 600 °C [114a].

2-Pyridylnitrenes undergo efficient ring expansion to 1,3-diazacyclohepta-1,2,4,6-tetraenes on FVP, e.g. **137–139** (Scheme 3.55).

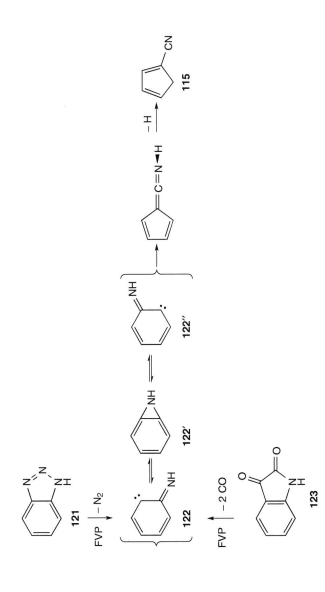

Scheme 3.52

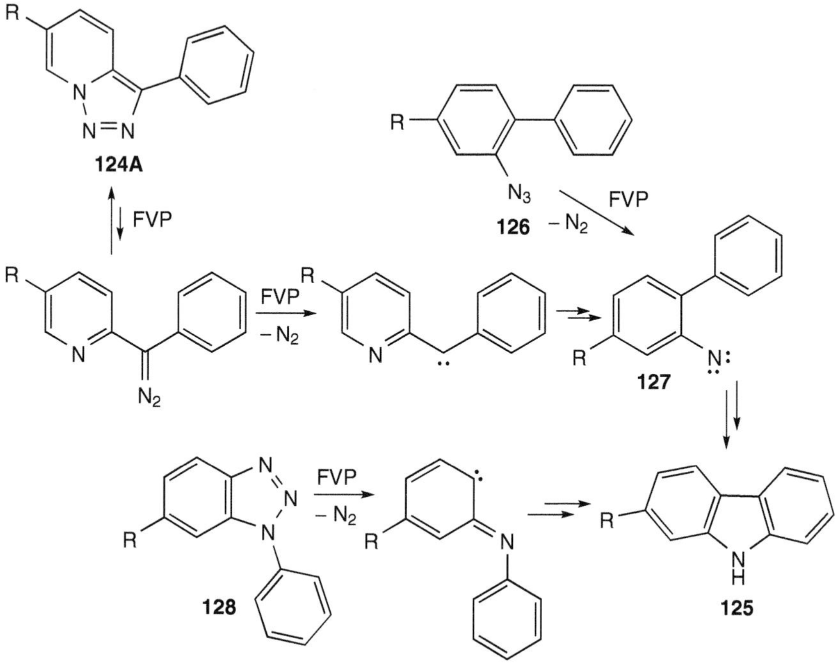

Scheme 3.53

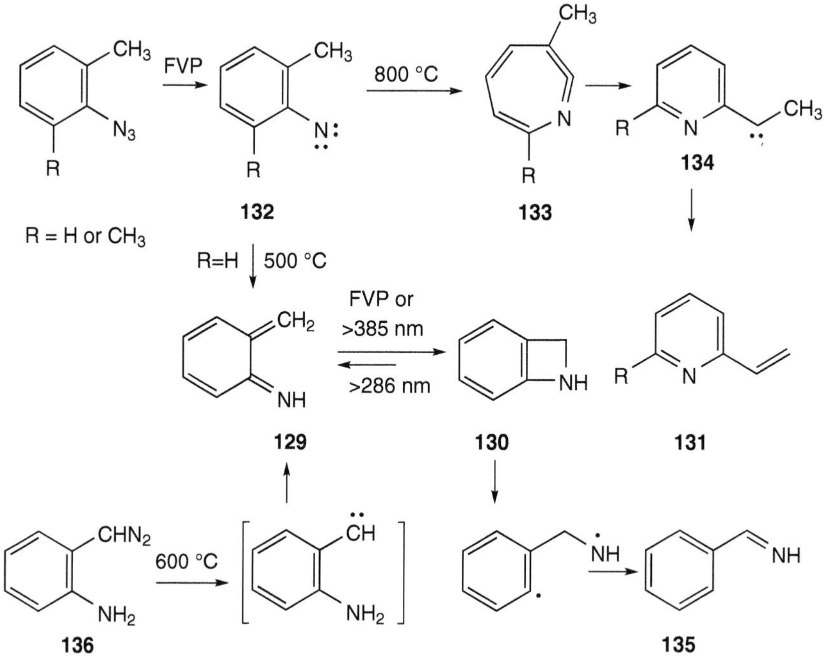

Scheme 3.54

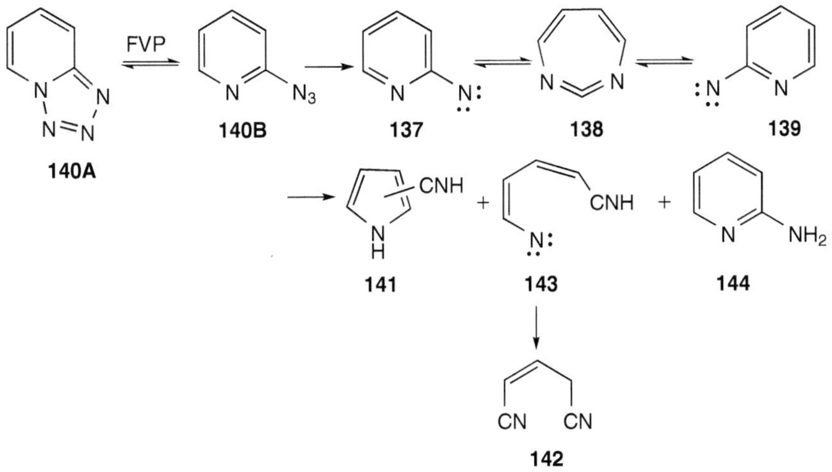

Scheme 3.55

Thus, FVP of tetrazolo[1,5-*a*]pyridine **140A** leads first to ring opening to the azide **140B**. Increased temperature produces the nitrene **137** (observable by ESR spectroscopy [107]) and the diazacycloheptatetraene **138** (observable by IR spectroscopy) [116]. The main product under preparative FVP conditions is a mixture of 2- and 3-cyanopyrroles **141**, which themselves interconvert on FVP. A small amount of glutacononitrile **142** was observed through ESR spectroscopy, and this is ascribed to ring opening to the cyanobutadienylnitrene **143**. 2-Aminopyridine **144** is usually formed in less than 10% yield due to H-abstraction by the triplet nitrene. Diazacycloheptatetraenes have been observed as products of rearrangement of several other aza-heterocyclic nitrenes. For example, the dibenzo-derivative **145** generated by FVP of tetrazolophenanthridine is stable until −40 °C, when it dimerizes (Scheme 3.56); the structure of the dimer **146** was determined by X-ray crystallography [116].

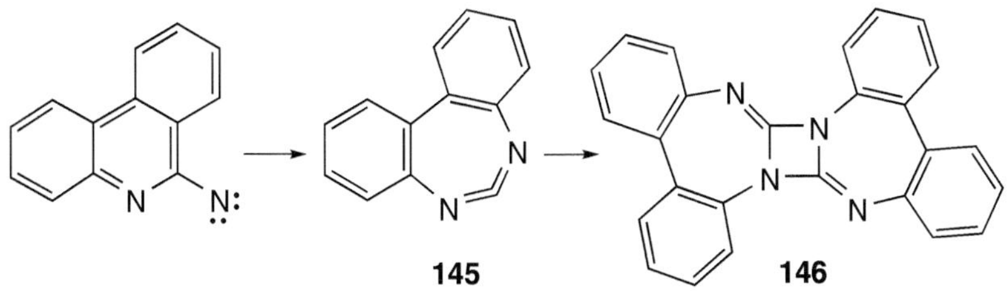

Scheme 3.56

The 2-, 3-, and 4-pyridyl azides all produce cyanopyrroles on FVP (Scheme 3.57), but the mechanisms are different. 2-Pyridyl azide **140B** was described earlier. 4-Pyridyl azide **147** is analogous to phenyl azide **112**. In 3-pyridyl azide **148**, the ring contraction to 3-cyanopyrrole takes place via a new type of ring opening to the nitrile ylide derivative **149** (X=N, Y=Z=CH) (Scheme 3.58). This ylidic ring opening takes place in several heteroarylcarbenes and nitrenes **149**, when there is a 1,3-relationship between a ring nitrogen atom and the carbene or nitrogen center [105, 106]. The ylides **150** have been observed directly by matrix-isolation IR and UV-vis spectroscopy in several cases, but under FVP conditions these cyclize to cyanopyrroles or cyanoimidazoles.

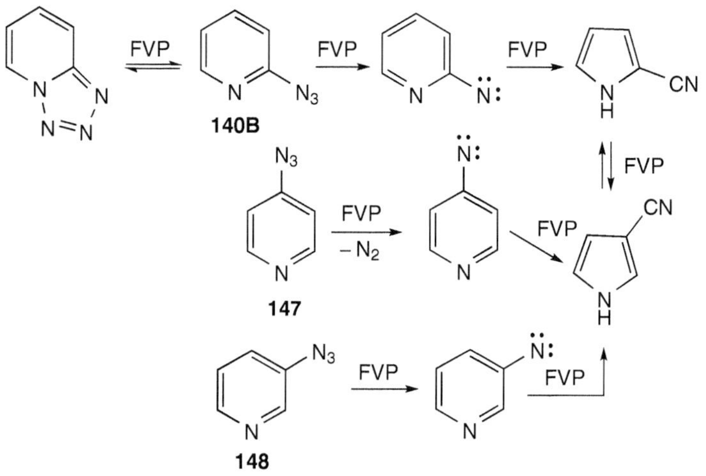

Scheme 3.57

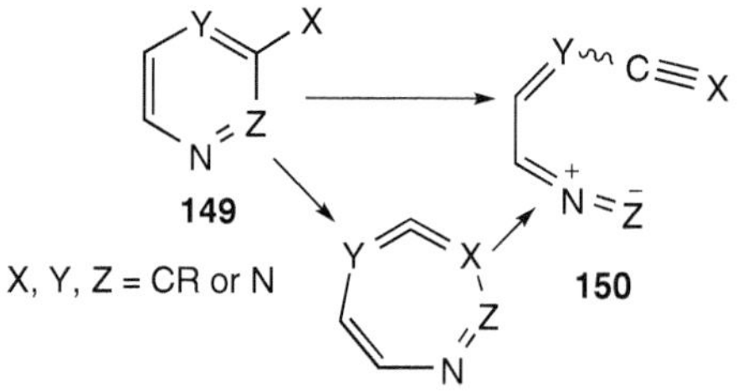

X, Y, Z = CR or N

Scheme 3.58

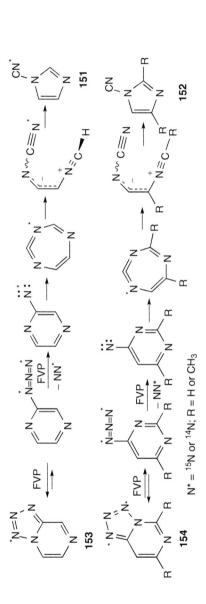

N* = ^{15}N or ^{14}N; R = H or CH$_3$

Scheme 3.59

Examples of the rearrangement to cyanoimidazoles **151** and **152** [105, 117] are shown in Scheme 3.59. The results of ^{15}N and methyl group labeling of the precursors **153** and **154** are in accord with the proposed mechanism.

References

1 Taylor, R. (1979, Chapter 15). Pyrolysis of acids and their derivatives. In: *The Chemistry of Functional Groups, Supplement B: Acid Derivatives*, vol. 2 (ed. S. Patai), 859–914. London: Wiley.

2 Inagaki, S., Fujimoto, H., and Fukui, K. (1976). *J. Am. Chem. Soc.* 98: 4693–5054.

3 Woodward, R.B. and Hoffmann, R. (1965). *J. Am. Chem. Soc.* 87: 2511–2513.

4 (a) Ripoll, J.L. and Vallee, Y. (1993). *Synthesis* 7: 659–677. (b) Paderes, G.D. and Jorgensen, W.L. (1992). *J. Org. Chem.* 57: 1904–1916. (c) Dubac, J. and Laporterie, A. (1987). *Chem. Rev.* 87: 319–334. (d) Oppolzer, W. and Snieckus, V. (1978). *Angew. Chem. Int. Ed. Engl.* 17: 476–486. (e) Hoffmann, H.M.R. (1969). *Angew. Chem. Int. Ed. Engl.* 8: 556–577. (f) Myers, A.G. and Zheng, B. (1996). *Tetrahedron Lett.* 37: 4841–4844. (g) Mikami, K. and Shimizu, M. (1992). *Chem. Rev.* 92: 1021–1050. (h) (2010). Retro-Ene Reaction. In: *Comprehensive Organic Name Reactions and Reagents*, vol. 534, 2373–2377. London: Wiley.

5 (a) Benn, F.R., Dwyer, J., and Chappell, I.J. (1977). *J. Chem. Soc. Perkin Trans.* 2: 533–535. (b) Stephenson, L.M. and Mattern, D.J. (1976). *J. Org. Chem.* 41: 3614–3619.

6 (a) Yu, Z.X. and Houk, K.N. (2003). *J. Am. Chem. Soc.* 125: 13825–13830. (b) Loncharich, R.J. and Houk, K.N. (1987). *J. Am. Chem. Soc.* 109: 6947–6952. (c) Thomas, B.E. IV, Loncharich, R.J., and Houk, K.N. (1992). *J. Org. Chem.* 57: 1354–1362.

7 (a) Price, J.D. and Johnson, R.P. (1985). *Tetrahedron Lett.* 26: 2499–2502. (b) Ward, H.R. and Karafiath, E. (1969). *J. Am. Chem. Soc.* 91: 7475–7480.

8 Viola, A., MacMillan, J.H., Proverb, R.J., and Yates, B.L. (1971). *J. Am. Chem. Soc.* 93 (25): 6967–6974.

9 (a) Trahanovsky, W.S. and Mullen, P.W. (1972). *J. Am. Chem. Soc.* 94: 5911–5913. (b) Riemann, J.M. and Trahanovsky, W.S. (1977). *Tetrahedron Lett.* 18: 1863–1866.

10 Zsindely, J. and Schmid, H. (1968). *Helv. Chim. Acta* 51: 1510–1513.

11 Woodward, R.B. and Hoffmann, R. (1969). *Angew. Chem.* 81: 797–869.

12 Schiess, P. and Heitzmann, M. (1977). *Angew. Chem. Int. Ed. Engl.* 16: 469–470.

13 Schiess, P. (1987). *Thermochim. Acta* 112 (1): 31–46.

14 (a)Koch, R., Finnerty, J., Murali, S., and Wentrup, C. (2012). *J. Org. Chem.* 77: 1749–1759.

15 Oppenheim, A. and Precht, H. (1876). *Chem. Ber.* 9: 323–325.

16 Holbrook, K. (1992). Vapour and gas-phase reactions of carboxylic acids and their derivative. In: *The Chemistry of Acid Derivatives, Supplement B: Vol 2* (ed. S. Patai), 703–746. London: Wiley.

17 Szware, M. and Murawski, J. (1951). *Trans. Faraday Soc.* 47: 269–274.

18 Allen, R.J.P., Forman, R.L., and Ritchei, P.D. (1955). *J. Chem. Soc.*: 2717–2725.

19 Brown, B.R. (1951). *Q. Rev. Chem. Soc.* 5: 131–146.

20 (a) Smith, G.C. and Blau, S.E. (1964). *J. Phys. Chem.* 68: 1231–1234. (b) Bigley, D.B. and May, R.W. (1967). *J. Chem. Soc. B.* (6): 577–579.

21 Smith, G.C. and Yates, B.L. (1965). *J. Chem. Soc.*: 7242–7246.

22 Taylor, R. (1975). *J. Chem. Soc. Parkin Trans.* 2: 1025–1029.

23 McEwen, I. and Taylor, R. (1982). *J. Chem. Soc. Perkin Trans.* 2: 1179–1183.

24 (a) Taylor, R. (1983). *J. Chem. Soc. Perkin Trans.* 2: 1157–1160. (b) August, R., McEwen, I., and Taylor, R. (1987). *J. Chem. Soc. Perkin Trans.* 2: 1683–1689.

25 Taylor, R. and Thorne, M.P. (1976). *J. Chem. Soc. Perkin Trans.* 2: 799–802.

26 Taylor, R. (1973). *J. Chem. Soc. Faraday Trans.* 2: 809–811.

27 Taylor, R. (1983). *J. Chem. Soc. Perkin Trans.* 2: 89–95.

28 Al-Awadi, N.A. and Taylor, R. (1988). *J. Chem. Soc. Perkin Trans.* 2: 177–182.

29 (a) Al-Awadi, N.A., Al-Bashir, R.F., and El-Dusouqui, O.M.E. (1989). *J. Chem. Soc. Perkin Trans.* 2: 579–581. (b) Al-Awadi, N.A., Al-Omran, F.A., and Mathew, T. (1995). *J. Kuwait University (Sci.)* 22: 53. (c) Al-Awadi, N.A. and Al-Omran, F.A. (1994). *Int. J. Chem. Kinet.* 26: 951–954.

30 Lossing, F.P. (1963, Chapter 11). Mass spectromety of free radicals. In: *Mass Spectrometry* (ed. C.A. McDowell), 442–505. New York: McGraw-Hill.

31 (a) Mardyukov, A. and Sander, W. (2009). *Chem. Eur. J.* 15: 1462–1467. (b) Mardyukov, A. and Sander, W. (2010). *Eur. J. Org. Chem.* 2010 (15b): 2904–2909.

32 Friderichsen, A.V., Radziszewski, J.G., Nimlos, M.R. et al. (2001). *J. Am. Chem. Soc.* 123: 1977–1988.

33 Cadogan, J.I.G., Hickson, C.L., and McNab, H. (1986). *Tetrahedron* 42: 2135–2165.

34 Vögtle, F., Fornell, P., and Löhr, W. (1979). *Chem. Ind.* (12): 416–418.

35 Vögtle, S., Rossa, F., and L. (1979). *Angew. Chem.* 91: 534–549.

36 Laufenberg, F., Pischel, N., Börsch, I. et al. (1997). *Liebigs Ann. Chem.-Recueil*: 1901–1906.

37 Corey, E.J. and Block, E. (1969). *J. Org. Chem.* 34: 1233–1240.

38 Vu, Y., Chrostowska, A., Huynh, T.K.X. et al. (2013). *Chem. Eur. J.* 19 (44): 14983–14988.

39 Justyna, K., Lesniak, S., Nazarski, R.B. et al. (2014). *Eur. J. Org. Chem.* (14): 3020–3027.

40 Abe, M. (2013). *Chem. Rev.* 113: 7011–7088.

41 Reyes, M.B. and Carpenter, B.K. (2000). *J. Am. Chem. Soc.* 122: 10163–10176.

42 Gassman, P.G. and Mansfield, K.T. (1969). *Org. Synth.* 49: 1.

43 Carpenter, B.K. (2013). *Chem. Rev.* 113: 7265–7286.

44 Khuong, K.S. and Houk, K.N. (2003). *J. Am. Chem. Soc.* 125: 14867–14883.

45 Chesick, J.P. (1961). *J. Phys. Chem.* 65: 2170–2173.

46 Flowers, M.C. and Frey, H.M. (1961). *J. Chem. Soc.*: 5550–5551.

47 Carpenter, B.K. (2009). *J. Phys. Chem. A* 113: 10557–10563.

48 Gilbert, J.C. (1969). *Tetrahedron* 25: 1459–1466.

49 (a) Hoffmann, R.W. (1968). *Dehydrobenzene and Cycloalkynes.* Weinheim: Verlag Chemie and Academic Press. (b) Gilchrist, T.L. and Rees, C.W. (1969). *Carbenes, Nitrenes and Arynes.* London: Nelson. (c) Hart, H. (1994, Chapter 18). Arynes and Heteroarynes. In: *The Chemisty of Triple-Bonded Functional Groups, Supplement C* (ed. S. Patai), 1017–1130. Chichester: Wiley. (d) Winkler, M., Wenk, H.H., and Sander, W. (2004, Chapter 16). *Arynes.* In: *Reactive Intermediate Chemistry* (eds. R.A. Moss, M.S. Platz and M. Jones Jr.), 741–796. Hoboken, NJ: Wiley. (e) Wentrup, C. (2010). *Aust. J. Chem.* 63: 979–986.

50 Kitamura, T. (2010). *Aust. J. Chem.* 63: 987–1001.

51 Radziszewski, J.G., Hess, B.A., and Zahradnik, R. (1992). *J. Am. Chem. Soc.* 114: 52–57.

52 (a) Brown, R.D., Godfrey, P.D., and Rodler, M. (1986). *J. Am. Chem. Soc.* 108: 1296–1297. (b) Robertson, E.G., Godfrey, P.D., and McNaughton, D. (2003). *J. Mol. Spectrosc.* 217: 123–126.

53 (a) Zhang, X. and Chen, P. (1992). *J. Am. Chem. Soc.* 114: 3147–3148. (b) Werstiuk, N.H., Roy, C.D., and Ma, J. (1994). *Can. J. Chem.* 73: 146–149. (c) Chrostowska, A., Pfister-Guillouzo, G., Gracian, F., and Wentrup, C. (2010). *Aust. J. Chem.* 63: 1084–1090.

54 (a) Fischer, I.P. and Lossing, F.P. (1963). *J. Am. Chem. Soc.* 85: 1018–1019. (b) Grützmacher, H.-F. and Lohmann, J. (1967). *Liebigs. Ann. Chem.* 705: 81–82. (c) Monsandl, T., Macfarlane, G., Flammang, R., and Wentrup, C. (2010). *Aust. J. Chem.* 63: 1076–1083.

55 Winkler, H. and Sander, W. (2010). *Aust. J. Chem.* 63: 1013–1047.

56 (a) Jagau, J.-C., Prochnow, E., Evangelista, F.A., and Gauss, J. (2010). *J. Chem. Phys.* 132: 144110–144119. (b) Li, X.-Z. and Paldus, J. (2010). *J. Chem. Phys.* 132: 114103–114110.

57 Wittig, G. and Ebel, H.F. (1960). *Angew. Chem.* 72: 564–564.

58 Ebel, H.F. and Hoffmann, R.W. (1964). *Justus Liebigs Ann. Chem.* 673: 1–12.

59 Grützmacher, H.-F. and Lohmann, J. (1967). *Liebigs. Ann. Chem.* 705: 81–90.

60 LeGoff, E. (1962). *J. Am. Chem. Soc.* 84: 3786–3786.

61 Günther, H. (1963). *Chem. Ber.* 96: 1801–1809.

62 (a) Brown, R.F.C. and Solly, R.K. (1965). *Chem. Ind.* (33): 1462–1483. (b) Brown, R.F.C. and Solly, R.K. (1966). *Aust. J. Chem.* 19 (6): 1045–1057. (c) Brown, R.F.C. (2010). *Aust. J. Chem.* 63: 1002–1006.

63 Fields, E.K. and Meyerson, S. (1965). *Chem. Commun.* (20): 474–476.

64 Cava, M.P., Mitchell, J., DeJongh, D.C., and Van Fossen, R.Y. (1966). *Tetrahedron Lett.* 7 (26): 2947–2951.

65 Wentrup, C., Blanch, R., Briehl, H., and Gross, G. (1988). *J. Am. Chem. Soc.* 110: 1874–1880.

66 Gerbaux, P. and Wentrup, C. (2012). *Aust. J. Chem.* 65: 1655–1661.

67 Sander, W. (1999). *Acc. Chem. Res.* 32: 669.

68 Wenthold, P.G. (2010). *Aust. J. Chem.* 63: 1091–1098.

69 (a) Diau, E.W.-G., Casanova, J., Roberts, J.D., and Zewail, A.H. (2000). *Proc. Natl. Acad. Sci.* 97: 1376–1379. (b) Moskaleva, L.V., Madden, L.K., and Lin, M.C. (1999). *Phys. Chem. Chem. Phys.* 1: 3967–3972.

70 Crawford, T.D., Kraka, E., Stanton, J.F., and Cremer, D. (2001). *J. Chem. Phys.* 114: 10638–10650.

71 (a) Jagau, J.-C., Prochnow, E., Evangelista, F.A., and Gauss, J. (2010). *J. Chem. Phys.* 132: 144110–144119. (b) Li, X.-Z. and Paldus, J. (2010). *J. Chem. Phys.* 132: 114103–114110.

72 (a) Bergman, R.G. (1973). *Acc. Chem. Res.* 6: 25–31. (b) Wenk, H.H., Winkler, M., and Sander, W. (2003). *Angew. Chem. Int. Ed. Engl.* 42: 502–528.

73 (a) Nikolaou, K.C. and Dai, W.-M. (1991). *Angew. Chem. Int. Ed. Engl.* 30: 1387–1415. (b) Nikolaou, K.C. and Smith, A.L. (1995). *Modern Acetylene Chemistry* (eds. P.J. Stang and F. Diederich), 203–228. Weinheim: VCH. (c) Poloukhtine, A., Karpov, G., and Popik, V.V. (2008). *Curr. Top. Med. Chem.* 8: 460–469. (d) Popik, V.V. (2010). *Aust. J. Chem.* 63: 1099–1107.

74 (a) Guo, X.-F., Zhu, X.-F., Shang, Y., and Zhang, S.-H. (2010). *Clin. Cancer Res.* 16: 2085–2094. (b) Roy, S. and Basak, A. (2010). *Chem. Commun.* 46: 2283–2285.

75 Kirmse, W. (1971). *Carbene Chemistry*, 2e, 1–634. New York: Academic Press.

76 Baron, W.J., DeCamp, M.R., Hendrick, M.E. et al. (1973). *Carbenes*, vol. 1 (eds. M. Jones Jr. and R.A. Moss), 4–151. New York: Wiley.

77 Brown, R.F.C. (1980). *Pyrolytic Methods in Organic Chemistry*. New York: Academic Press.

78 Richardson, D.C., Hendrick, M.E., and Jones, M. Jr. (1971). *J. Am. Chem. Soc.* 93: 3790–3791.

79 Wentrup, C., Bibas, H., Kuhn, A. et al. (2013). *J. Org. Chem.* 78: 10705–10717.

80 Kvaskoff, D., Lüerssen, H., Bednarek, P., and Wentrup, C. (2014). *J. Am. Chem. Soc.* 136: 15203–15214.

81 (a) Crow, W.D. and Paddon-Row, M.N. (1973). *Aust. J. Chem.* 26: 1705–1723. (b) Wentrup, C. and Wilczek, K. (1970). *Helv. Chim. Acta.* 53: 1459–1463. (c) Schissel, P., Kent, M.E., McAdoo, D.J., and Hedaya, E. (1970). *J. Am. Chem. Soc.* 92: 2147–2149.

82 Wentrup, C. and Müller, P. (1973). *Tetrahedron Lett.* 14 (31): 2915–2918.

83 Wiersum, U.E. and Nieuwenhuis, T. (1973). *Tetrahedron Lett.* 14 (28): 2581–2584.

84 (a) Baron, W.J., Jones, M. Jr., and Gaspar, P.P. (1970). *J. Am. Chem. Soc.* 92: 4739–4740.

85 (a)Chapman, O.L., McMahon, R.J., and West, P.R. (1984). *J. Am. Chem. Soc.* 106: 7973–7974. (b) Chapman, O.L., Uh-Po, T., and E. (1984). *J. Am. Chem. Soc.* 106: 7974–7976. (c) Chapman, O.L., Abelt, C.J., Johnson, J.W. et al. (1988). *J. Am. Chem. Soc.* 110: 501–509.

86 Trahanovsky, W.S. and Scribner, M.E. (1984). *J. Am. Chem. Soc.* 106: 7976–7978.

87 (a) Becker, J. and Wentrup, C. (1980). *J. Chem. Soc. Chem. Commun.* (4): 190–191. (b) Wentrup, C., Mayor, C., Becker, J., and Lindner, H.J. (1985). *Tetrahedron* 41: 1601–1612. (c) Engler, T.A. and Shechter, H. (1999). *J. Org. Chem.* 64: 4247–4254.

88 Wentrup, C., Becker, J., and Winter, H.-W. (2015). *Angew. Chem. Int. Ed.* 54: 5702–5704.

89 Becker, J., Diehl, M., and Wentrup, C. (2015). *J. Org. Chem.* 80: 7144–7149.

90 Kvaskoff, D., Becker, J., and Wentrup, C. (2015). *J. Org. Chem.* 80: 5030–5034.

91 (a) Brown, R.F.C., Harrington, K.J., and McMullen, G.L. (1974). *J. Chem. Soc. Chem. Commun.* (4): 123–124. (b) Brown, R.F.C., Eastwood, F.W., and Jackman, G.P. (1977). *Aust. J. Chem.* 30: 1757–1767.

92 Wentrup, C., Winter, H.-W., and Kvaskoff, D. (2015). *J. Phys. Chem. A* 119: 6370–6376.

93 Brown, R.F.C., Eastwood, F.W., Harrington, K.J., and McMullen, G.L. (1974). *Aust. J. Chem.* 27: 2391–2402.

94 Becker, J., Wentrup, C., Katz, E., and Zeller, K.-P. (1980). *J. Am. Chem. Soc.* 102: 5110–5112.

95 Scott, L.T., Hashemi, M.M., Meyer, D.T., and Warren, H.B. (1991). *J. Am. Chem. Soc.* 113: 7082–7084.

96 (a) Neilen, R.H.G. and Wiersum, U.E. (1996). *Chem. Commun.* (2): 149–150. (b) Banciu, M.D., Brown, R.F.C., Coulston, K.L. et al. (1996). *Aust. J. Chem.* 49: 965–976.

97 McIntosh, G.J. and Russell, D.K. (2014). *J. Phys. Chem. A* 118: 12205–12220.

98 Jing, W., Zheng, S., Xinjiang, Z. et al. (2001). *Angew. Chem. Int. Ed.* 40: 3055–3057.

99 Chen, C.C. and McQuaid, M.J. (2012). *J. Phys. Chem. A* 116: 3561–3576.

100 (a) Gieseler, G. and König, W. (1964). *Z. Phys. Chem.* 227: 81–92. (b) Stepanov, R.S., Kruglyakova, L.A., and Buka, E.S. (1986). *Kinet. Katal.* 27: 479–482.

101 Zeng, Y., Sun, Q., Meng, L. et al. (2004). *Chem. Phys. Lett.* 390: 362–369.

102 (a) Bock, H. and Dammel, R. (1988). *J. Am. Chem. Soc.* 110: 5261. (b) Wentrup, C. (2013). *Aust. J. Chem.* 66: 852–863.

103 Wentrup, C. (2015). *J. Phys. Chem. A* 119: 8256–8257.

104 Pritzkov, W. and Timm, D. (1966). *J. Prakt. Chem.* 32: 178–189.

105 Wentrup, C. (2011). *Acc. Chem. Res.* 44: 393–404.

106 Wentrup, C. (2013). Matrix Studies on Aromatic and Heteroaromatic Nitrenes and their Rearrangements. In: *Nitrenes and Nitrenium Ions* (eds. D.E. Falvey and A.D. Gudmundsdottir) Chapter 8. Hoboken, NJ: Wiley.

107 Kuzaj, M., Lüerssen, H., and Wentrup, C. (1986). *Angew. Chem. Int. Ed. Engl.* 25: 480–482.

108 Kvaskoff, D., Bednarek, P., George, L. et al. (2006). *J. Org. Chem.* 71: 4049–4058.

109 Crow, W.D. and Wentrup, C. (1968). *Tetrahedron Lett.* 9 (59): 6149–6152.

110 Aylward, N., Winter, H.-W., Eckhardt, U., and Wentrup, C. (2016). *J. Org. Chem.* 81: 667–672.

111 Wentrup, C. and Crow, W.D. (1970). *Tetrahedron* 26: 3965–3981.

112 Mayor, C. and Wentrup, C. (1975). *J. Am. Chem. Soc.* 97: 7467–7480.

113 Graebe, C. and Ullmann, F. (1896). Ueber eine neue Carbazolsynthese. *Justus Liebigs Ann. Chem.* 291: 16–17.

114 (a) Morawitz, J., Sander, W., and Träubel, M. (1995). *J. Org. Chem.* 60: 6368–6378. (b) Wentrup, C. (1969). *Chem. Commun.* (23): 1386–1387.

115 Wentrup, C. (1976). *Top. Curr. Chem.* 62: 173–251.

116 Wentrup, C. and Winter, H.-W. (1980). *J. Am. Chem. Soc.* 102: 6159–6161.

117 Addicott, C., Wong, M.W., and Wentrup, C. (2002). *J. Org. Chem.* 67: 8538–8546.

4

Structure/Reactivity Correlation

Mechanisms that were earlier based only on analysis of products of reaction, analogy, and intuition can now be substantiated by kinetic studies. This chapter explains how accruing information on physical and structural properties made it possible to explain and correlate molecular reactivities essential for a better understanding of the mechanism in the gas-phase pyrolysis reaction.

4.1 Diketones

Products obtained from the gas-phase elimination reaction of pentane-2,4-dione 1 were identified as acetone and ketene, which are formed via a six-membered transition state [1]. Kinetic data were used to compare the mechanism suggested for gas-phase pyrolysis of the dione with those adopted from structurally related acetic anhydride [2], diactyl sulfide [3], and diacetamide [4], as pyrolysis of these compounds involves a concerted but asynchronous six-membered transition state (Scheme 4.1).

Gas-Phase Pyrolytic Reactions: Synthesis, Mechanisms, and Kinetics,
First Edition. Nouria A. Al-Awadi.
© 2020 John Wiley & Sons, Inc. Published 2020 by John Wiley & Sons, Inc.

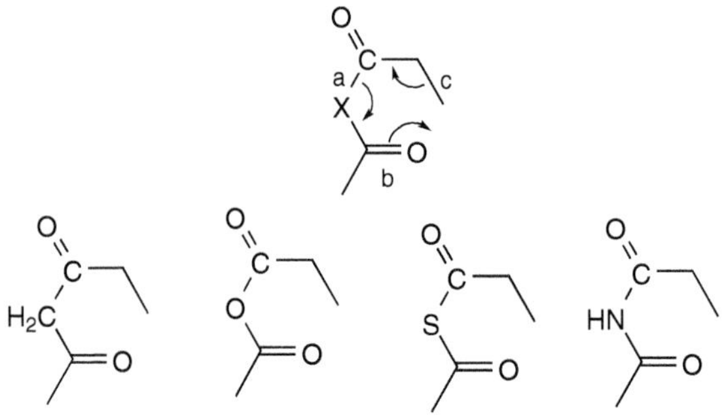

Scheme 4.1

Increase in the polarity of the C—X bond (a) from the methylene unit in pentane-2,4-dione to O in the anhydride, to S in the sulfide, and to NH in the diacetamide has resulted in rate ratios for pentane-2,4-diones of $1 : 3 \times 10^2$ relative to the anhydride, $1 : 6 \times 10^3$ relative to the sulfide, and $1 : 5 \times 10^4$ relative to the diamide.

In an attempt to increase the molecular reactivity of the dione system, one CH_3 group was replaced by OCH_3 in order to increase the polarity of bond (a) and hence the reactivity of the molecule. This acetylacetone **2** was expected to give ketone and methylacetate. The products identified were methanol and dehydroacetic acid **3**. Evidently, this structural modification resulted in a change of mechanism from that suggested for the 2,4-dione, which seems to be highly affected by the substituent at the carbonyl carbon. The methoxy group has obviously encouraged intermolecular cyclization in the acetylacetone with loss of alcohol molecules (Scheme 4.2). The stability of the product **4**, which has an aromatic character, gives support to this mechanism.

To confirm this finding, the gas-phase thermal elimination reaction of ethylacetoacetate was also studied, and analysis of the pyrolysate shows formation of the same product – dehydroacetic acid **4** – which supports the suggested mechanism as shown in Scheme 4.2.

We then investigated the substituent electronic effect on the methylene carbon by replacing the two hydrogens with a phenylhydrazone group in 3-phenylhydrazopentane-2,4-dione **5**. The only products detected from pyrolysis of **5** were ketene and compound **6** (Scheme 4.3), which means the arylhydrazono (ArNHN=) group has restored the elimination pathways of the dione reaction shown in Scheme 4.1 with

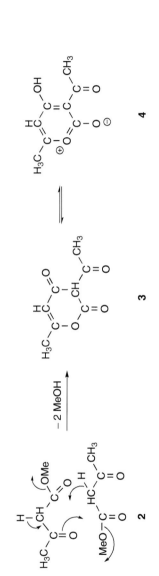

Scheme 4.2

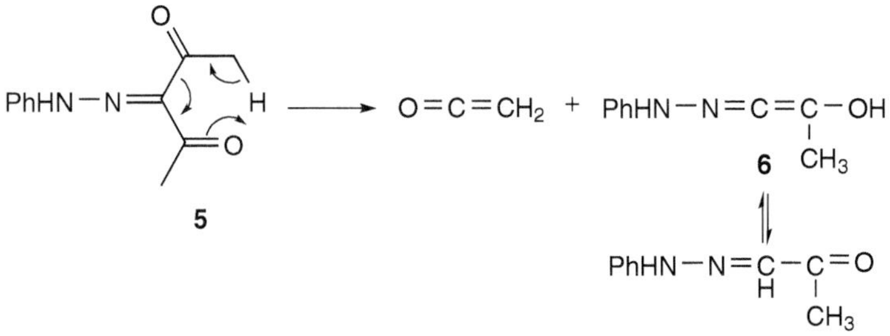

Scheme 4.3

a rate coefficient of 1.27×10^{-3} at 227 °C. This is 240 times faster than the corresponding pentene-2,4-dione (Scheme 4.3).

The two protophilic carbonyl functions of the diketone **1** are equivalent, and both are equally available for interaction with the incipient H from either of the two equivalent terminal CH_3 moieties. Incorporation of an arylhydrazono-2,4-propanedione **5** provides an acidic hydrogen site that augments the positive charge on the incipient H involved in the elimination pathway through extended delocalization of the methyl σ-electrons bond (c), as shown in Scheme 4.4.

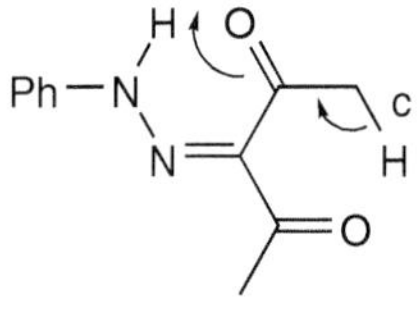

Scheme 4.4

The effect of this important structural modification is reflected in a rate enhanced by a factor of 240 in compound **5** over **1**.

The effect of electronic factors on the direction of elimination is shown in the pyrolysis of methyl trans-2-acetoxycyclohexancarboxylate **7**, which gives 97% of the conjugated ester **8** and only 3% of the unconjugated isomer **9** [5] (Scheme 4.5).

A steric effect also operates in the pyrolysis of 1-methylcyclohexyl-acetate **10** at 450 °C, which yields 24% exo-methylene cyclohexene **11** and 76% endo isomer **12** [6] (Scheme 4.6). On the other hand, the only product from 1-methylcyclopentylacetate **13** was 1-methylcyclopentene **14** (Scheme 4.6).

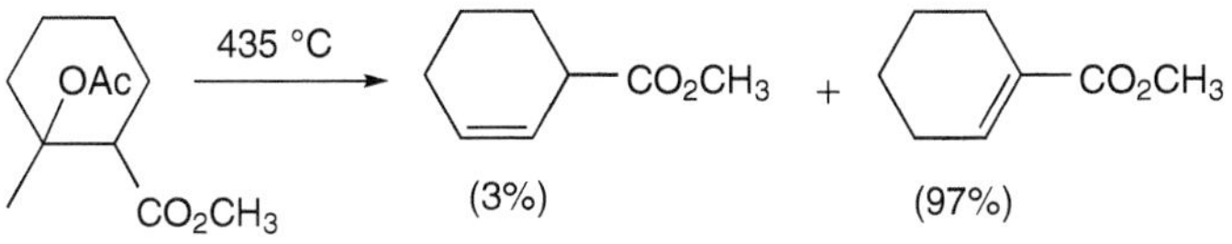

Scheme 4.5

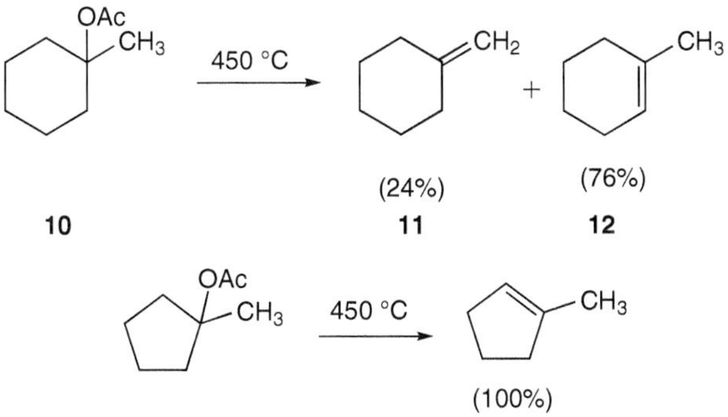

Scheme 4.6

The structure–reactivity relationship in the gas-phase pyrolysis of diketones and cyclic 2,4-diketones shown in Scheme 4.7 have been examined [7].

The rate of gas-phase elimination reactions was measured over a temperature range of 50 °C. The kinetic data and product analysis were used to assess the influence of the ring size and the heteroatom on the reactivity of cyclic 2,4-diketones **15–20**. The rate coefficient of these diketones at 467 °C, shown in Scheme 4.7, reflects the effect of structural modification in enhancing and retarding the rate of the gas-phase elimination reaction of cyclic 2,4-diketones. Product analysis suggests a cyclic six-membered transition state involving proton transfer to protophilic/nucleophilic carbonyl, as shown in Scheme 4.1. According to this transition model, the structural factors upon which the overall substrate molecular reactivity depends include the following:

i. Protophilicity/nucleophilicity of bond (b), the effect of the polarity of bond (a), and the H-bond donor-acidity of bond (c)
ii. Nature and magnitude of charge at each carbon atom involved in the transition state

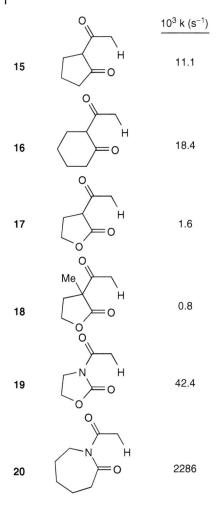

Scheme 4.7

iii. Thermodynamic stability of the incipient fragments, and any subsequent rearrangement products therefrom

iv. Conformational preorganization, and any possible statistical factors [1, 8–15].

From the first-order rate constants of the unimolecular gas-phase thermal elimination reactions of the cyclic 2,4-diketones, it is evident that 2-acetylcyclohexanone **16** is 1.7 times more reactive than its cyclopentanone counterpart **15**. Its reactivity can

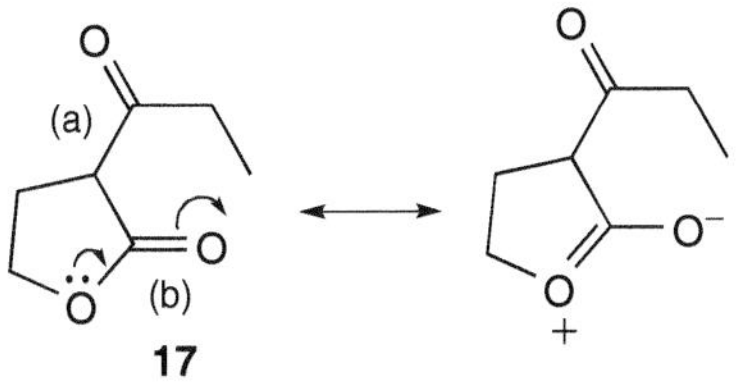

Scheme 4.8

be explained in terms of the relative thermodynamic stability of the 1-cyclohexanol to 1-cyclopentanol formed from **16** and **15**, respectively. 2-Acetylbutyrolactone **17** would be much more reactive than **15** and **16** due to the resonance structure shown in Scheme 4.8, which will increase the protophilicity of the oxygen of the carbonyl group involved in the six-membered transition state; but it shows less reactivity than **15** and **16**, which can be explained by the fact that the ester group in **17** has less of an electron-withdrawing effect than the keto groups in **15** and **16**. This will make bond (a) of **17** less polar, thus rendering it more difficult to break. This means that breaking bond (a) is more important than the protophilic attack of bond (b) on the hydrogen atoms in these cyclic diketones.

Introducing a methyl group in 2-acetyl-2-methylbutyrolactone **18** reduced the rate as a result of the methyl group's affect on the alignment of the atoms involved in the transition state. The enhanced reactivity of 3-acetyl-2-oxazolidinone **19** over compounds **15–18** is a result of increasing the polarity of bond (a) C—N relative to the C—C bond in **15–18**. The outstanding reactivity of N-acetylcaprolactam **20** is not only due to the polarity of bond (a) C—N. Another factor that might contribute to the acceleration of the reaction rate of **20** is the development of a double bond into a seven-membered ring, which requires less energy than the same in a five membered ring. The thermodynamic stability of the imidole formed from **20** over that of the enols formed from **15–18** will also enhance the molecular reactivity of the diketone **20**.

4.2   Cyanoketones

Product analysis and kinetic data for the gas-phase elimination reactions of cyanoketones **21–25** suggest the elimination pathway shown in Scheme 4.9 [1].

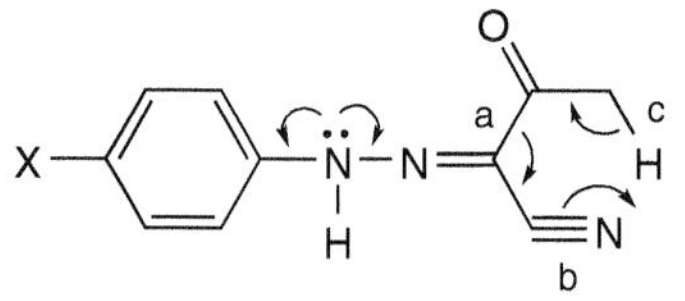

Scheme 4.9

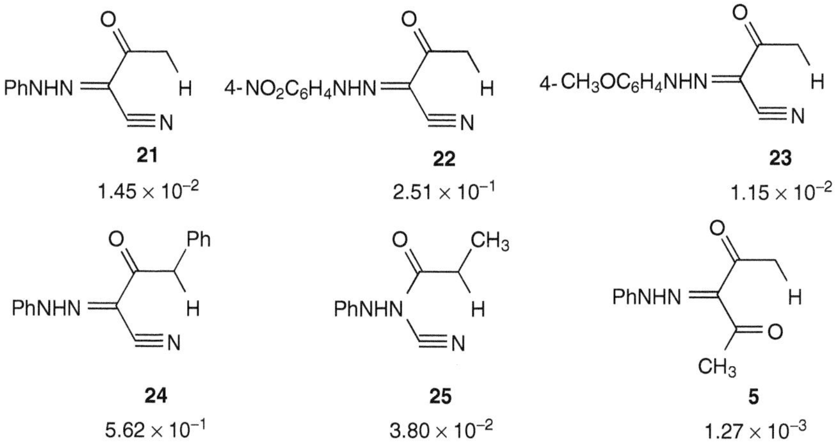

Scheme 4.10

From the rate data reported in Scheme 4.10, the effect of the hydrazone substituent is reflected in the rate of reaction of compound **21** compared to **22** and **23**: the p-NO$_2$ group in **22** has increased reactivity by a factor of 17, and the p-OMe group in **23** has slightly decreased reactivity by a factor of 0.79 due to its mesomeric electron-donating. The moderate electronic effect observed by the NO$_2$ and OMe groups on the reactivity of **22** and **23** could be explained by the cross-over conjugation of the lone pair of electrons on the NH moiety of the hydrazono substituent shown in Scheme 4.9. The electron-withdrawing effect of the phenyl group in compound **24** on the acidity of the proton involved in the elimination reaction enhanced reactivity 39 times more than unsubstituted compound **21**; accordingly, the methyl substituent in **25** would be expected to reduce the acidity of the incipient proton, but it slightly enhanced reactivity by a factor of 3.93 relative to unsubstituted compound **21**. This could be explained in terms of the greater thermodynamic stability of the methyl ketene resulting from the pyrolysis of **25** relative to the ketene produced from **21**.

Diketone **5** is 11 times less reactive than cyanoketone **21**. This rate-enhancing effect of —C≡N could be explained in terms of the two labile π bonds of C≡N together with their protophilic character compared to only one π bond of C=O in diketone **5**.

4.3 Ketoamides

The two carbonyl groups in 3-oxo-N-phenylbutanamide **26** are non-equivalent: an amide C=O group and a keto C=O function. The gas-phase pyrolysis of **26** involves a transition state shown in Scheme 4.11, with the participation of the ketone (C=O) group as the protophilic center, and the amino (NH) function for provision of the incipient proton fragment.

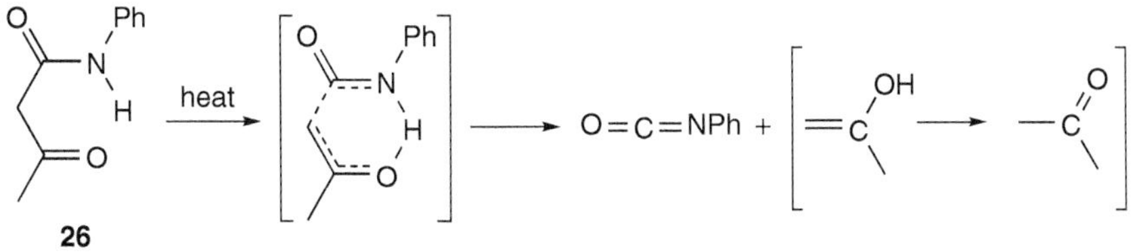

Scheme 4.11

Evidence for this mechanism includes (i) accepted higher donor acidity of the NH bond compared to the C—H bond of the terminal CH_3 group; (ii) Hammett correlation of the values of σ-constants for groups substituted in the phenyl moiety at the amino function ($\rho = 1.22 \pm 0.3$); (iii) identification and characterization of the elimination fragments [16].

Incorporation of an arylhydrazono group in the oxobutanamide **26** frame produces **27**. Substrate **27** was found to be 2.33×10^3 times less reactive than **26**. In contrast, the arylhydrazono group increased the rate of thermal gas-phase elimination of the analogous diketone (2,4-pentanedione) by a factor of 2.4×10^2.

The changeover in the character of the arylhydrazone group from a strongly activating diketone and cyanoketone to a strongly deactivating substituent is justified by the fact that the arylhydrazono group in compound **27** has engaged the ketone (C=O) protophilic center in a competitive intramolecular cyclization process (Scheme 4.12).

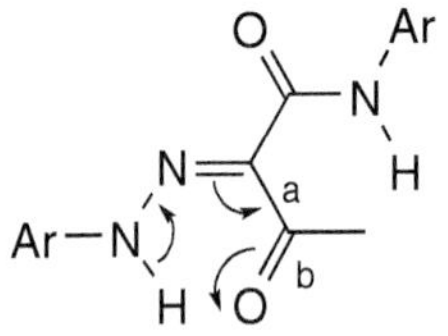

27

Scheme 4.12

Intramolecular cyclization is, on the other hand, a dead-end process; it affects reaction kinetics without being directly involved in any fragmentation. It is of considerable interest to note the experimental value of activation energy barriers measured by the thermal gas-phase elimination reaction of the parent 3-oxo-N-phenylbutanamide **26** and its 3-oxo-2-phenylhydrazono-N-phenylbutanamide derivative **27**: 106.1 kJ mol^{-1} for **26** and 158.t kJ mol^{-1} for **27** [1].

4.4 Benzotriazoles

Pyrolysis of benzotriazole derivatives often results in homolytic cleavage of the N—N bond, leading to the formation of reactive intermediates followed by either rearrangement or cyclization involving

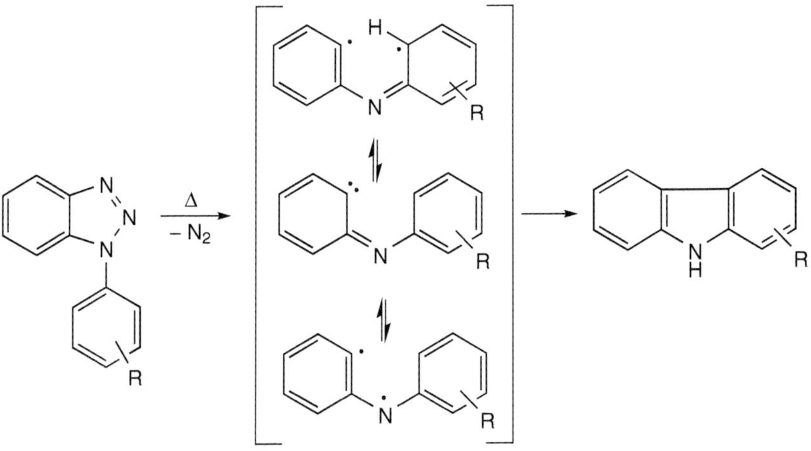

Scheme 4.13

either carbon/nitrogen or carbon/carbon biradicals or an iminocarbene intermediate shown in Scheme 4.13. This behavior of benzotriazoles in pyrolytic reaction has received considerable attention in synthesis and mechanistic investigations based on product analysis and the intermediates involved [17–25]. On the other hand, molecular reactivities of benzotriazole and its derivatives have received less attention. Molecular reactivities require kinetic investigations to measure the rate of reactions over a reaction temperature range of not less than 50 °C, Arrhenius parameters, energy, and entropy of reaction are required for correlation of the structure of the benzotriazoles with their reactivities and to assess the electronic effect of substituents on the molecular reactivities. The first kinetic investigation of the thermal gas-phase elimination reactions of benzotriazole compounds was published in 2003 [26].

An interesting comparative structure/reactivity is obtained from the nine selected benzotriazole derivatives 28–36, shown in Table 4.1.

Kinetic studies performed on the gas-phase pyrolysis of benzotriazole (28–36), rate coefficients, and Arrhenius parameters were measured using a static reactor following the procedure described in Chapter 1. Compounds 28–36 all behaved well kinetically and gave reproducible first-order rate constants; the kinetic data are provided in Table 4.2.

Reaction products from complete gas-phase pyrolysis of substrates 28–36 were obtained by sealed-tube static and flash vacuum pyrolysis (FVP), applying the procedures explained in Chapter 1. The products were separated by column chromatography and characterized using GC–MS, LCMS, and ^{1}H, ^{13}C NMR spectroscopies [25].

Table 4.1 shows the major products obtained from both modes of pyrolysis with their percent yields. It is noteworthy that the differences in products from the pyrolysates from static and flash vacuum pyrolysis are a result of the difference in substrate residence time in the heating tube of each reactor; in static mode, it is in the range of minutes, whereas it is in milliseconds in a flow system in FVP.

Benzotriazoles 29–31 pyrolyzed to give 37a,b and 38; a suggested mechanism is given in Scheme 4.14 to account for the formation of compounds 37a,b.

Table 4.1 Products obtained for pyrolysis of **28–36** and percentage yields.

Substrate	Structure	Condition	Pyrolysis product (percentage yield)
28		FVP	Charring (products could not be identified)
		STP	Charring (products could not be identified)
29		FVP	**37** (28%)
		STP	**37** (15%), **43** (13%)
30		FVP	**37** (18%)
		STP	**38** (15%), **43** (10%)
31		FVP	**37** (16%), **39** (9%)
		STP	**38** (18%), **43** (8%)
32		FVP	**40** (25%), **41** (5%), **42** (5%), **43** (12%)
		STP	**38** (18%), **41** (15%), **42** (8%), **43** (8%)

Table 4.1 (Continued)

Substrate	Structure	Condition	Pyrolysis product (percentage yield)
33		FVP	**38** (15%), **45** (43%), CH_3CN (10%)
		STP	**38** (35%), CH_3CN (10%)
34		FVP	**44** (19%), **47** (37%)
		STP	**44** (5%), **46** (10%), **47** (48%)
35		FVP	**38** (5%), **47** (65%)
		STP	**38** (10%), **47** (55%)
36		FVP	**37** (30%), CH_3CN (15%)
		STP	**37** (28%), CH_3CN (14%)

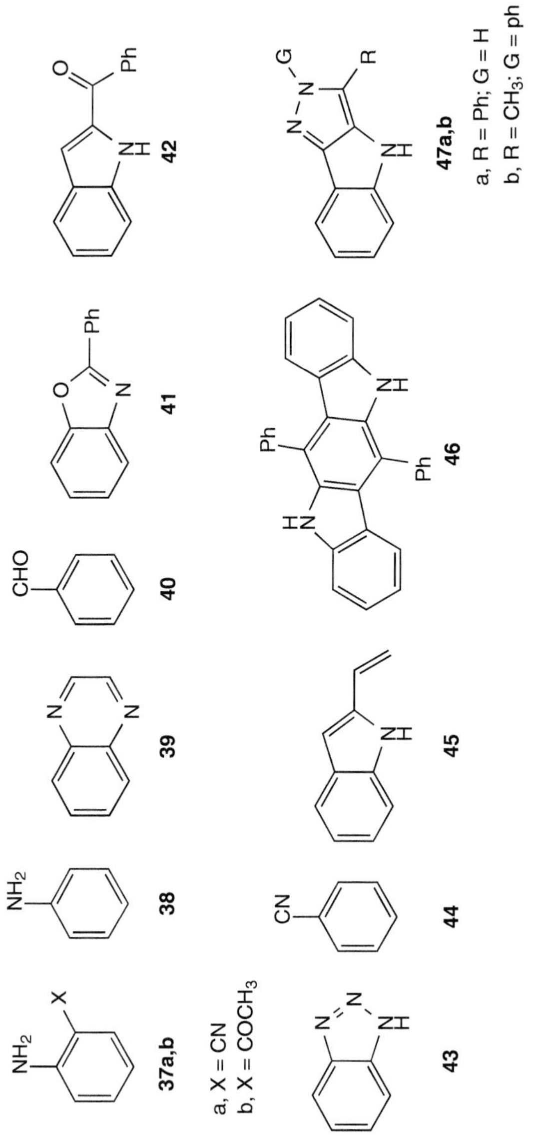

37a,b

a, X = CN
b, X = COCH₃

38

39

40

41

42

43

44

45

46

47a,b

a, R = Ph; G = H
b, R = CH₃; G = ph

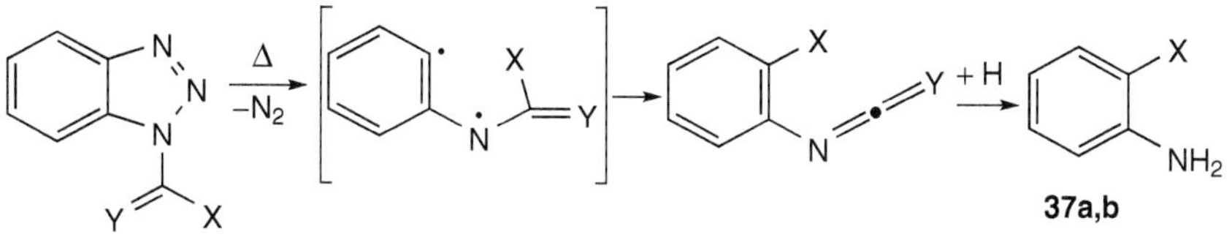

	X	Y
29	CN	2H
30	CN	CHNMe$_2$
31	COMe	CHNMe$_2$

	X
37a	CN
37b	COMe

Scheme 4.14

Table 4.2 Kinetic data and Arrhenius parameters of compounds **28–36** at 227 °C.

Substrate	Log A s^{-1}	E_a kJ mol^{-1}	k s^{-1}
28	15.39 ± 0.44	163.60 ± 3.85	7.14×10^{-1}
29	13.58 ± 0.12	206.10 ± 1.43	1.05×10^{-6}
30	15.34 ± 0.69	190.90 ± 7.38	1.66×10^{-3}
31	13.15 ± 0.19	156.30 ± 2.19	2.04×10^{-4}
32	5.25 ± 0.12	87.79 ± 1.31	8.22×10^{-4}
33	13.50 ± 0.13	192.40 ± 1.46	1.70×10^{-5}
34	11.52 ± 0.31	168.20 ± 3.57	3.55×10^{-5}
35	9.95 ± 0.15	148.30 ± 1.67	1.82×10^{-5}
36	9.95 ± 0.15	148.30 ± 1.67	7.41×10^{-5}

It is instructive to note that both FVP and static pyrolysis (STP) reactions of benzotriazoles and related macrocycles have been reported to yield small amounts of aniline derivatives (similar in nature to compounds **37** and **38**), through abstraction of hydrogen from the reaction mixture and formation of reactive diradical or carbene intermediates [19, 21, 26–28]. The presence of quinoxaline **39** among the products of the FVP reaction of substrates **31** is shown in Scheme 4.15.

The loss of the ring substituent from the reaction intermediate, and the subsequent cyclization of the resulting N/N diradical, would help to explain the presence of benzotriazole **43** among the products of pyrolysis of **31**. Benzotriazole **43** is also detected among

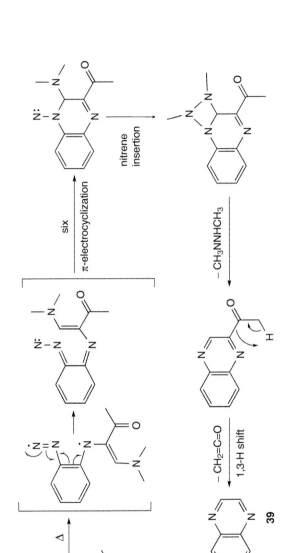

Scheme 4.15

the pyrolysates of substrates **32** and **33**, possibly due to the loss of analogous substituents and their recyclization into the benzotriazole.

Products **41** and **42** are believed to result from the STP and FVP reactions of compound **32**, according to the two pathways shown in Scheme 4.16.

The routes that outline the formation of 2-benzoylindole **42** are believed to commence with the extrusion of molecular nitrogen and formation of a C/N biradical intermediate: one route in Scheme 4.16 involves the carbonyl π bond in the subsequent cyclization step leading to **41**, whereas the other cyclization to **42** engages the alkene π bond. Benzaldehyde **40** is formed in a separate step as a result of the fragmentation of 2-benzoylindole **42**; the fragmentation is formally an extrusion of the PhCO moiety only from substrate **32**.

Product analysis of the pyrolysis reactions of benzotriazoles **32–34** is a representative example of how gas-phase pyrolysis reactions determine the effect of substituents (CH_3/Ph) on the nature of the products formed. The only structural difference between **33** and **34** is that the substituent on the pyrazole ring is methyl in **33** and phenyl in **34**. A mechanism suggested for the formation of 2-vinyl-1-H-indole **45** from FVP of **33** through a carbene intermediate by its insertion is shown in Scheme 4.17. Its counterpart **34** pyrolyzed in a static reactor to give the condensed heterocyclic **45** resulting from dimerization of the C/C biradical followed by aromatization.

The same bicyclic framework **47b** resulted from the pyrolysis of benzotriazole derivative **35** (Scheme 4.18) because of the common structural feature between benzotriazoles **34** and **35** in which a phenyl substituent varies its position in the pyrazole ring and the replacement of H in the NH moiety of the pyrazole ring of **35** with the phenyl group, which changes the π bond's position in the same ring.

Aniline **37b** and acetonitrile were detected from the pyrolysis of benzotriazole **36**.

The rates of the gas-phase elimination reactions of substrates **28–36** were measured over a wide temperature range of 237–387 °C, a prerequisite for obtaining reliable activation parameters in gas-phase pyrolytic reactions [29, 30]. The values of the Arrhenius log A (10.0–15.3 s^{-1}) and E_a (148–209 kl mol^{-1}) are of the same order, as expected for homogeneous unimolecular thermal gas-phase elimination processes [15, 31], and are listed in Table 4.2. The first-order rate constants (k/s^{-1}) at 277 °C for substrates **28–36** are given in Scheme 4.19.

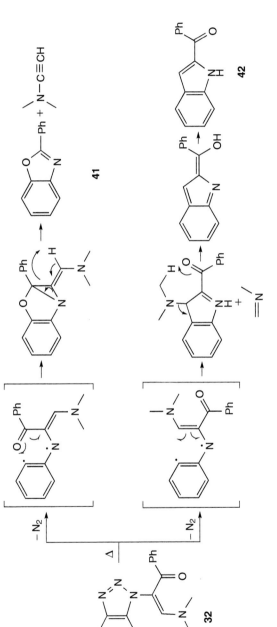

Scheme 4.16

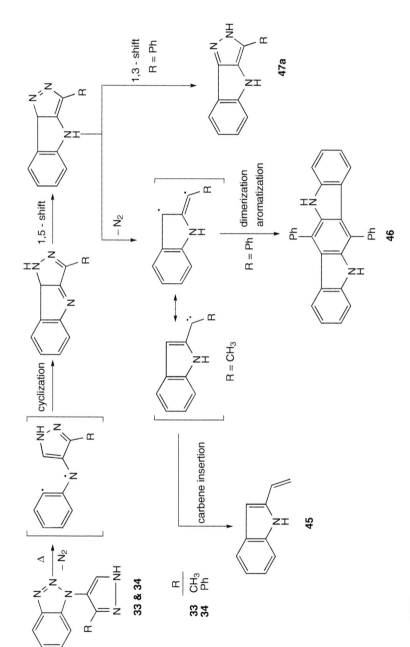

Scheme 4.17

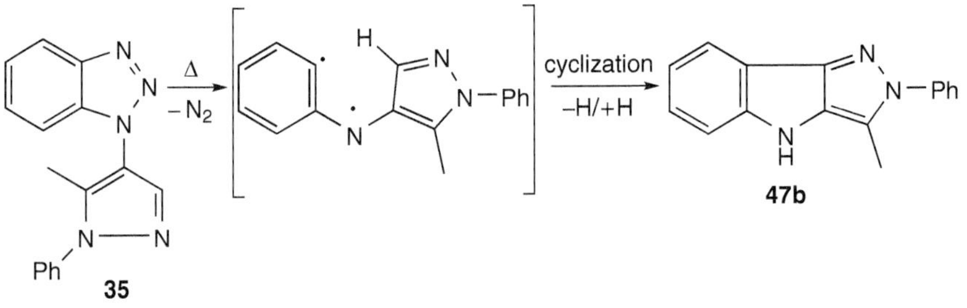

Δ
$-N_2$
H
N–Ph
cyclization
–H/+H
N–Ph
N H
47b
Ph
35

Scheme 4.18

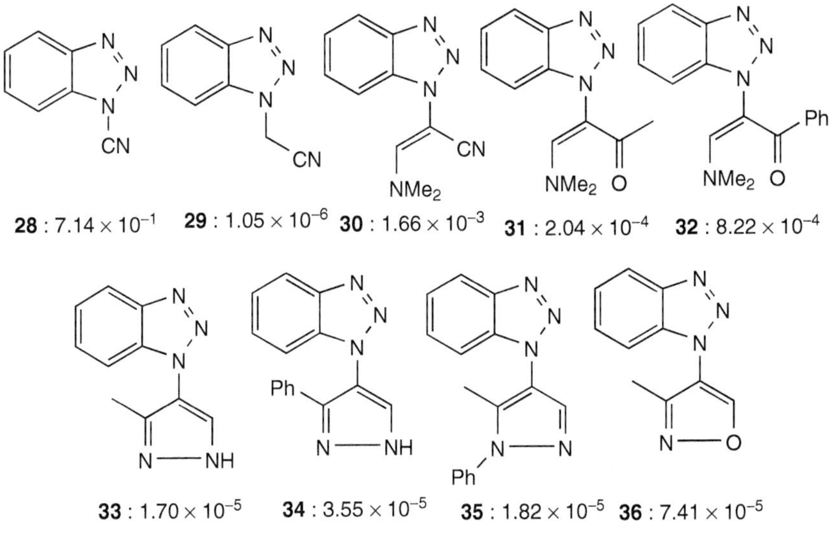

CN
CN
CN
NMe2
NMe2 O
Ph
NMe2 O
28 : 7.14 × 10^−1
29 : 1.05 × 10^−6
30 : 1.66 × 10^−3
31 : 2.04 × 10^−4
32 : 8.22 × 10^−4
N—NH
Ph
N—NH
Ph
N—O
33 : 1.70 × 10^−5
34 : 3.55 × 10^−5
35 : 1.82 × 10^−5
36 : 7.41 × 10^−5

Scheme 4.19

Rationalization of the rate constants and relative molecular reactivities of substrates **28–36** and related benzotriazoles is based on the mechanism of thermal gas-phase elimination that involves the initial formation of C/N and N/N biradical intermediates, following in each case the loss of a molecular nitrogen fragment. This pathway allows several valuable molecular structure/reactivity correlations to be drawn:

i) It is evident that the stabilization of the reactive intermediates, and hence the promotion of molecular reactivity in gas-phase pyrolysis of benzotriazoles, is affected considerably by conjugative interactions. This is clearly evident from a comparison of the rates of pyrolysis (Scheme 4.19) of (benzotriazol-1-yl)acetonitrile **29** and 1-cyanobenzotriazole **28**. The latter, benzotriazole **28**, is ca. 7×10^5 times more reactive than the former substrate as it has a reactive intermediate in which the C≡N multiple bond is directly conjugated with the biradical/carbene centers of the intermediate (Scheme 4.14). This structural relationship is absent in **29**. Greater stability of biradicals due to π conjugation has previously been deduced from measurements of the lifetimes of the biradicals and has been observed in gas-phase pyrolysis of compatible benzotriazoles [17–19, 26, 27].

ii) Likewise, π conjugation involving the enamine moiety of benzotriazole **30** increased this substrate reactivity over **29** by a factor of 1.6×10^3. Two structural features related to this rate factor (compared to the previously mentioned factor of 7×10^5) are noteworthy: (i) unlike the CN group in 1-cyanobenzotriazole **28**, the enamine moiety is an electron-donating group; (ii) the enamine moiety of **30** is also cross-conjugated with the neighboring cyano group, a feature that reduces the conjugative interaction of the enamine group with the radical center.

iii) Benzotriazole **30** is 8 and 2 times more reactive than compounds **31** and **32**, respectively. This rate enhancement is attributed to a more electron-withdrawing CN group **30**, compared to the keto group in **31** and **32**. Compound **32** is 4 times more reactive than benzotriazole **31**, a rate factor associated with the electron-withdrawing phenyl group in the former and the electron-donating methyl group in the latter.

iv) Substituent replacement effects associated with heteroaryl moieties adequately explain the reactivity order for **31/33**, **33/34**, **34/35**, and **33/36**. Compound **33** is 12 times less reactive than **31**, a rate factor associated with the ring (C=N) moiety of the pyrazole group in **33** formally replacing the electron-withdrawing C=O function of **31**. Besides, the π-excessive pyrazolyl group as a whole is itself an electron-donating substituent. Compound **34** is more reactive than both **33** and **35** due to the position and electron-withdrawing nature of the phenyl group. Replacement of N by the more-electronegative O atom explains the greater reactivity of **36** over **33**.

4.5 Hammett Correlation in Gas-Phase Pyrolysis

The effects of substituents in a gas-phase pyrolysis reaction of 1-arylethyl acetate **48** were investigated through a plot of $\log k_{rel}$ ($\log k/k_0$) at 277 °C against σ^- constants, which gave a poor correlation [32], whereas a very satisfactory correlation was obtained when the values of $\log k_{rel}$ plotted against σ^+ constants [33], giving $\rho = -0.66$ at 377 °C. The literature emphasizes that the reaction that follows a ρ/σ^+ relationship must involve an electron-deficient center, such as a fully or partially formed carbonium ion [33–36] (Scheme 4.20). The ρ factor for the effect of the substituents in the pyrolysis reaction of 2-arylethyl acetate **49** at 650 is ca. 0.3; this means substituents in the phenyl group that will withdraw electrons and hence increase the acidity of the β-hydrogen increase the rate of pyrolysis. This ρ factor is numerically less than that obtained (−0.66) for the same effect in the pyrolysis of 1-arylethyl acetate **48**; this finding makes it apparent that the strength of the C—O bond, which is reflected in the stability of the partially formed carbonium ion, is more important in determining the stability of the esters than the C—H bond's strength.

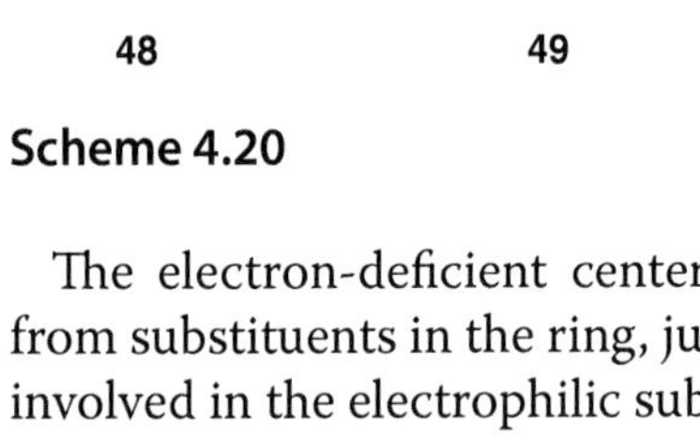

Scheme 4.20

The electron-deficient center is thus stabilized by electron supply from substituents in the ring, just as these stabilize the transition states involved in the electrophilic substitution reaction [37].

The electrophilic reactivity of heterocyclic nitrogen compounds cannot be determined by measuring the rate at which they react with external electrophiles, since protonation will occur, and reaction takes place only under extreme conditions, e.g. pyridinium ion.

It is of interest to note that gas-phase pyrolysis reactions have the advantage of being effectively free from solvation, hydrogen bonding, and protonation effects common in solution reactions.

The precise correlation between log k/k_0 values and σ^+ constants in the pyrolysis of 1-arylethyl acetates exists because of the electronic (mesomeric and inductive) stabilization factor of the incipient carbonium ion by the substituents in the gas-phase pyrolytic reaction (Scheme 4.21). Pyrolysis of this ester is, therefore, allied to electrophilic aromatic substitution, and it follows that substituent effects can be related to each other. Taylor has elegantly used the ρ factor of -0.66 at $227\,^{\circ}C$ for 1-arylethyl acetates **48** to determine quantitatively the electrophilic reactivity of the 2-, 3-, and 4-positions of pyridine under conditions whereby protonation of the nitrogen atom could not occur [38].

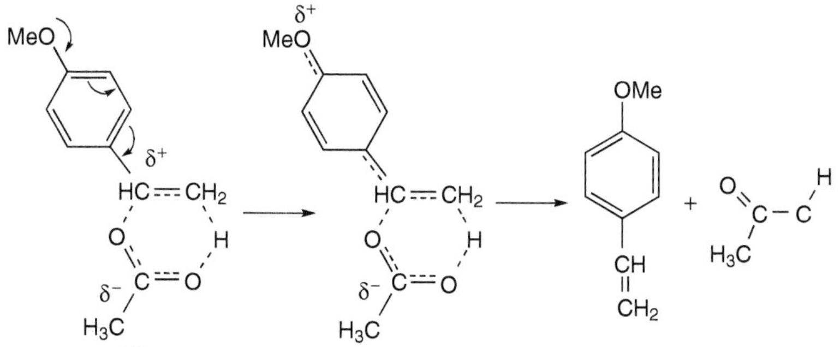

Scheme 4.21

The ability of the aryl substituent to stabilize an adjacent carbonium ion was found to be in the order Ph > 3-Py > 2-Py > 4-Py, which means all nuclear positions in pyridines will be deactivated toward electrophilic substitution; the 3-position is the least deactivated. These results are in accordance with the well-known reactivity of pyridine in electrophilic substitution reactions.

From $\rho = -0.66$ at $227\,^{\circ}C\,K$ and $\rho = -0.63$ at 625 K for 1-arylethyl acetate, and from the log k/k_0 value calculated for the 2-, 3-, and 4-pyridyl substituents, σ^+ constants for positions 2, 3, and 4 in pyridine were calculated and found to be $+0.80$, $+0.30$, and -0.87, respectively. This approach was extended to the measurement of electrophilic aromatic reactivities from gas-phase pyrolysis of 1-arylethyl acetate for the 2- and 3-positions of furan and thiophene [39]. The σ^+ values were obtained by dividing the log k/k_0 values at $227\,^{\circ}C$ by the ρ factor (-0.66 at $227\,^{\circ}C$). These results with σ^+ values, shown in Table 4.3, reveal that

Table 4.3 σ^+ values of the 2- and 3-positions of thiophene, furan, and pyridine.

Compound	σ_2^+	σ_3^+	σ_4^+
Thiophene	−0.79	−0.38	−
Furan	−0.885	−0.415	−
Pyridine	0.80	0.30	0.87

Table 4.4 σ^+ values of the 2- to 8-positions of quinoline.

Quinoline	$\sigma+$
2-	0.73
3-	0.08
4-	0.75
5-	−0.11
6-	0.065
7-	0.15
8-	0.065

the 3-position in furan is slightly more reactive than the 3-position of thiophene, and the 2-position of furan is more reactive than the 2-position of thiophene.

Evidently, the gas-phase pyrolysis reaction provides an excellent model for studying electrophilic substitution effects in heterocyclic compounds. Quantitative electrophilic reactivity of neutral quinoline molecules has also been evaluated [40]: the electrophilic substituent constants σ^+ for the seven positions in quinoline were measured and are given in Table 4.4.

With the exception of the 2- and 4-positions, the reactivity of all other positions in quinoline is deactivated toward electrophilic substitution.

The electronegativity of the nitrogen makes all the positions in quinoline less reactive in electrophilic substitution than their counterpart positions in naphthalene. On the other hand, all positions of the

N-substituted ring of quinoline were found to be more reactive than the counterpart positions in pyridine as a result of the activating effect of a benzo-substituent.

The σ^+ values for all positions of quinoline, naphthalene, and pyridine are provided in Scheme 4.22. The ρ values of the Hammett ρ/σ correlation are mainly used as indicators of the degree of polarity of the transition state in organic reactions, and the sign and magnitude of the charge developed in atoms involved in transition-state formulation.

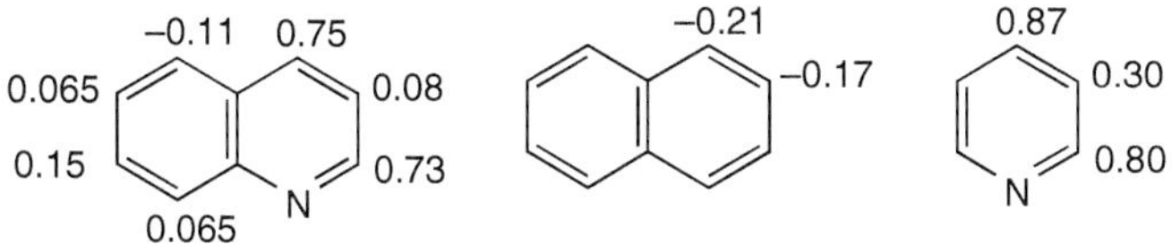

Scheme 4.22

The reaction ρ constant at 227 °C for tert-butyl carboxylate ester **50** substituted at the carbonyl carbon atom was found to be 0.58, whereas it is 0.39 for tert-butyl ethanoate ester **51** [41, 42] (Scheme 4.23).

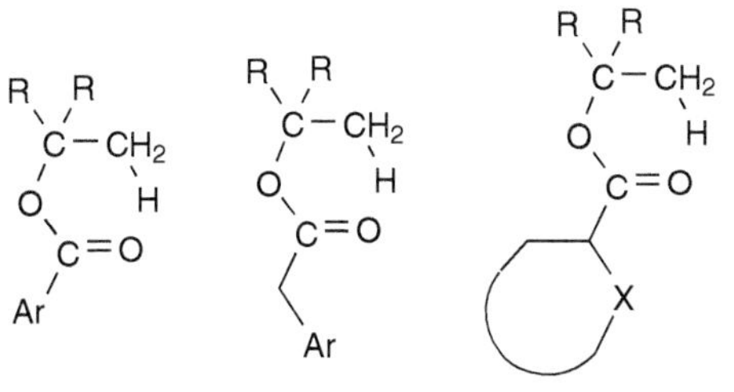

Scheme 4.23

Interestingly, the hetero atoms (X) of the heteroaryl groups of carboxylate esters **52** could be viewed as substituents in their own right rather than as complementary units in heterocyclic rings [43]; accordingly, these can be considered part of molecular forms closer in structure to aryl ethanoate system **51** than to aryl carboxylate **50**. To support this view, the Hammett ρ constants of the gas-phase pyrolytic elimination reactions of tert-butyl heteroarylcarboxylate esters **53a–g** (Scheme 4.24) were measured [44].

The rates and relative values of esters **53a–g** at 227 °C are given in Table 4.5.

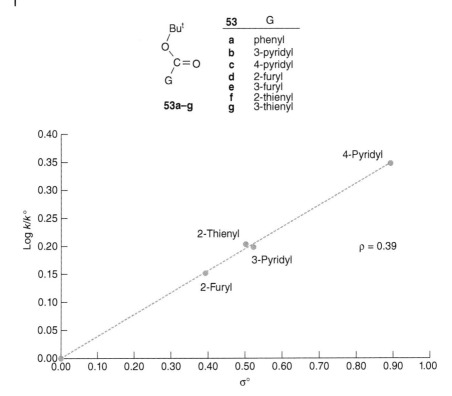

Scheme 4.24

The log k_{rel} values were plotted against the σ° values of heterocyclic groups from the data compiled by Exner, Blanck, and Otsuji et al. [45–47] from the hydrolysis of substituted carboxylate esters and phenylethanoic acids.

The Hammett ρ value calculated from the plot of the gas-phase pyrolysis of tert-butyl heteroarylcarboxylate esters **53a–g** shown in Scheme 4.24 is 0.39 (227 °C). This value of the ρ parameter is in complete agreement with the ρ value of the tert-butyl arylethanoate ester **51** in Scheme 4.23 and notably different from the ρ value obtained for corresponding tert-butyl arylcarboxylate esters **50**.

This finding suggests that Hammett σ/ρ parameters reflect the structural resemblance of heteroarylcarboxylate frame **52** to aryl ethanoate frame **51** as in esters **52**, with at least one carbon atom inserted between the ester moiety and the heteroatom functioning as an individual substituent.

Table 4.5 Rate data at 227 °C for pyrolysis of **53a–g**.

53	Rate const. $10^3\ k\ s^{-1}$	Log k_{rel}
a	54.1	–
b	85.3	0.198
c	120.5	0.348
d	76.9	0.153
e	43.2	−0.098
f	86.3	0.203
g	46.1	−0.069

Substituent σ^0 constant values for various heterocyclic groups have further been investigated by Al-Awadi and her group [48].

4.6 Alkoxy versus Amino Group

The structural and electronic effects of replacing the oxygen atom of the alkoxy heterocycles with the NH moiety of the amino analogues were assessed by a detailed kinetic study of the pyrolysis gas-phase reaction of ethyl, isopropyl, and tert-butyl pyrazine (**54–59**) and pyrimidines (**60–65**) [49].

2-N-Alkylaminopyrazines and their pyrimidine analogues undergo unimolecular first-order thermal elimination to give alkenes and aminopyrazines and pyrimidines, respectively, through a six-membered transition state, according to the mechanism shown in Scheme 4.25.

Rate data for the alkylaminopyrazines and pyrimidines (Scheme 4.26) show that replacing the oxygen atom with the NH moiety decreases the reactivity consistently by factors of $4.2 \times 10^2 – 4.2 \times 10^4$.

This is to be expected, since cleavage of the C—O bond in the alkoxy compounds proceeds faster than the cleavage of the less-polar C—N bond in the amino compounds. The relative reactivities of 1°: 2°: 3° N-alkylaminopyrazine are 1 : 14 : 38, which are smaller than those of N-alkylaminopyrimidines (1 : 21 : 163). This rate factor of alkoxy versus alkylamino heterocycles $k_{(O)}/k_{(NH)}$ increases with branching in the alkyl group. Scheme 4.26 shows a ratio of 424, 795, 41 640 for pyrazines

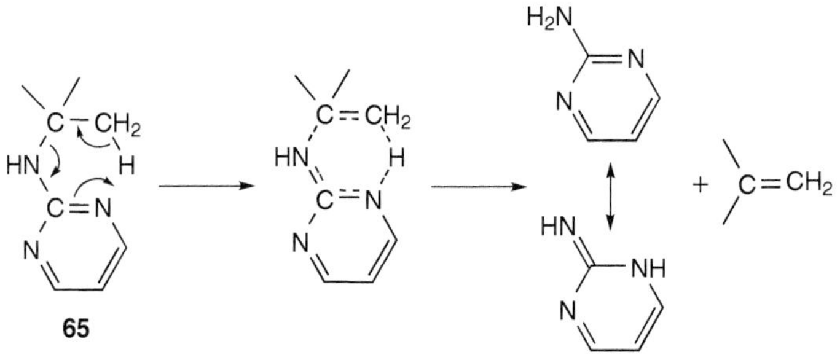

Scheme 4.25

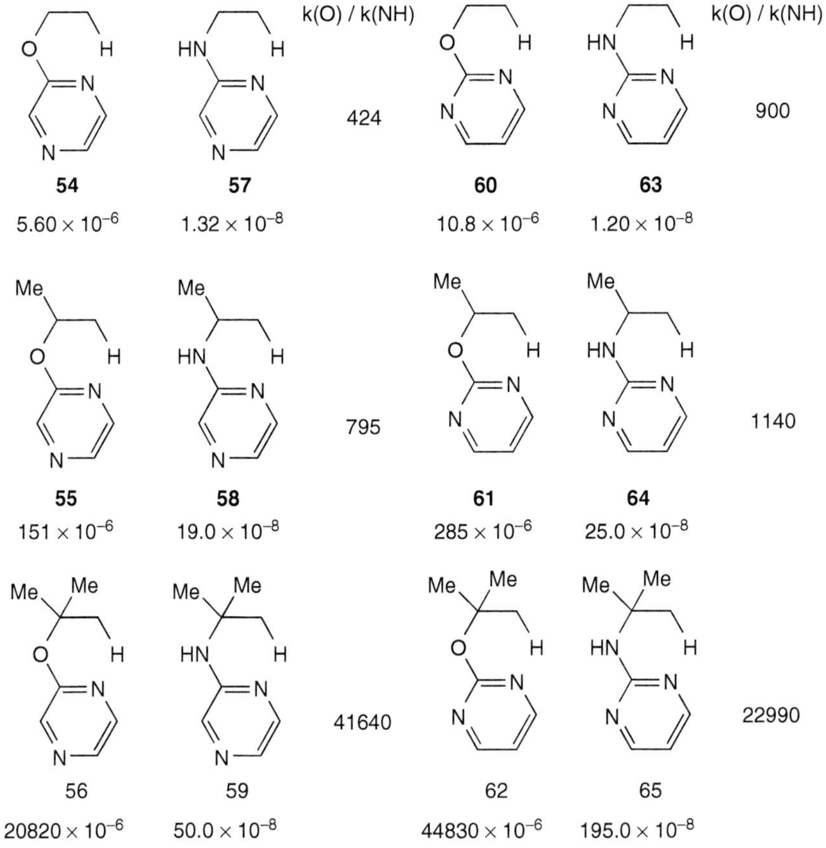

Scheme 4.26

and 900, 1 140, 22 990 for pyrimidines. This is due to the relative stabilizing effect of the alkyl groups on the carbocation transition state, and the electron-withdrawing effect of the two ortho N-replacement substituents of the electron-deficient pyrimidine nucleus in comparison with the ortho- and meta-nitrogens of the pyrazine isomer. This will facilitate C—N bond cleavage in the pyrimidine as the most rate-controlling step of the reaction. Moreover, the polarity of the transition state should be greater for pyrimidines, as appears to be the case.

4.6.1 Neighboring Group Participation

Alan Maccoll 1956 [50] suggested neighboring group participation in the gas-phase pyrolysis reaction of alkyl halide via an intimate ion-pair mechanism. Later, Chuchani and his group showed that ethyl chlorides substituted with a polar group on the β position are much more active than the unsubstituted ethylchloride, and that has been confirmed by a detailed kinetic study [51]. Several examples of the gas-phase pyrolysis reactions of 2-substituted ethyl chlorides are described in Table 4.6 [52–55]; it is assumed that the higher rate ratio relative to the unsubstituted chlorides is a result of the significant effect of H_3CS and $(H_3C)_2N$ groups in stabilizing and polarizing the C—Cl bond in the transition state of the reaction (Scheme 4.27).

It is thought that substituents Cl, OH, and OCH_3 operate by their inductive effect (—I), which results in a low relative rate value, while the substituents thiomethyl SMe and N-dimethyl $N(Me)_2$ have a

Table 4.6 Relative rates for ZCH_2CH_2Cl pyrolysis at 400 °C.

Z	Relative rate
H	1
Cl	4.3
OH	2.12
OCH_3	2.39
SCH_3	558.7
$N(CH_3)_2$	560.0

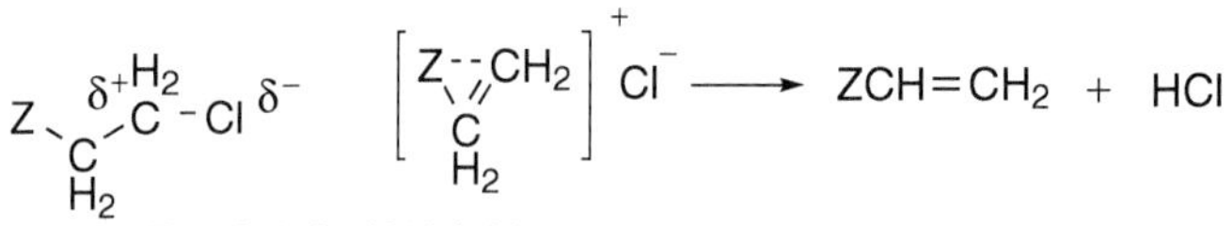

$$Z\underset{H_2}{\overset{\overset{+}{\delta}H_2}{\underset{C}{\overset{C}{\diagdown}}}}\overset{C}{\underset{}{}}-Cl\;\delta^- \quad \left[\underset{\overset{C}{\underset{H_2}{}}}{Z\cdots CH_2}\right]^+ Cl^- \longrightarrow ZCH=CH_2 \;+\; HCl$$

Z = CH₃S, (CH₃)₂N

Scheme 4.27

much larger effect on reactivity. The pronounced effect on reactivity is explained in terms of neighboring group participation exerted by the SMe and N(Me)₂ substituents, as these groups will be polarized by the effect of the partial positive carbon atom resulting from C—Cl bond polarization in the transition state (Scheme 4.28).

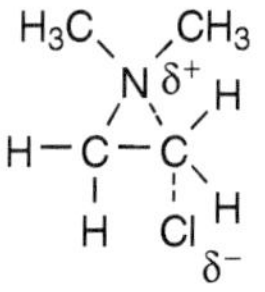

$$\underset{H}{\overset{H_3C}{\diagdown}}\underset{}{\overset{}{N}}\overset{\delta^+}{\underset{}{}}\underset{}{\overset{CH_3}{\diagup}}H$$

H—C—C
 | |
 H   Cl   H
      δ⁻

Scheme 4.28

It is emphasized that the order of reactivity shown in Table 4.6 is similar to that reported in solution, where the neighboring group participation of the sulfur and nitrogen atoms is in a fully ionized transition state, enhancing the reactivity of the substrates in the reaction [56].

The favorable effect of SMe and N(Me)₂ groups in solution reaction for the five-membered cyclic transition state relative to the three-membered ring has been assessed in gas-phase pyrolysis reactions by Chuchani's group, who studied the effect of SMe on the pyrolysis reaction of 1-chloro-4-(methylthio)-butane **66** through kinetic investigation [57]. This effect is evidenced by the fact that only a cyclic molecule is obtained, suggesting an intimate ion-pair type intermediate that will enhance the departure of the leaving chloride ion by intramolecular solvation or outosolvolysis to give tetrahydrothiophene **67** and methylchloride (Scheme 4.29).

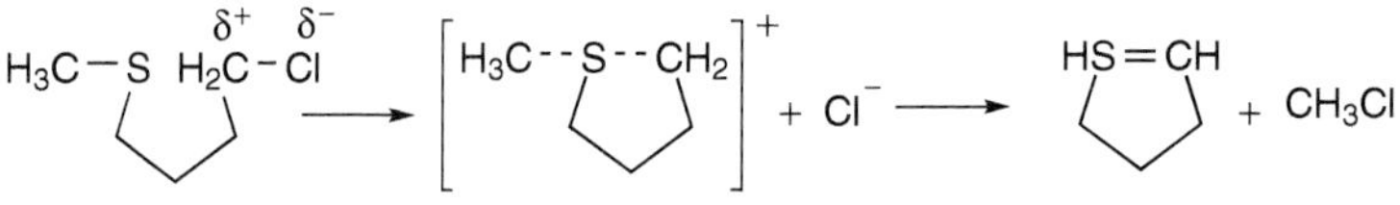

$$H_3C-S\;\overset{\delta^+}{H_2}C\overset{\delta^-}{-Cl} \longrightarrow \left[\underset{}{H_3C\cdots S\cdots CH_2}\right]^+ + Cl^- \longrightarrow HS=CH + CH_3Cl$$

66   **67**

Scheme 4.29

It was proposed that neighboring group participation is most effective in a gas-phase pyrolysis reaction when the transition state is polar [58a, b]. This phenomenon is assessed by investigating the effect of the SMe group on the pyrolysis reaction of butylacetate **68** and 1-chlorobutane **69**; the latter has a more polar transition state. Kinetic data in Scheme 4.30 show that the (SMe) substituent has a much more pronounced effect in accelerating the pyrolysis rate in polar 1-chlorobutane than in butyl-4-(methylthio)-acetate **70** on unsubstituted butylacetate **68**.

	Relative rate
$CH_3(CH_2)_3$ — Cl (**69**)	1
MeS —$(CH_2)_4$— Cl (**66**)	720
$CH_3(CH_2)_3$—OAc (**68**)	1
MeS —$(CH_2)_4$—OAc (**70**)	25

Scheme 4.30

From the kinetic data obtained in Scheme 4.31, it is evident that the anchimeric assistance of the carbomethoxy group($COOCH_3$) [59] is much more effective in the stabilization of the more-polar allylic C—Br bond in the transition state of methyl-(4-bromocrotonate) **71** relative to the same effect on the less-polar C—Br bond of the transition state for the saturated compound 4-bromobutyrate **72**.

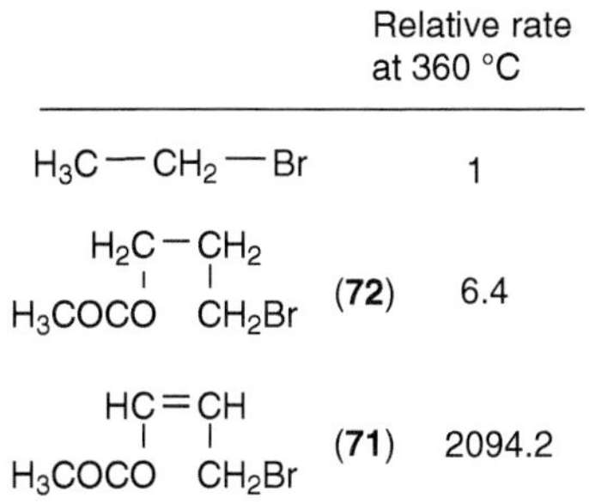

	Relative rate at 360 °C
H_3C — CH_2 — Br	1
H_2C — CH_2 $\quad\mid\qquad\mid$ $H_3COCO\quad CH_2Br$ (**72**)	6.4
HC = CH $\quad\mid\qquad\mid$ $H_3COCO\quad CH_2Br$ (**71**)	2094.2

Scheme 4.31

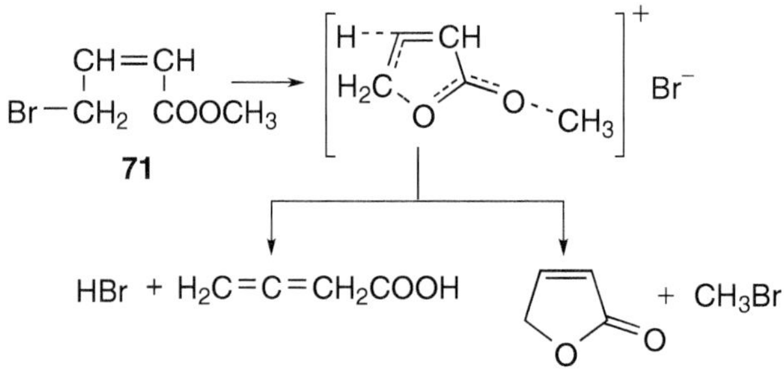

Scheme 4.32

The involvement of the carbomethoxy group is also supported by the nature of the products formed (Scheme 4.32) during dehydrobromination upon the pyrolytic reaction of compound **71** to give lactone.

4.6.2 Amino Esters

The same approach in assessing the intimate ion-pair mechanism resulting from neighboring group participation has also been applied to gas-phase pyrolysis reactions for amino esters. The product analysis and relative rate of reaction support such a reaction mechanism. 3-(Dimethylamino)-1-propylacetate **73** and 3-(dimethylamino)-1-butylacetate **74** were studied and compared with their unsubstituted counterpart propyl- and butyl-acetates [60].

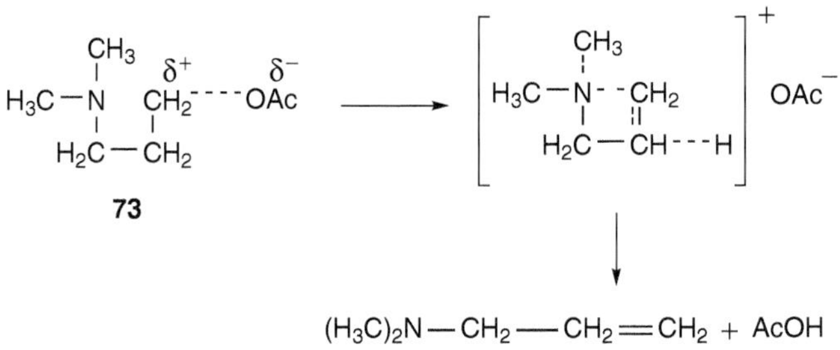

Scheme 4.33

The mechanism suggested and shown in Scheme 4.33 for the gas-phase pyrolysis of **73** is based on product formation. The cyclic four-membered transition state shown in Scheme 4.33 was also observed by Grob and Jenny [61] in solvolysis of 3-(dimethylamino)-

1-chloropropane, which had a much higher rate of solvolysis than unsubstituted chloropropane.

This finding was supported by the product analysis and kinetic data for the butyl esters (Scheme 4.34).

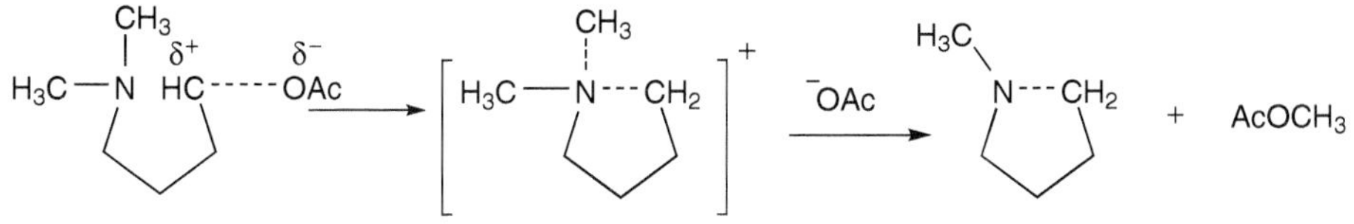

74

Scheme 4.34

Scheme 4.35 shows the relative rate data for substituted acetates **73** and **74** relative to their unsubstituted counterparts.

	Rel. Rate
AcO—(CH$_2$)$_2$—CH$_3$	1
AcO—(CH$_2$)$_2$—CH$_2$—N(Me)$_2$ **(73)**	2.3
AcO—(CH$_2$)$_3$—CH$_3$	1
AcO—(CH$_2$)$_3$—CH$_2$—N(Me)$_2$ **(74)**	158

Scheme 4.35

To assess the nucleophilic effect of the amino substituents in neighboring group participation, 4-(N-phenylamino)-1-butylacetate **75** and 4-(N-methyl-N-phenylamino)-1-butylacetate **76** were pyrolyzed and their kinetic data investigated [62]. Products analysis of compounds **75** and **76** suggests two different mechanisms, as shown in Scheme 4.36.

Path A proceeds via anchimeric assistance by the methylphenyl amino substituent to form an intimate ion-pair intermediate, which leads to intramolecular solvation of the acetate anion. Path B involves a six-membered cyclic transition state well known for ester pyrolysis.

From the comparative kinetic data for butylacetate and the substituted butylacetates **75** and **76** shown in Scheme 4.37, it becomes evident that the nucleophilicity of the amino substituent has an important role in the mechanism of pyrolysis reactions for butyl esters due to its anchimeric assistance in stabilizing the polarized C—O bond in the transition state.

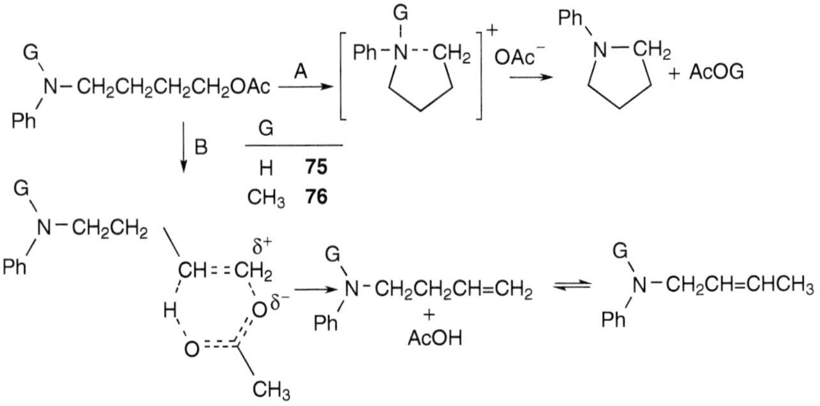

Scheme 4.36

This effect is consistent with the reactivity order reflected in the rate coefficients $(CH_3)_3N > Ph(CH_3)N > PhNH$ (Scheme 4.37).

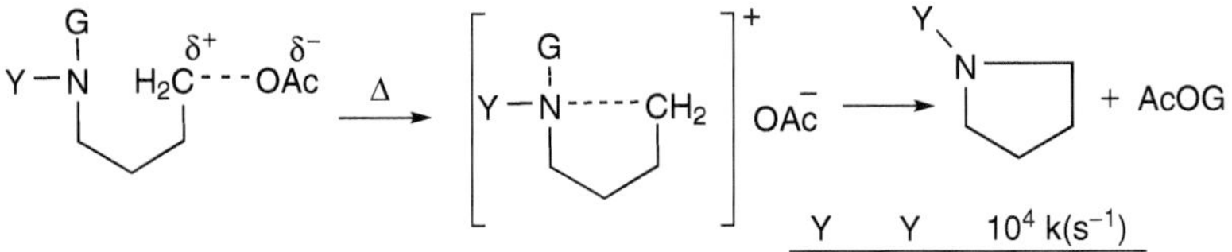

Y	Y	$10^4\ k(s^{-1})$
CH_3	CH_3	678.19
Ph	CH_3	7.76
Ph	H	6.22

Scheme 4.37

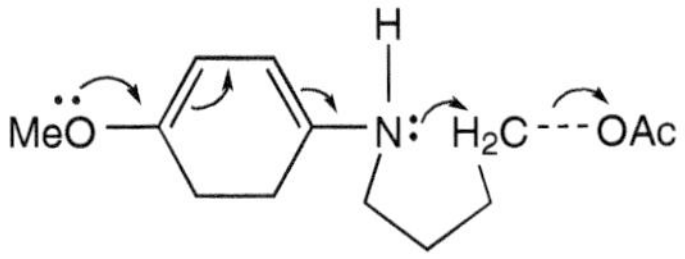

77a

Scheme 4.38

This order of reactivity suggests that decreasing the nucleophilicity of the nitrogen atom of the amino moiety in substituted butylacetates **75** and **76** results in decreasing its anchimeric assistance in stabilizing the transition state of the reaction. This reactivity order is

Table 4.7 Rate constants at 227 °C and relative rates of gas-phase pyrolysis reactions of compounds **77a–c** (p-X C$_6$H$_4$[CH$_2$]$_4$OAc).

X	$10^4\,k\,s^{-1}$	k_{rel}
H	0.29	1
OMe	133.4	460.0
Me	78.9	272.1
Cl	15.6	53.8

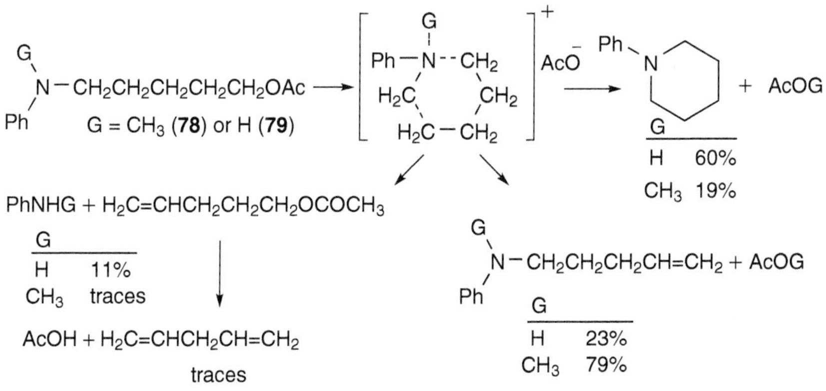

Scheme 4.39

further confirmed by product and kinetic analysis investigation for 5-(N-arylamino)-1-butylacetates **77a–c**, in which the aryl group of the amino moiety is para-substituted with a group with the opposite electronic effect (**a**: OMe, **b**: Me, and **c**: Cl) [63]. Characterization of the pyrolysis reaction products reveals the formation of the expected N-arylpyrolidines. Moreover, the kinetic data indicate that the reactivity of these substituted butyl acetates **77a–c** has increased due to the positive mesomeric electron-donating effect of the OMe substituent as a result of increasing the nucleophilicity of the nitrogen atom involved, thus resulting in large anchimeric assistance in stabilizing the C—O bond (Scheme 4.38).

This result is well justified by the opposite electronic effect of the chloro substituent of **77c** and the modest positive inductive effect of the methyl group in **77b** (Table 4.7).

The effect of a cyclic six-membered structure versus a five-membered structure in neighboring group participation for the gas-phase pyrolysis reactions of 5-(N-methyl-N-phenylamino)-1-pentyl acetate **78** and 5-(N-phenylamino)-1-pentyl acetate **79** has also been studied [64], where compounds **78** and **79** resulted in the formation of N-phenylpiperidine (Scheme 4.39), suggesting an intimate ion-pair mechanism.

References

1 Al-Awadi, N.A., Elnagdi, M.H., and Mathew, T. (1995). *Int. J. Chem. Kinet.* 27: 517–523.

2 Szware, M. and Marawski, J. (1951). *Trans. Farady Soc.* 47: 269–274.

3 (a) Bailey, W.J. and Bayloung, R.A. (1959). *J. Am. Chem. Soc.* 81: 2126–2129. (b) Bailey, W.J. and King, C. (1956). *J. Org. Chem.* 21: 858–861.

4 Al-Awadi, N.A., Al-Bashir, R.F., and El-Dusouqui, O.M.E. (1989). *J. Chem. Soc. Perkin Trans.* 2: 579–581.

5 Bailey, W.J. and Hale, W.F. (1959). *J. Am. Chem. Soc.* 81: 651–655.

6 Froemsdorf, D.H., Collins, C.H., Hammoud, G.S., and Depuy, C.H. (1959). *J. Am. Chem. Soc.* 81: 643–647.

7 Al-Awadi, N.A. and Mathew, T. (1995). *Int. J. Chem. Kinet.* 27: 843–848.

8 Taylor, R. (1975). *J. Chem. Soc. Perkin Trans.* 2: 1025–1029.

9 Smith, G.G. and Kelly, F.W. (1971). *Progress in Physical Organic Chemistry*, vol. 8 (eds. A. Streitwieser Jr., and R.W. Taft), 75. New York: Wiley.

10 Holbrook, K.A. (1992). *In the Chemistry of Functional Groups: Supplement B* (ed. S. Patai), 703, Chapter 12. Chichester: Wiley.

11 (a)Taylor, R. (1991). *Int. J. Chem. Kinet.* 23: 247–250. (b) Taylor, R. (1983). *J. Chem. Soc. Perkin Trans.* 2: 809–811.

12 Al-Awadi, N.A. and El-Dusouqui, O.M.E. (1997). *Int. J. Chem. Kinet.* 29: 295–298.

13 Al-Juwaiser, I.A., Al-Awadi, N.A., and El-Dusouqui, O.M.E. (2002). *Can. J. Chem.* 80: 499–503. and references therein.

14 Bigley, D.B. and Clarke, M.J. (1982). *J. Chem. Soc. Perkin Trans.* 2: 1–6.

15 Al-Awadi, N.A., Elnagdi, M.H., and El-Dusouqui, O.M.E. (1998). *Int. J. Chem. Kinet.* 30: 457–462.

16 Malhas, R.N., Al-Awadi, N.A., and El-Dusouqui, O.M.E. (2007). *Int. J. Chem. Kinet.* 39: 82–91.

17 Wentrup, C. and Crow, W.D. (1970). *Tetrahedron* 26: 3965–3981.

18 Kvaskoff, D., Lüerssen, H., Bednarek, P., and Wentrup, C. (2014). *J. Am. Chem. Soc.* 136: 15203–15214.

19 Wentrup, C. and Freiermuth, B. (2016). *J. Anal. Appl. Pyrolysis* 121: 67–74.

20 Brown, R.F.C. (1980). *Pyrolytic Methods in Organic Chemistry*. New York: Academic Press.

21 Crow, W.D. and Wentrup, C. (1968). *Chem. Commun.*: 1026–1027.

22 Ohashi, M., Tsujimoto, K., Yoshino, A., and Yonezawa, T. (1970). *Org. Mass Spectrom.* 4: 203–210.

23 Mehta, L.K., Patrick, J., and Payne, F. (1993). *J. Chem. Soc. Perkin Trans.* 1: 1261–1267.

24 Fan, W.-Q. and Katritzky, A.R. (1996). *Comprehensive Heterocyclic Chemistry*, vol. 4 (eds. A.R. Katritzky, C.W. Rees, E.F.V. Scriven and R.C. Storr), 1–126. Oxford: Elsevier.

25 Dib, H.H., John, E., El-Dusouqui, O.M.E. et al. (2017). *Anal. Appl. Pyrolysis* 124: 403–408.

26 Dib, H.H., Al-Awadi, N.A., Ibrahim, Y.A., and El-Dusouqui, O.M.E. (2003). *Tetrahedron* 59: 9455–9464.

27 Dib, H.H., Al-Awadi, N.A., Ibrahim, Y.A., and El-Dusouqui, O.M.E. (2004). *J. Phys. Org. Chem.* 17: 267–272.

28 Al-Awadi, N.A., Geroge, J.B., Dib, H.H. et al. (2005). *Tetrahedron* 61: 8257–8263.

29 Johnston, L.J. and Scaiano, J.C. (1989). *Chem. Rev.* 89: 521–547. and references therein.

30 Al-Awadi, N.A., Ibrahim, Y.A., Patel, M. et al. (2007). *Int. J. Chem. Kinet.* 39: 59–66.

31 Al-Awadi, N.A., Kaul, K., and El-Dusouqui, O.M.E. (2000). *J. Phys. Org. Chem.* 13: 499–504.

32 Smith, G.G., Bagley, F.D., and Taylor, R. (1961). *J. Am. Chem. Soc.* 83: 3647–3653.

33 Brown, H.C. (1958). *J. Am. Chem. Soc.* 80: 4979–4987.

34 Bartlett, P.D. and Rüchardt, C. (1960). *J. Am. Chem. Soc.* 82: 1756–1762.

35 Harrison, A.G., Kebarle, P., and Lossing, F. (1961). *J. Am. Chem. Soc.* 83: 777–780.

36 Pearson, D.E. and Baxter, J.F. (1952). *J. Org. Chem.* 17: 1511–1518.
37 Norman, R.O. and Taylor, R. (1965). *Electrophilic Substitution in Benzenoid Compounds*, 283. Amsterdam: Elsevier.
38 Taylor, R. (1962). *J. Chem. Soc.*: 4881–4888.
39 Taylor, R. (1968). *J. Chem. Soc. B*: 1397–1401.
40 Taylor, R. (1971). *J. Chem. Soc. B*: 2382–2387.
41 van Bekkum, H., Verkade, P.E., and Wepster, B.M. (1959). *Rec. Trav. Chim.* 78: 815–863.
42 Taft, R.W. (1960). *J. Phys. Chem.* 84: 1805–1815.
43 ten Thije, P.A. and Janssen, M.J. (1965). *Rec. Trav. Chim.* 84: 1169–1176.
44 Al-Awadi, N.A., Al-Bashir, R.F., and El-Dusouqui, O.M.E. (1990). *Tetrahedron* 46: 2903–2910.
45 Exner, O. (1978). *Correlation Analysis in Chemistry: Recent Advances* (eds. N.B. Chapman and J. Shorter), Chapter 10. New York: Plenum Press.
46 Blanch, J.H. (1966). *J. Chem. Soc. B*: 937–939.
47 Otsuji, Y., Koda, Y., Kubo, M. et al. (1959). *Nippon Kagaku Zasshi* 80: 1300–1307.
48 Al-Awadi, N.A., Al-Bashir, R.F., and El-Dusouqui, O.M.E. (1990). *Tetrahedron* 46: 2911–2916.
49 Al-Awadi, N.A., El-Dusouqui, O.M.E., Kaul, K., and Dib, H.H. (2000). *Int. J. Chem. Kinet.* 32: 403–407.
50 Maccoll, A. and Thomas, P.J. (1955). *J. Chem. Soc.*: 979–986.
51 Chuchani, G., Martin, I., and Martin, G. (1979). *Int. J. Chem. Kinet.* 11: 109–115.
52 Tsang, W. (1964). *J. Chem. Phys.* 41: 2487–2494.
53 Holbrook, K.A., Walker, R.W., and Watson, W.R. (1971). *J. Chem. Soc. B*: 577–581.
54 Skingle, D.C. and Stimson, V.R. (1976). *Aust. J. Chem.* 29: 609–615.
55 Chuchani, G., Martin, I., and Bigley, D.B. (1978). *Int. J. Chem. Kinet.* 10: 649–652.
56 Winstein, S. and Grunwald, E. (1948). *J. Am. Chem. Soc.* 70: 828–837.
57 Chuchani, G., Martin, I., Hernandez, J.A. et al. (1980). *J. Phys. Chem.* 84: 944–948.
58 (a) Jagau, J.-C., Prochnow, E., Evangelista, F.A., and Gauss, J. (2010). *J. Chem. Phys.* 132: 144110. (b) Li, X.-Z. and Paidus, J. (2010). *J. Chem. Phys.* 132: 144103.
59 Chuchani, G. and Martin, I. (1988). *Int. J. Chem. Kinet.* 20: 1–8.

60 Chuchani, G., Rotinov, A., Dominguez, R.M., and Gonzalez, N. (1984). *J. Org. Chem.* 49: 4157–4160.

61 Grob, C.A. and Jenny, F.A. (1960). *Tetrahedron Lett.* 23: 25–29.

62 Chuchani, G., Al-Awadi, N.A., Dominguez, R.M. et al. (2000). *J. Phys. Org. Chem.* 13: 266–271.

63 Al-Awadi, N.A., Elnagdi, M.H., Kaul, K., and Chuchani, G. (2000). *J. Phys. Org. Chem.* 13: 675–678.

64 Chuchani, G., Al-Awadi, N.A., Dominguez, R.M., and Kaul, K. (2001). *J. Phys. Org. Chem.* 14: 180–186.

5

Functional Group and Structural Frame Interconversions

This chapter presents several illustrative examples of functional and structural frame interconversion. An extensive discussion of thermal retro-ene reactions as valuable synthetic routes to various organic compounds is included.

5.1   Functional Group Interconversion

There is considerable scope for the development of flash vacuum pyrolysis (FVP) methods for functional group transformations, particularly where the advantage of chemistry without reagents provides for convenient work-up and isolation.

5.1.1   Thermal Retro-Ene Reactions

The formation of an alkene unit is a common FVP transformation [1, 2], which is often accomplished by retro-ene reactions. The synthetic applications of retro-ene reactions have been reviewed by Hoffmann [1] and Ripoll et al. [2]. A comparative study of acetate, xanthate, and tosylate pyrolysis has been used to optimize the preparation of 6-chlorohex-1-ene [3]. The main problem here is avoiding secondary elimination of HCl to give hexa-1,5-diene. Yields of up to 80% of the

Gas-Phase Pyrolytic Reactions: Synthesis, Mechanisms, and Kinetics,
First Edition. Nouria A. Al-Awadi.
© 2020 John Wiley & Sons, Inc. Published 2020 by John Wiley & Sons, Inc.

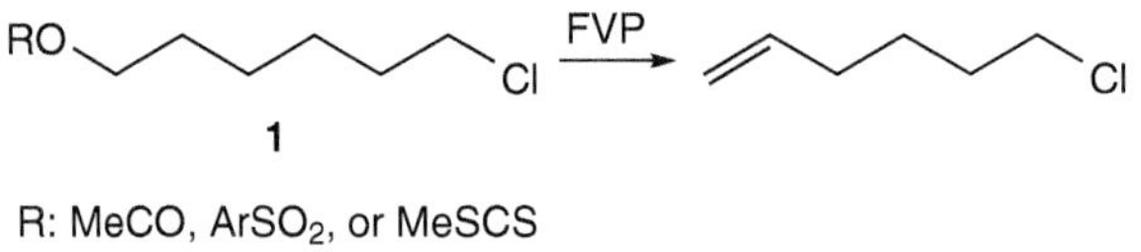

RO
Cl
FVP
Cl
1
R: MeCO, $ArSO_2$, or MeSCS

Scheme 5.1

required product were obtained from compound **1** (Scheme 5.1), which requires a lower pyrolysis temperature than the alternative precursors.

A functional group transformation of unsaturated alcohols to unsaturated ketones was obtained in the gas-phase pyrolysis reaction of 4-vinylhepta-1,6-dien-4-ol **2**, which undergoes a retro-ene reaction to give hexadiene-3-one **3** and a competing oxy-Cope-rearrangement giving 1,8-nonadiene-4-one **4** [4] (Scheme 5.2).

O−H
2
retro-ene
oxy-Cope
O +
3
O−H
O
4

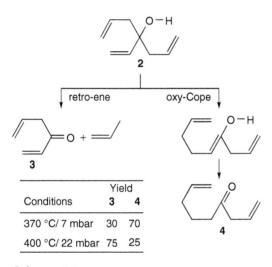

Conditions	Yield 3	Yield 4
370 °C/ 7 mbar	30	70
400 °C/ 22 mbar	75	25

Scheme 5.2

The gas-phase pyrolysis of unsaturated acetals **5** represents an interesting example of functional group interconversion, where carbonyl compounds, alkenes, and unsaturated ethers are formed in thermal retro-ene reactions [5] (Schemes 5.3 and 5.4).

The thermal interconversion of the functional groups of vinyl ethers **6** to ketones takes place quantitatively by extrusion of an ethene molecule [6] (Scheme 5.5).

Similarly, pyrolysis of n-butyl vinyl ether **7** at 480 °C produces 1-butene and acetaldehyde (Scheme 5.6)

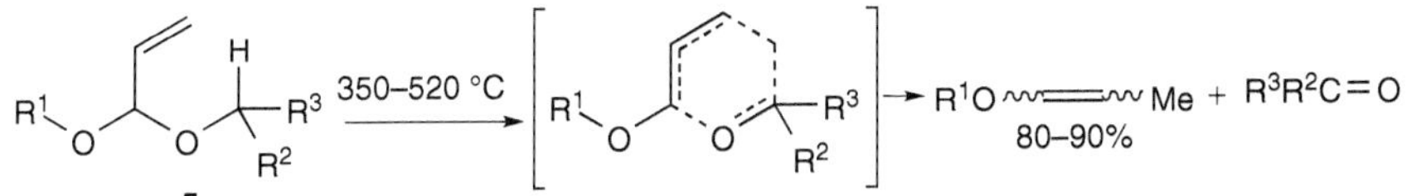

Scheme 5.3

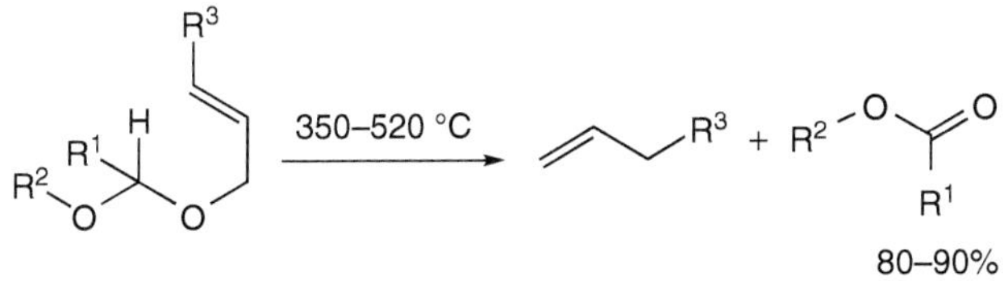

R^1-R^3 = H or alkyl

Scheme 5.4

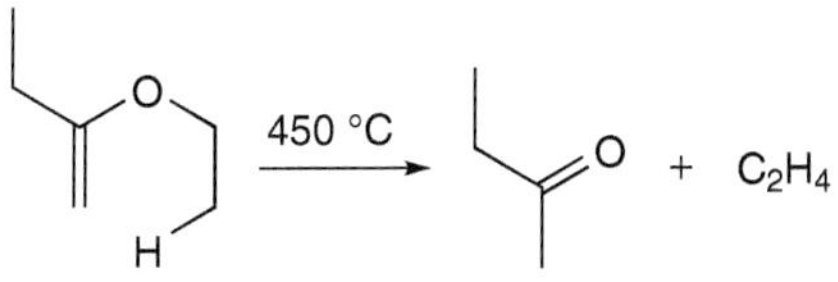

Scheme 5.5

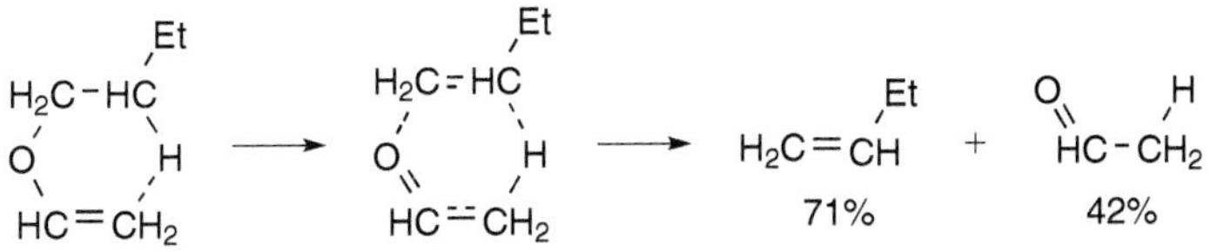

Scheme 5.6

Pyrolysis of 2-ethoxy-1-butene **8** at 460 °C followed the same approach to yield a quantitative conversion to methyl ethyl ketone in 94% yield (Scheme 5.7).

A direct route to vinyl ethers **9** was achieved [6] from the pyrolysis of α-acetoxy ether **10** at 380 °C (Scheme 5.8).

Several amine-imines interconvert by a retro-ene reaction of β-aminoalkenes **11a–c** and β-aminoalkynes **12a,b** [7] (Scheme 5.9).

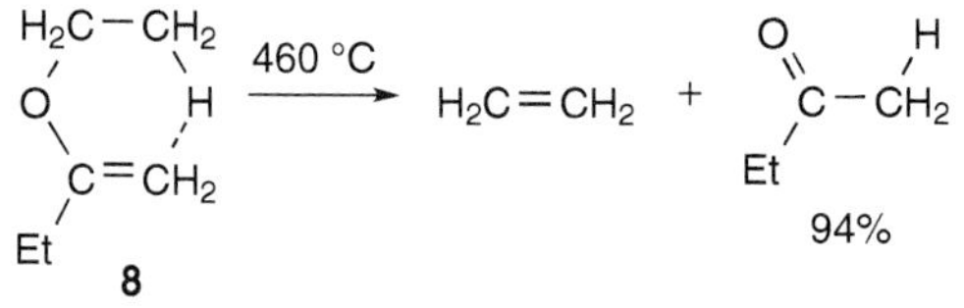

$$H_2C=CH_2 + \underset{Et}{\overset{O}{\underset{|}{C}}}-CH_2 \quad \text{(structure with 94%)}$$

Compound **8** with 460 °C reaction arrow yields $H_2C=CH_2$ + product at 94%

Scheme 5.7

Compound **10** with 380 °C reaction arrow yields $CH_3-CH_2-CH_2-CH_2-O-CH=CH_2$ (48%) + **9**

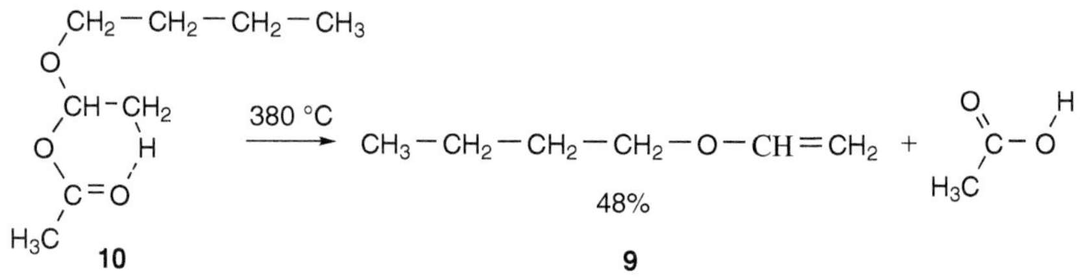

Scheme 5.8

3-Hydroxypropionitrile **13** eliminates to unsubstituted ketenimine **14** by a retro-ene reaction that involves cyclic rearrangement and elimination [8–10] (Scheme 5.10).

Allyl sulfide derivatives have been successfully converted to unsaturated thiocarbonyl compounds [11, 12]. Allyl cyano sulfide **15a** undergoes a retro-ene reaction at 800 °C by FVP to yield propene and 1,3,5-trithiane **16** [13], which were observed in the flow thermolysis of thioformyl cyanide **17** [14] (Scheme 5.11). On the other hand, sulfide **15b**, upon FVP under the same conditions, produced monothiogloxal **18** [13] (Scheme 5.11).

Diallyl sulfide **19**, allyl benzyl sulfide **20**, and 3-allylthio-2-butanone **21** at 615–660 °C resulted in thioacrolein **22**, thiobenzaldehyde **23**, and monothiobiacetyl **24**, respectively, with the elimination of propene [14] (Scheme 5.12).

The cycloreversions of the retro Diels-Alder reactions of cyclohexene and, more specifically, hetero cyclohexenes provided an easy access for the synthesis of conjugated dienic and, more importantly, heterodienic systems [15, 16].

Accordingly FVP at 800 °C coupled with low-temperature IR spectroscopy for heterocycle **25** resulted in the detection of thioxoethanol **26** and ethene [17a, b]. The characterization of these products was further established by photoelectron spectroscopy (Scheme 5.13).

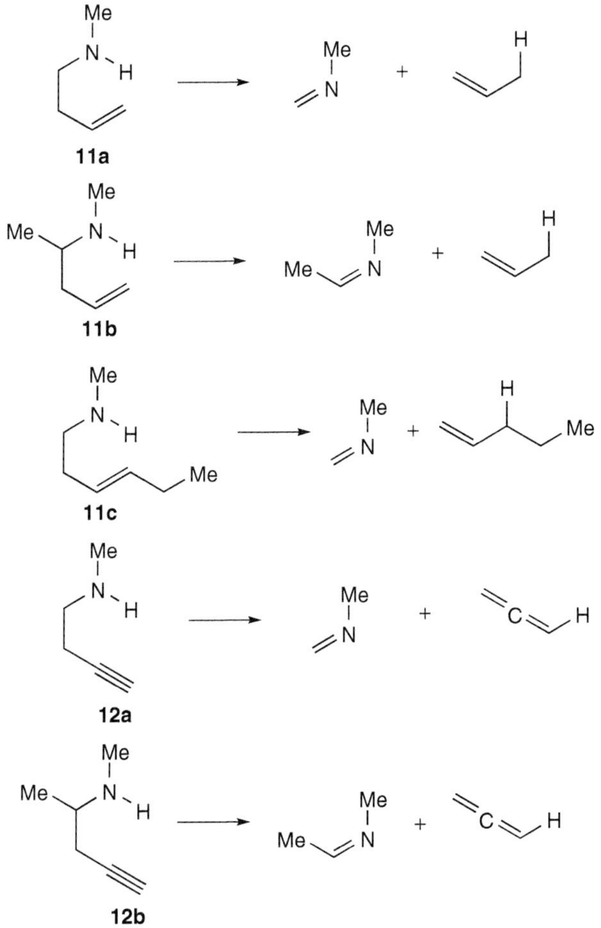

Scheme 5.9

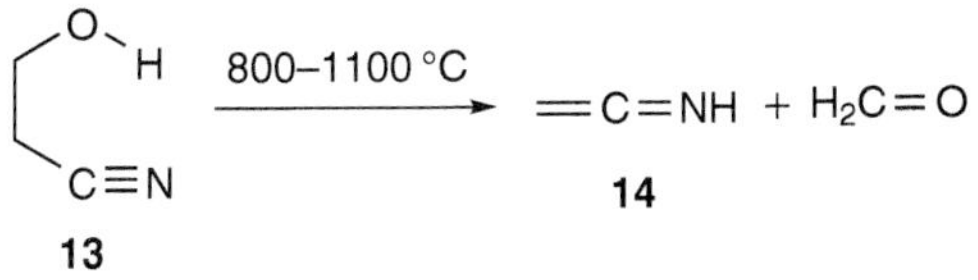

Scheme 5.10

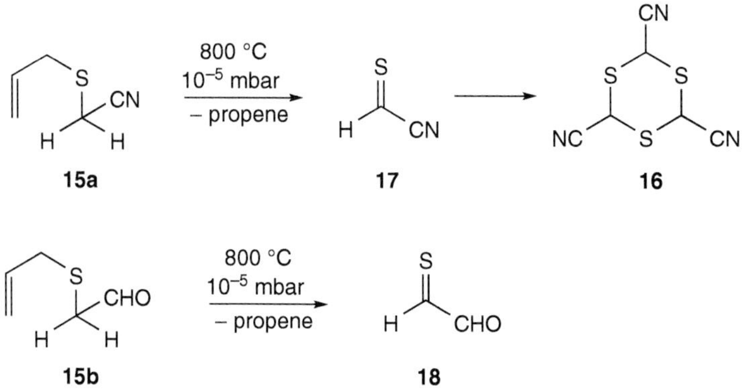

Scheme 5.11

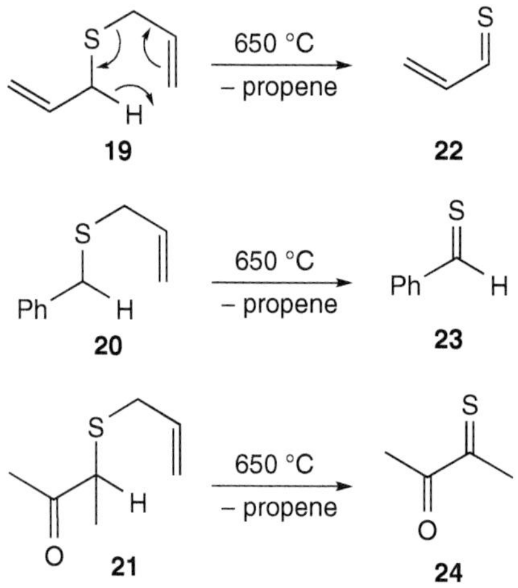

Scheme 5.12

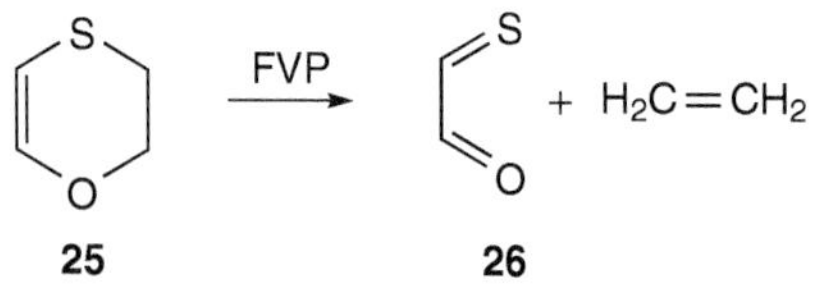

Scheme 5.13

A one-step retro-ene reaction converts diprop-2-ynyl ether **27** under FVP of 750 °C into propynal **28** [18] (Scheme 5.14).

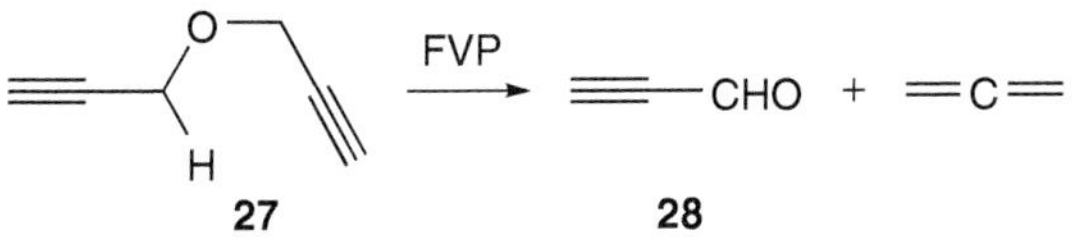

Scheme 5.14

Effective functional group transformation of propargyl thiocyanates [19] and their seleno [20] analogous **29** into isothiocyanates **30** takes place at 350–400 °C by gas-phase pyrolysis, which results in 98% yield (Scheme 5.15). Both products are useful substrates for heterocyclization reactions by the treatment with nucleophilic reagents.

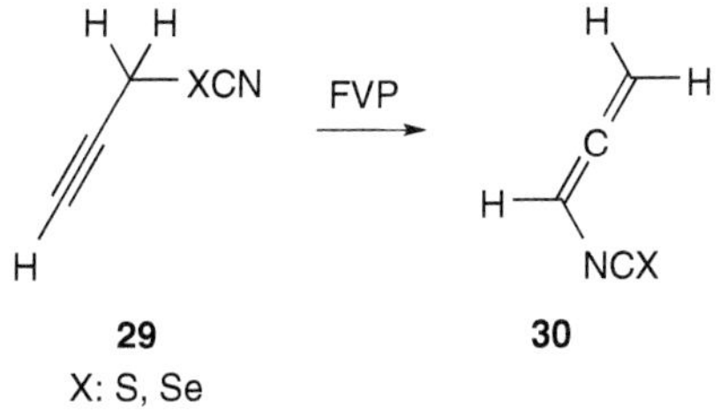

X: S, Se

Scheme 5.15

Functional group interconversion of isonitrile-nitrile rearrangement (Scheme 5.16) is an example that can be carried out with great efficiency under FVP conditions [21]. Yields obtained are high, and a wide range of structurally diverse isonitrile derivatives undergo this rearrangement under standard conditions.

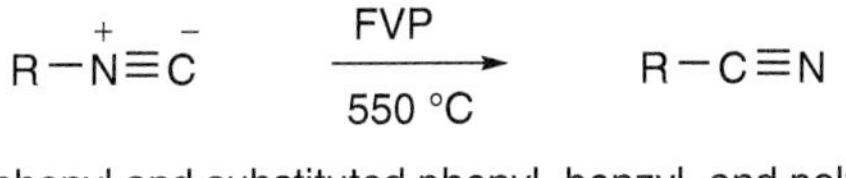

R = alkyl, phenyl and substituted phenyl, benzyl, and polycyclic

Scheme 5.16

5.1.1.1 α-Substituted Carboxylic Acids

An extensive study by Chuchani and co-workers led to the elucidation of the mechanism of the gas-phase pyrolysis reaction of α-substituted carboxylic acids and their alkyl esters [22–26]. The transition state for the reaction is believed to be concerted, polar, and five-membered, and it involves the elimination of a hydrogen atom from the carboxylic acid group (Scheme 5.17).

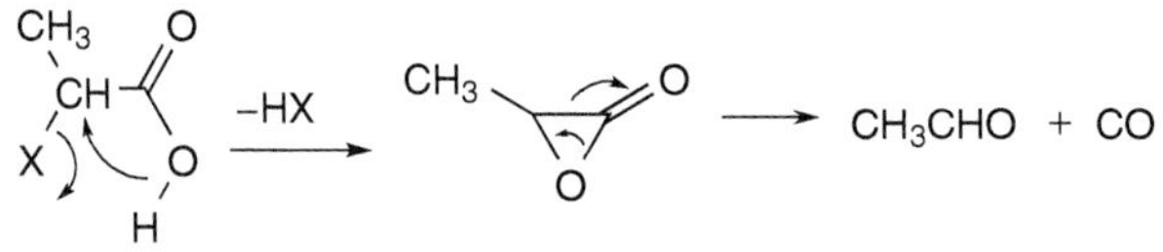

X = Cl, Br, OH, OR, OPh, NHPh, SPh

Scheme 5.17

The reaction shown in Scheme 5.17 can be applied to various α-substituted carboxylic acids, where X is a halogen, hydroxyl, alkoxy, or aryloxy, or its amino or thio analogues.

Elimination of HCl from 2-chloropropionic acid **31** and that of HBr from its bromo analogue takes place via a polar five-membered cyclic transition state [23] in which the hydrogen atom of the COOH group is involved (Scheme 5.18).

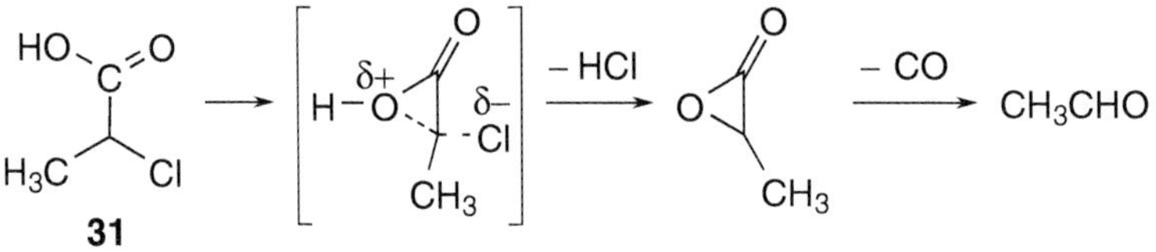

Scheme 5.18

In the absence of this hydrogen atom, as in methyl 2-halopropionate **32** (Scheme 5.19), elimination of HX proceeds via a four-centered transition state that involves hydrogen of the β-carbon. It is noteworthy that the rate of elimination of HCl from methyl ester **32** is about 610 times slower than from acid **31** (Table 5.1); the same behavior is observed with the bromo compounds.

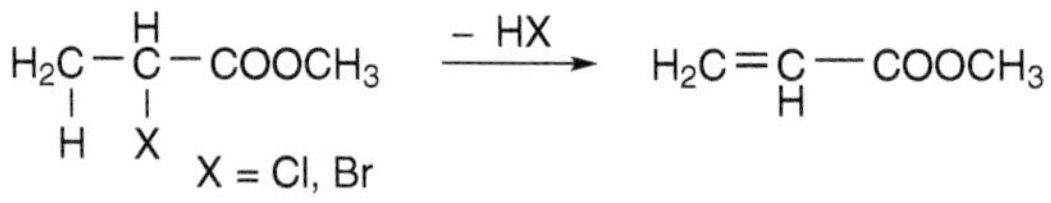

X = Cl, Br

Scheme 5.19

The kinetic results support the fact that a Br atom is a better leaving atom than a Cl atom in the molecular process of dehydrohalogenation of organic halogen compounds in the gas phase [23] (Scheme 5.19).

Table 5.1 Rates and relative rates of α-chloro and α-bromo carboxylic acids and esters at 360 °C.

Substrate	$10^4 \, k \, s^{-1}$	Relative rate
$CH_3CHClCOOCH_3$	0.02	1
$CH_3CHClCOOH$	12.80	610
$CH_3CHBrCOOCH_3$	0.37	1
$CH_3CHBrCOOH$	33.88	92

5.1.1.2 2-Hydroxycarboxylic Acids

Intermolecular assistance of the hydrogen in COOH for the elimination of a halogen atom in a polar five-membered cyclic TS has been demonstrated in the molecular dehydration process in the gas-phase pyrolysis of primary, secondary, and tertiary-α-hydroxycarboxylic acids [24, 25] (Scheme 5.20).

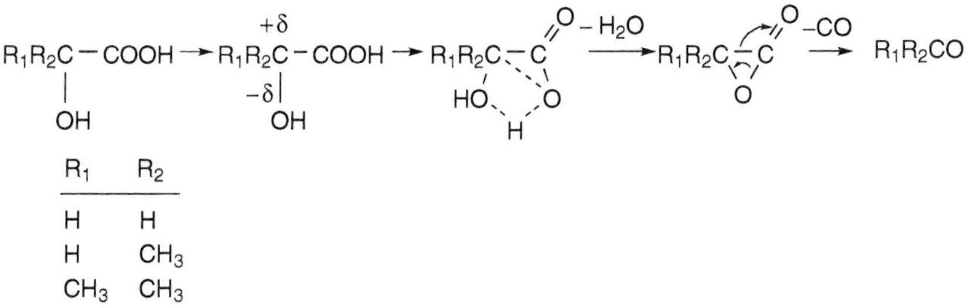

R_1	R_2
H	H
H	CH_3
CH_3	CH_3

Scheme 5.20

The greater the basicity of the OH group of the alcohol moiety in the gas phase [26], the faster its elimination through the assistance of the acidic hydrogen of the COOH group, and consequently the observed sequence of rates from 1° to 3° α-hydroxycarboxylic acid. Chuchani has suggested that the elimination of H_2O in Scheme 5.20 proceeds via a semipolar five-membered cyclic TS, where the C-OH bond polarization, in the direction of $C^{\delta+}\ldots^{\delta-}OH$, is the rate-determining factor. The intermediate α-propiolactone is unstable under the experimental pyrolytic conditions and decomposes rapidly into acetaldehyde and carbon monoxide.

5.1.1.3 2-Alkoxycarboxylic Acids

According to the pyrolysis products of 2-methoxy, 2-ethoxy, and 2-isopropoxy acetic acid, the gas-phase elimination reaction is suggested to follow the same type of mechanism described in Scheme 5.21 for α-hydroxy carboxylic acid, in which the acidic hydrogen of the COOH group assists the leaving of the alkoxy substituent through an intramolecular displacement reaction.

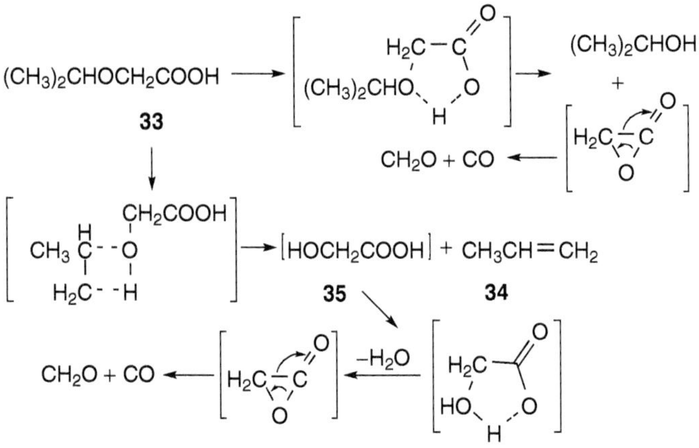

Scheme 5.21

Propene was detected as one of the pyrolysis products from pyrolysis of α-isopropoxy-acetic acid **33**, which suggests a different, parallel mechanism (Scheme 5.21), viz. polarization of $(CH_3)_2^{\delta+}CH \ldots^{\delta-}O$ to yield propene **34** and glycolic acid **35**; the latter further decomposes into formaldehyde, water, and CO. The temperature required for isopropoxy-acetic acid **33** to pyrolyze is higher than that needed for glycolic acid **35** to decompose [27], so the secondary decomposition is not avoidable in this case.

A combined experimental/theoretical study [28] of 2-methoxy-, 2-ethoxy-, and 2-isopropoxy-propanoic acids supports the concerted molecular two-step mechanism; the first rate-limiting step is the elimination of the alkoxy substituent via a five-membered cyclic transition structure associated with an intramolecular hydrogen transfer from the COOH group. The resulting α-lactone intermediate undergoes ring opening in the second step to yield acetaldehyde and CO.

Table 5.2 Rates and relative rates for the pyrolysis of 2-phenoxy carboxylic acids at 300 °C.

Substrate	$10^4\, k\, s^{-1}$	Relative rate
2-Phenoxyacetic acid	0.054	1.0
2-Phenoxypropionic acid	3.31	61.3
2-Phenoxylbutyric acid	4.37	80.9
2-Phenoxyisobutyric acid	257.04	4760.0

5.1.1.4 2-Phenoxycarboxylic Acids

The rates of gas-phase pyrolysis of the phenoxyalkanoic acids (Table 5.2) increase on going from a primary to a tertiary carbon bearing the phenoxy substituent [29].

The mechanism is rationalized in terms of a moderately polar, bicyclic transition structure, where an α-lactone intermediate is formed with the assistance of the acidic hydrogen of MeCOOH followed by a nucleophilic attack of the carbonyl oxygen (Scheme 5.22).

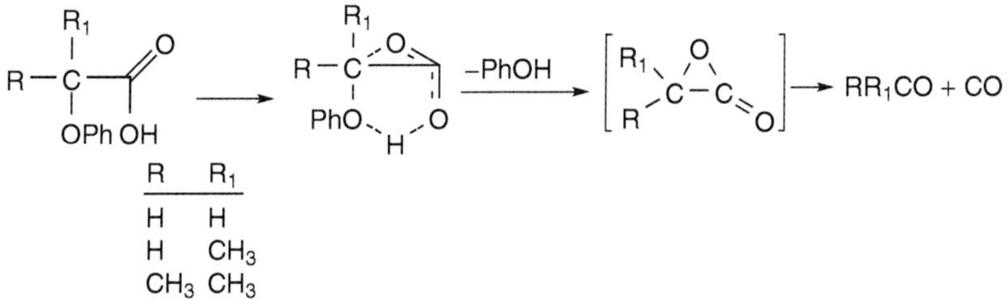

R	R_1
H	H
H	CH_3
CH_3	CH_3

Scheme 5.22

The formation of methacrylic acid **36** from the pyrolysis of 2-phenoxyisobutric acid **37** is explained in terms of unimolecular elimination of phenol via a polar four-centered cyclic TS (Scheme 5.23).

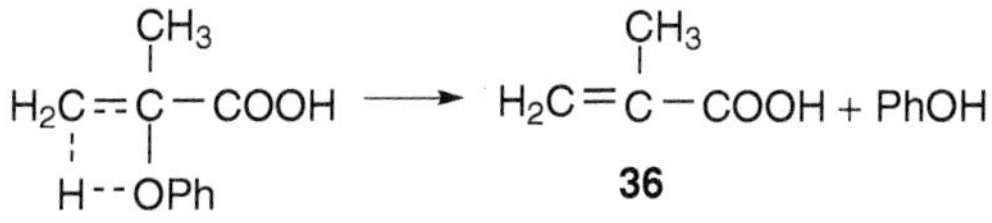

Scheme 5.23

Theoretical calculations by Andres and his group [30–33] on the pyrolysis reactions of α-substituted carboxylic acids have confirmed the experimental results of Chuchani and co-workers and provided further insight into the nature of the molecular transition structures along the reaction coordinate, which has placed the mechanism of these unimolecular thermal gas-phase elimination processes on a firm theoretical and experimental basis.

The rate-limiting step was considered to involve a transition structure [Scheme 5.24, TS (**a**)] in which fragmentation and α-lactone formation is assisted by the oxygen atom of the carbonyl moiety from the carboxyl group. A second transition structure [TS (**b**)] was identified, associated with the decomposition of the α-lactone intermediate to give carbon monoxide and the aldehyde as other products of the reaction. Accordingly, the molecular transition structures along the reaction coordinate follow the Scheme 5.24.

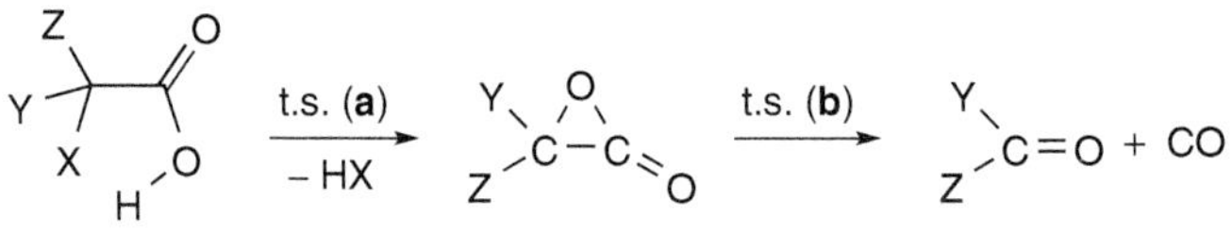

Scheme 5.24

The thermal decomposition of 2-phenoxyacetic acid **38**, 2-phenoxypropionic acid **39**, 2-phenoxybutyric acid **40**, and 2-phenoxyisobutyric acid **37** in the gas phase was studied computationally using several ONIOM combinations [34]. In all cases, for the reaction channel to form phenol, CO, and the corresponding carbonyl compound (Scheme 5.25), one intermediate and two transition states were identified and characterized.

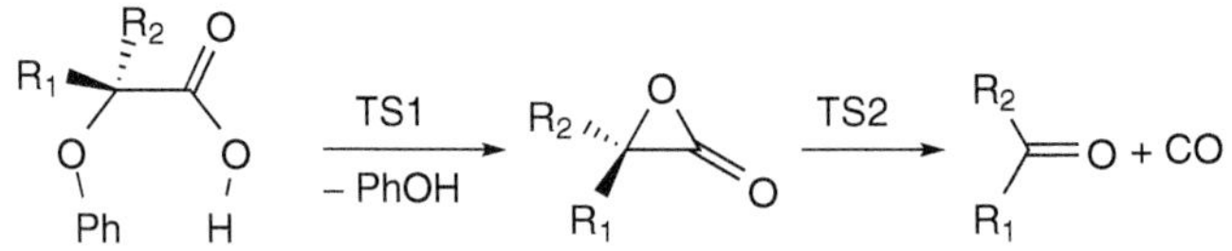

	R_1	R_2
37	CH_3	CH_3
38	H	H
39	CH_3	H
40	CH_3CH_2	H

Scheme 5.25

Table 5.3 Rates and relative rates of α-substituted propanoic acid at 327 °C.

X:	Cl	Br	OH	HNPh	OPh	SPh
$10^4\,k\,\mathrm{s}^{-1}$	1.81	5.17	2.11	3150	155	97.4
k_{rel}	1	2.86	1.17	1740	85.6	53.8

The relative reactivities of α-substituted propanoic acids, CH_3—CH(X)—COOH, with different groups X, are given in Table 5.3. The effect of the α-substituent (X) is related to what has been described as the basicity of X or its capacity to accept a proton.

The aryl substituents (—NHPh, —OPh, —SPh) show comparatively larger rates due to their pronounced resonance effects. The trend among the aryl groups reflects the relative acid strengths of the incipient phenol, thiophene, and aniline formed due to protonation by the carboxylic acid hydrogen, as well as the relative stability and proton affinity of the conjugate base of each of the departing aromatic molecules [35].

The influence of the aryl substituents on the negative charge being developed during the transition state is supported by the electronic effects of the substituents at phenoxy moieties in α-phenoxy propanoic acids (Table 5.4).

The rate differences are nevertheless systematic and consistent with the expected electron-withdrawing and electron-donating effects of substituents at the *meta* and *para* positions. Cyclization associated with the formation of the lactone intermediate (Scheme 5.25) and H-bonding between X and the acid proton might have a combined effect to reduce the negative charge developing on the oxygen of the phenoxy moiety and thus moderate the effects of the substituents.

A large drop in reactivity is observed when the acidity of the proton donor is reduced by replacing the carboxylic acid proton with a

Table 5.4 Rates and relative rates of α-phenoxy propanoic acids at 327 °C.

Group:	H	4-NO$_2$	4-Cl	3-Cl	4-CH$_3$	4-OCH$_3$
$10^4\,k\,\mathrm{s}^{-1}$	155	260	233	187	157	132
k_{rel}	1	1.68	1.50	1.21	1.01	0.85

hydroxyl proton. A relative rate factor of 2000 applies to the pyrolysis of 2-phenoxypropanoic acid **39** and 2-phenoxy-1-propanol **41** (Scheme 5.26). The rate factor is much greater (3.3×10^7) when the comparison involves the more basic 2-(N-phenylamino) moiety of acid **42** to that of the corresponding alcohol **43** [35] (Scheme 5.26).

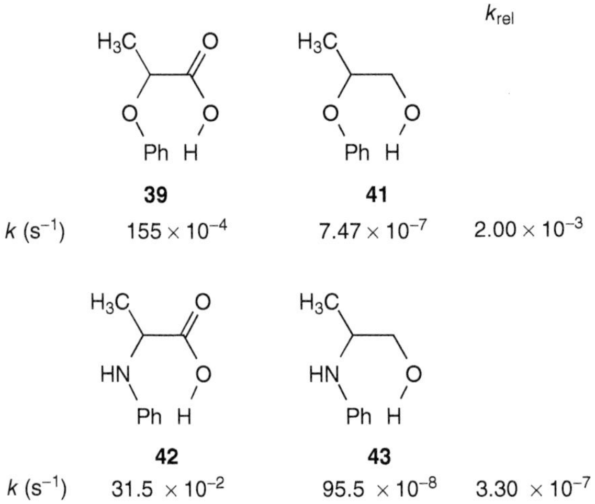

k_{rel}

	39	**41**
k (s^{-1})	155×10^{-4}	7.47×10^{-7}	2.00×10^{-3}

	42	**43**
k (s^{-1})	31.5×10^{-2}	95.5×10^{-8}	3.30×10^{-7}

Scheme 5.26

5.1.1.5 2-Aminocarboxylic Acids

The reactivity correlations for α-substituents in propanoic acid were originally shown to be in the order –NHPh > –OPh > –SPh [35]. We later re-evaluated the rate constant for 2-(N-phenylamino)propanoic acid, and the order now is –OPh > –SPh > –NHPh [36]. This order corresponds more closely to the trend expected from the acidity of the incipient phenol, thiphenol, and aniline molecules.

The noteworthy observation in these kinetic studies is that there seems to be no opposing substituent effect from the *p* position of the aryl moiety [37] of the amino group consistent with the electron-donating character of the methyl and the electron-withdrawing character of the chloro substituent (Table 5.5).

Chuchani et al. noted the difficulty in investigating the kinetics of pyrolysis of α-aminocarboxylic acids as neutral molecules in their static pyrolysis system owing to the insolubility of the solid, low-molecular-weight amino acids in organic solvents and their tendency to form zwitterionic species in aqueous solution [37].

Table 5.5 Rate coefficients for pyrolysis of 2- and 3-(arylamino)propanoic acids at 327 °C.

Compound	$10^3 \, k \, s^{-1}$
2-Phenylaminopropanoic acid	3.68
2-(p-Chlorophenylamino)propanoic acid	6.63
2-(p-Methylphenylamino)propanoic acid	5.47
3-(Phenylamino)propanoic acid	4.64
3-(p-Chlorophenylamino)propanoic acid	3.46
3-(p-Methylpphenylamino)propanoic acid	6.21

Despite these limitations, the kinetics of gas-phase elimination of N,N-dimethylglycine **44** [38], picoline acid **45** [39], and N-phenylglycine **46** [40] and their ethyl ester **47** were examined, and mechanisms for their gas-phase pyrolytic reactions have been suggested as shown in Schemes 5.27–5.29, respectively.

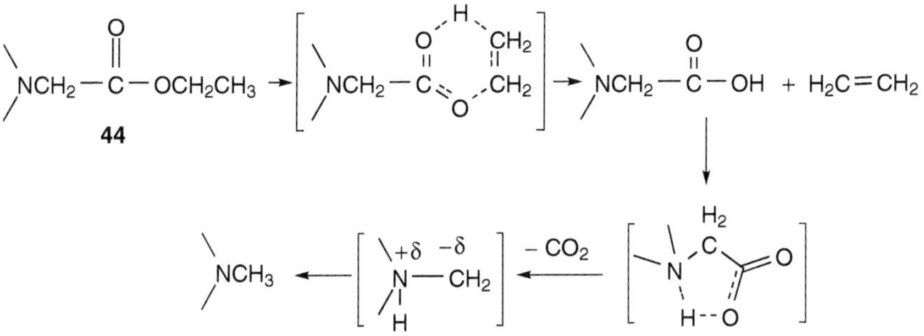

Scheme 5.27

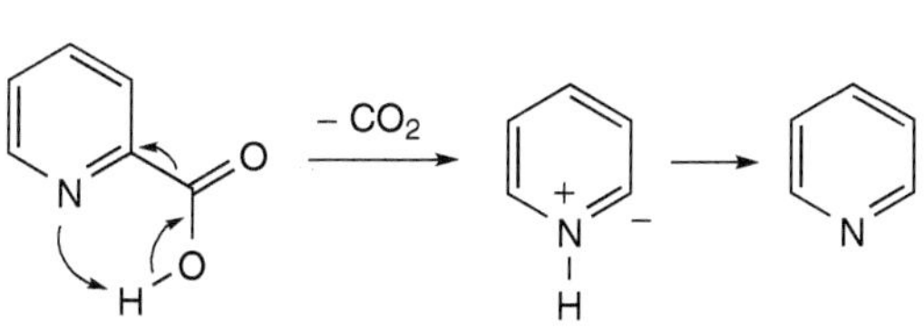

Scheme 5.28

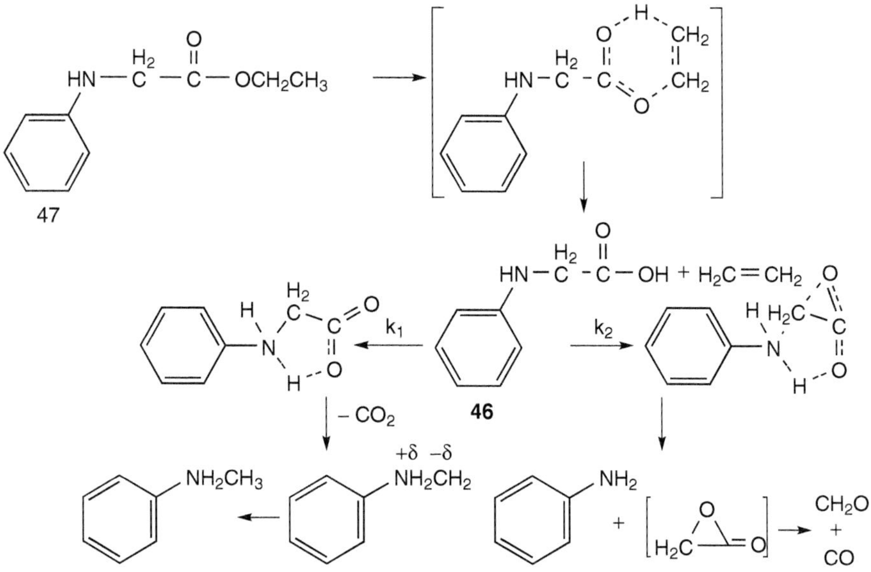

Scheme 5.29

Chuchani et al. further examined the gas-phase pyrolysis of N-benzylglycine ethyl ester **48**. They suggested the most probable mechanism for the elimination as given in Scheme 5.30, where the first step is an ester pyrolysis with formation of the corresponding carboxylic acid and ethene. The acid intermediate then undergoes a very rapid decarboxylation process through a five-membered cyclic transition state to give mainly the corresponding substituted N-benzyl-N-methylamine **49** and CO_2 [41].

5.1.1.6 2-Acetooxycarboxylic Acids

Product analysis of the gas-phase elimination reaction of 2-acetoxycarboxylic acids **50** suggests a mechanism proceeding through a moderately semipolar bicyclic transition structure [42], as shown in Scheme 5.31, where an α-lactone is produced with the assistance of the hydrogen of the COOH group followed by nucleophilic attack of the carbonyl oxygen of the acid moiety on the α-carbon atom. The lactone intermediate then fragments into the carbonyl compound and CO.

The kinetic data obtained from the reactions indicate an increase in the elimination rate from primary to tertiary carbon bearing the acetoxy substituent (Table 5.6) [42].

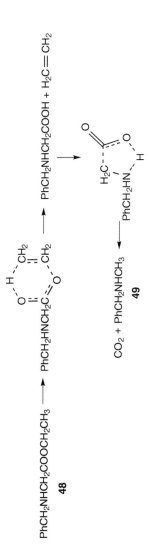

$PhCH_2NHCH_2COOCH_2CH_3 \longrightarrow PhCH_2HNCH_2C \underset{\underset{O}{\parallel}}{\overset{CH_2 \cdots H}{}} \underset{O}{\overset{CH_2}{\parallel}} \longrightarrow PhCH_2NHCH_2COOH + H_2C{=}CH_2$

48

$\longrightarrow$ H_2C\overset{O}{\underset{PhCH_2HN}{\diagdown}}\underset{H}{\overset{O}{}}

$CO_2 + PhCH_2NHCH_3$

49

Scheme 5.30

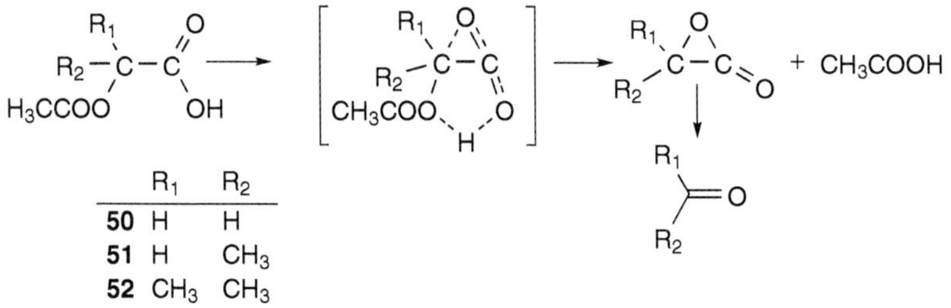

	R$_1$	R$_2$
50	H	H
51	H	CH$_3$
52	CH$_3$	CH$_3$

Scheme 5.31

Table 5.6 Rates and relative rates of α-lactone formation at 300 °C.

Substrate	$10^4\ k\,s^{-1}$	Relative rate
2-Acetoxyaceticacid acid **50**	2.88	1.0
2-Aceoxypropionic acid **51**	19.05	6.6
2-Acetoxy-2-methylpropionic acid **52**	371.54	129.0

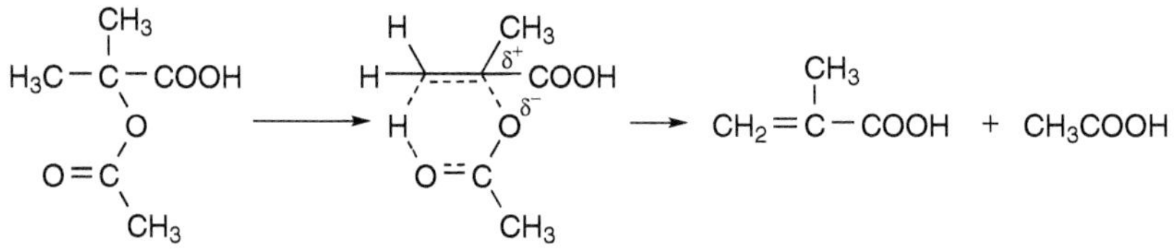

53

Scheme 5.32

Product analysis for the gas-phase pyrolysis of 2-acetoxy-2-methylpropionic acid **52** indicates the formation of methacrylic acid together with acetic acid and acetone. This parallel elimination pathway was explained in association with the gas-phase pyrolysis of carboxylic esters **53**, whereby a unimolecular elimination to acetic and methacrylic acids involves a semipolar six-membered cyclic transition state (Scheme 5.32).

5.1.1.7 2-Ketocarboxylic Acids

The gas-phase pyrolysis of oxalic acid **54**, the simplest 2-keto carboxylic acid, was found to proceed with elimination of CO_2 and formation of formic acid [43] via a five-membered cyclic transition state (Scheme 5.33).

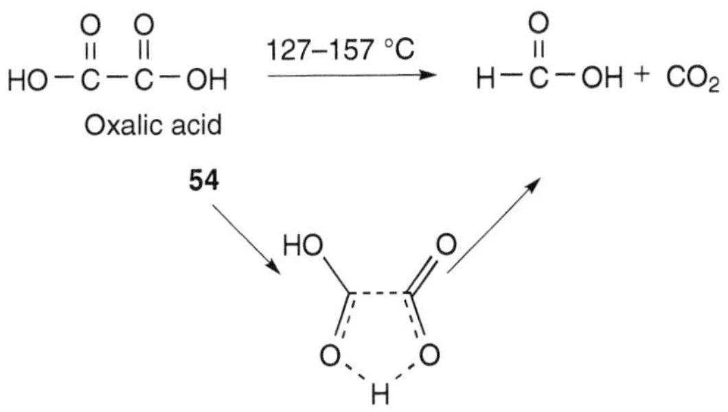

Scheme 5.33

Pyruvic acid **55** fragments thermally into carbon dioxide and acetaldehyde [42] (Scheme 5.34).

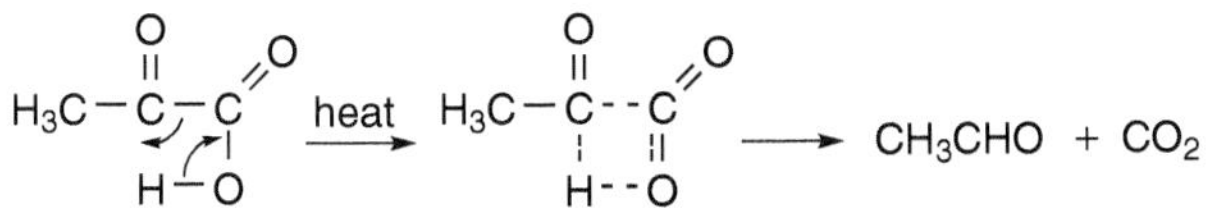

Scheme 5.34

This is believed to proceed via a semiconcerted reaction with a four-center cyclic transition state involving partial development of a negative charge (relative to the ground state) at the α-carbonyl carbon [44].

This mechanism accounts for the fact that oxalic acid (in which OH replaces Me) is much more reactive, as the −I inductive effect of OH stabilizes the adjacent negative charge better than Me. There is also a statistical factor of two involved with oxalic acid, but this is much lower than the 290-fold rate difference observed at 600 K.

According to this suggested mechanism, benzoylformic acid **56** should be more reactive than pyruvic acid **55** in view of the −I inductive effect of phenyl relative to methyl. The kinetic data show that benzoylformic acid is ca. 46 times more reactive than pyruvic acid. It is, however, much less reactive than oxalic acid [43], consistent with the greater −I effect of OH relative to Ph. A plot of logarithms of the statically corrected relative rates at 600 K of pyruvic **55**, benzoylformic

56, and oxalic acids **54** [45a] (1 : 19 : 145) against σ values of the group adjacent to the α-carbonyl group (HO, Ph, Me) provides a reasonable correlation. This result supports the proposal that negative charge develops at the α-carbon in the transition state and is stabilized by inductive electron withdrawal by the neighboring group.

A joint theoretical and experimental study of the homogeneous, unimolecular gas-phase elimination kinetics of methyl oxalyl chloride was reported [45b]. The reaction was found to proceed via a concerted, semipolar mechanism, as described in Scheme 5.34.

5.1.1.8 α-Substituted Esters

Replacing the OH group in benzoylformate **56** with a methoxy group results in a 2-ketoester **57**. The gas-phase elimination reaction of methyl benzoylformate involves a five-membered cyclic transition state. This is consistent with the product formation and analogous to the four-center transition state for the elimination pathway of the α-keto acids proposed in Scheme 5.34, in which a partial negative charge is located on the α-carbonyl carbon.

The kinetic data (Scheme 5.35), reaction products, and proposed elimination pathways suggest that the elimination of benzoylformic acid **56** is only ca. 46 times faster than acetyl formic acid, due to the electronic effect of the phenyl group in neutralizing the partial negative charge on the α-carbonyl carbon [45a]. However, benzoylformic acid (α-keto acid) **57** eliminates 10^6 times faster than methyl benzoylformate **56** (α-keto ester). This million-fold drop in reactivity reflects a decrease in the acidity of the incipient hydrogen on replacing the hydrogen of

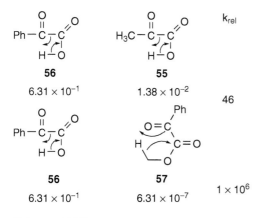

Scheme 5.35

the hydroxyl (O—H) moiety of the acid function with the methoxy (OCH$_2$—H) hydrogen of the ester group [46].

Different reaction mechanisms for the decomposition of 2-oxo-esters have been postulated. These pathways are depicted in Scheme 5.36. Compounds with electron-releasing groups, such as CH$_3$ (ethyl 2-oxo-propionate) **58** and (CH$_3$)$_2$CH (3-methyl-2-oxo-butyrate) **59**, decompose as organic esters to produce the corresponding carboxylic acid and ethene (Scheme 5.36a), where the elongation and subsequent polarization of the C—O bond, in the sense C$_\alpha^{\delta+}$...O$^{\delta-}$, are the rate-determining step. The corresponding 2-oxo-acids and ethene are formed via six-membered cyclic transition states. These oxo-acids formed are unstable under the reaction conditions and rapidly decarboxylate [31, 32] to give the corresponding aldehyde. The intermediate product, 2-oxo-propionic acid or pyruvic acid, from the elimination of ethyl 2-oxo-propionate (Scheme 5.36b), fragments rapidly in the gas phase [47] into acetaldehyde and CO$_2$ at temperatures of 284–334 °C.

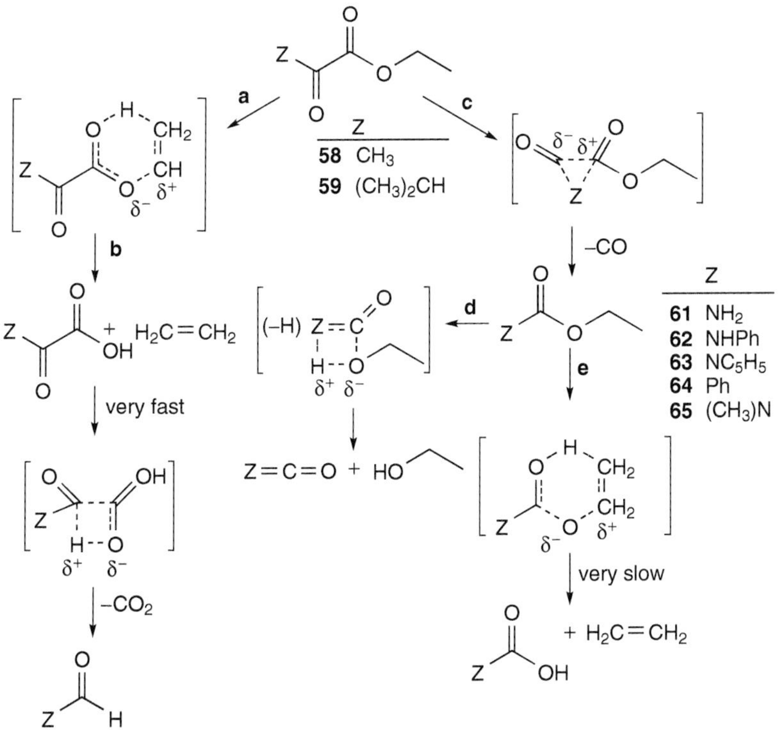

Scheme 5.36

Thus, the greater the electron release of the alkyl group R, the greater the nucleophilicity of the carbonyl oxygen; and with more stabilization of the positive C, the faster the elimination rate. Electron-withdrawing substituents like Cl in methyl chlorooxalate **60** (Scheme 5.37), NH_2 in ethyl oxamate **61**, C_6H_5NH in ethyl oxanilinate **62**, C_5H_5N in ethyl piperidine glyoxylate **63**, and C_6H_5 in ethyl benzoyl formate **64** lead to decarbonylation [33, 35] via Scheme 5.36c and Scheme 5.37. The hydrogen atom attached to the nitrogen substituent may lead to subsequent elimination of ethanol (Scheme 5.36d) or ethene. An exception is $Z = (CH_3)_2N$ in ethyl N,N-dimethyl oxamate **65** (Scheme 5.36), where no migration of this substituent to the carboethoxy group is observed.

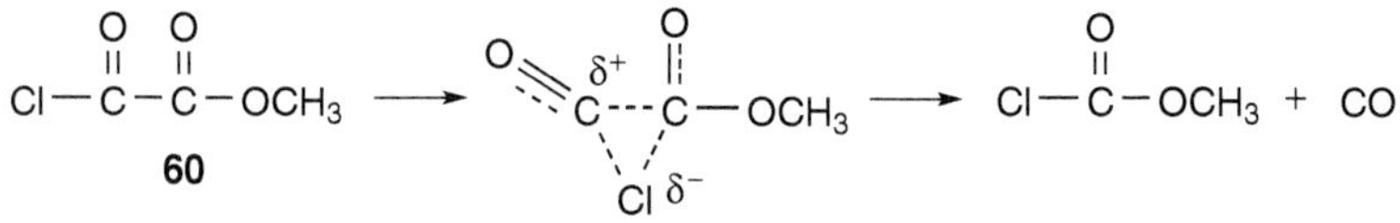

Scheme 5.37

The mechanism of pyrolysis of methyl esters of 2-hydroxy carboxylic acids – that is, methyl 2-hydroxypropionate and methyl 2-hydroxyisobutyrate – in the gas phase was reported [48] and has been examined computationally at the B3LYP/6–31 + G**, MP2/6–31 + G**, and MP2/6–31 + G** levels with the conclusion that the reaction proceeds via a concerted, asynchronous, five-membered cyclic transition state structure. The extension of the O—H bond with subsequent migration of the OH proton appears to be the driving force for the elimination process. Calculated thermodynamic and kinetic parameters were found to be in good agreement with the experimental results.

A significant difference in rate of elimination has been found for 2-hydroxypropionic acid (lactic acid) and 2-hydroxyisobutyric acid, while little difference in rates was found for their corresponding methyl esters [24, 49]. In addition to this fact, mandelic acid, a secondary hydroxy acid, decarbonylates faster than 2-hydroxypropionic acid [49].

5.1.1.9 β-Substituted Carboxylic Acids

The pyrolysates of β-substituted carboxylic acid derivatives **66–69** (Scheme 5.38) were analyzed and identified as PhOH, PhSH, $PhNH_2$, and $PhCH_3$ [50]. Two alternative mechanistic pathways were suggested

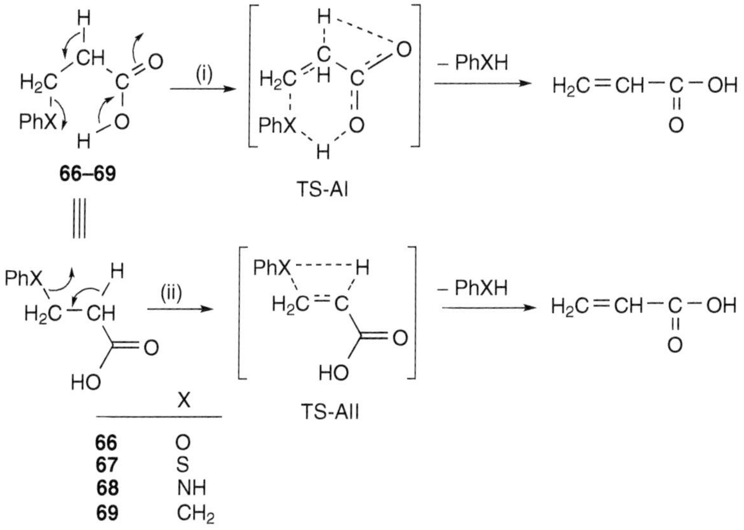

Scheme 5.38

to explain the formation of the products, as shown in Scheme 5.38; one is through a six-membered transition state pathway that involves the proton of the carboxylic acid group, and the other is through a four-membered transition state pathway that involves the less acidic proton of the α-carbon atom.

Examination of the two competitive reaction pathways for 3-(N-phenylamino)propionic acid **68** carried out computationally using the ab initio SCF method [36] shows that the single cyclic four-membered ring transition state, TS-AII, has a lower free-energy barrier than the cyclic four- and six-membered TS-AI shown in Scheme 5.38. The calculated free energies of activation were 89.59 and 66.41 kJ mol^{-1} for the reaction via the transition states TS-AI and TS-AII, respectively. The overall process was exergonic, with reaction free energy equal to −92.89 kJ mol^{-1} (Figure 5.1).

Based on the products of the pyrolysis reaction of β-substituted ethyl esters **70** and **71**, it is suggested that the pyrolysis of these esters involves two steps; it starts with elimination of ethene and formation of the β-substituted acids via a cyclic six-membered transition state, and the formed acid then fragments into PhXH and acrylic acid (Scheme 5.39). This justifies the slower rate of pyrolysis of β-substituted esters relative to their corresponding acids, shown in Scheme 5.40. On the other hand,

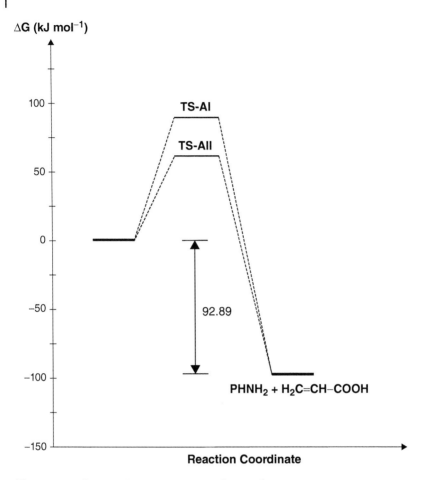

Figure 5.1 Competitive reaction pathways for 3-(N-phenylamino)propionic acid **68** using the ab initio SCF method.

pyrolysis of methyl esters **72** and **73** produced methyl acrylate and not acrylic acid. These results seem to substantiate a methyl ester reaction pathway involving a four-membered cyclic TS analogous to that shown for the acids (Scheme 5.41).

It is of interest to note that the results of the ab initio calculations show the gas-phase thermodynamic stabilities of the arene fragments to be in the order $PhNH_2 \gg PhCH_3 \sim PhOH > PhSH$ [51]. The results appear to correlate with the expected gas-phase acidities of these compounds and the relative proton affinities of their incipient conjugate bases.

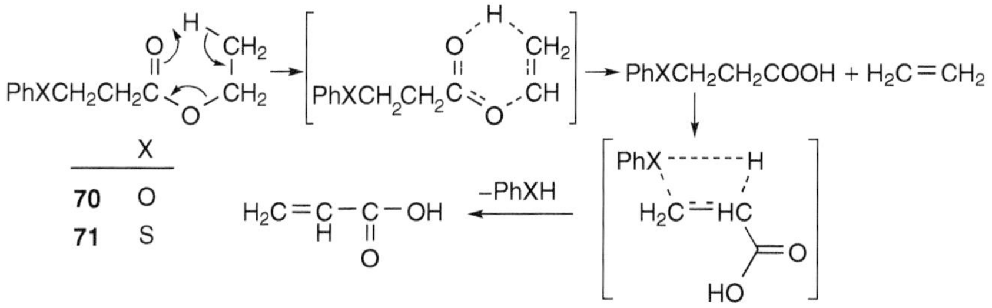

Scheme 5.39

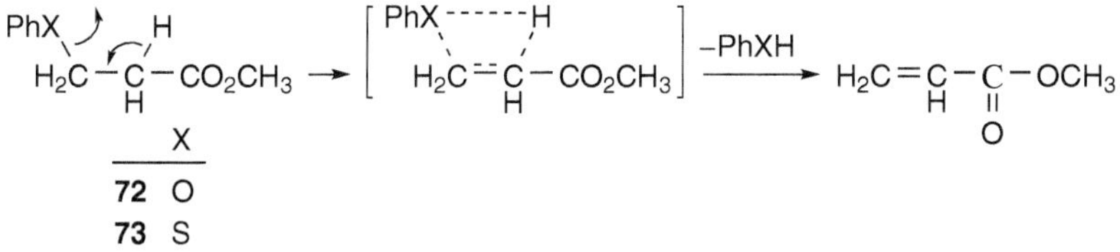

Scheme 5.40

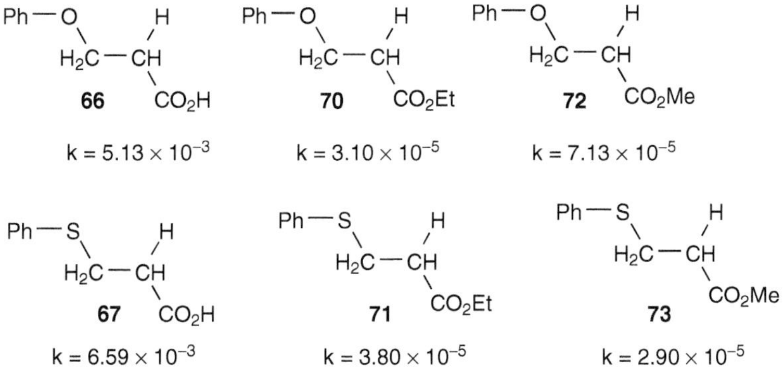

Scheme 5.41

5.2 Structural Frame Interconversion

A structural framework interconversion of allyl ethers **74–81** with a range of substituents on either the benzene ring or the propenolate chain at 650 °C is an effective method for synthesis of 2-substituted benzofurans **82–88** [52] (Scheme 5.42).

Scheme 5.43 explains the mechanism for the FVP of ether **89** that yields furan **90** with no byproducts. The main feature of the mechanism is that the ester group COOR acts as a leaving group.

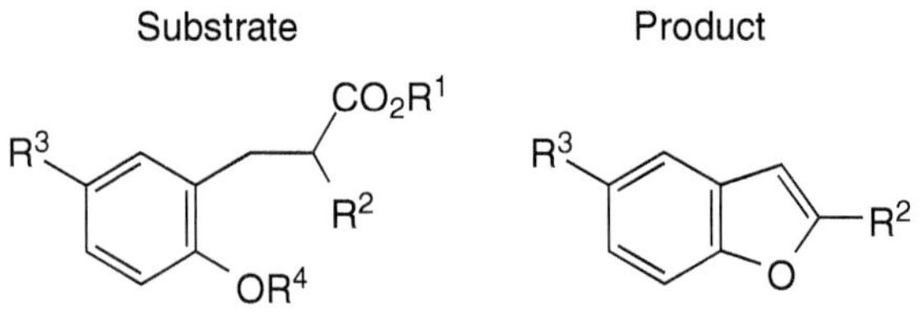

Substrate   Product

	R1	R2	R3	R4		R2	R3	%Yield
74	Me	H	H	allyl	82	H	H	65
75	Me	H	Cl	allyl	83	H	Cl	60
76	Me	H	NO$_2$	allyl	84	H	NO$_2$	55
77	Me	Me	H	allyl				
					85	Me	H	75
78	Et	Me	H	allyl				
79	Et	Me	Cl	allyl	86	Me	Cl	85
80	Me	COOMe	H	allyl	87	COOMe	H	95
81	Me	CN	H	allyl	88	CN	H	52

Scheme 5.42

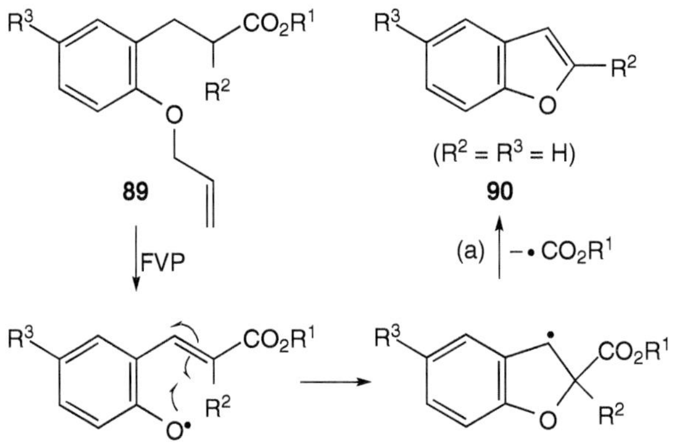

Scheme 5.43

Annelation of furan onto a thiophene ring is also possible by this method (Scheme 5.44), but lower yields were obtained in such pyrolysis [53].

Examination of the products and kinetic investigation of the gas-phase pyrolysis reactions for compounds **91–97** shown in Scheme 5.45 suggest similar thermal retro-ene reactions via a cyclic six-membered

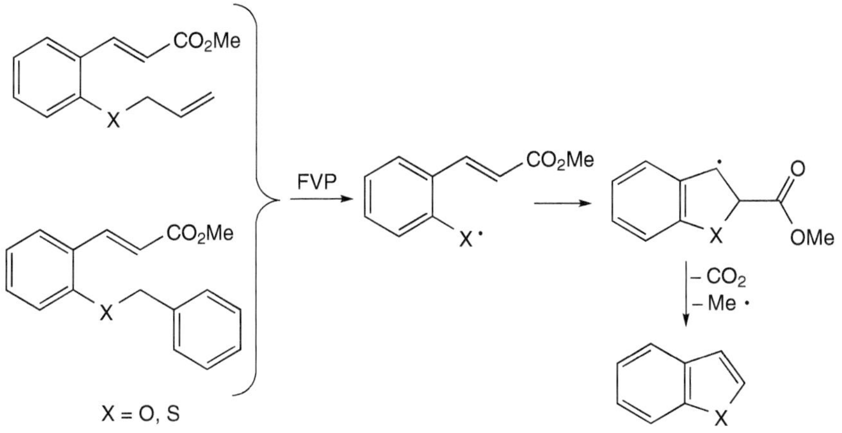

Scheme 5.44

transition state involving C=C of different systems [54]: acyclic in vinyl ether **91**, cyclic in 3,4-dihydro-2-H-pyrans **92**, aromatic in 2-ethoxybenzene **93**, and an aromatic fused ring **94**. It seems that these structural frame differences do not alter the mechanistic approach of their fragmentation in gas-phase pyrolysis, resulting in different products of different structures.

Compounds of group I in Scheme 5.45 follow a similar mechanistic pathway with participation of C=C that is part of the acyclic unsaturated systems with a different environment: vinyl ether **91**, allyl ether **95**, and hydroxy alkene **97**. This supports the idea that structure frame interconversion of substrates does not necessarily alter the reaction mechanism. This also applies to compounds in groups II, III, and IV in Scheme 5.45, as these all feature retro-ene reactions via a six-membered cyclic transition state with C=C participation under a different environment. Needless to say, the molecular reactivity of each compound depends primarily on the electronic environment of each substrate.

Product analysis of FVP of chroman-1-one **98** shows that this compound does not eliminate ethene by a retro-ene reaction like chroman **94**. Rather, it dehydrogenates to produce the aromatic coumarin, which upon extrusion of a CO molecule forms benzofuran **99** as a major product (Scheme 5.46); for this, the driving force might be the gain in aromaticity [55].

Isochroman-3-one **100** was converted to benzocyclobutene [56] by electrocyclic ring closure: extrusion of CO_2 from isochroman-3-one

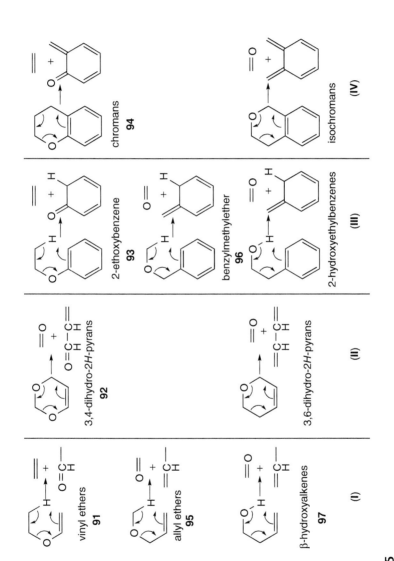

vinyl ethers **91**

3,4-dihydro-2*H*-pyrans **92**

chromans **94**

(I)

(II)

(IV)

allyl ethers **95**

2-ethoxybenzene **93**

benzylmethylether **96**

β-hydroxyalkenes **97**

3,6-dihydro-2*H*-pyrans

2-hydroxyethylbenzenes

(I)

(II)

(III)

isochromans

Scheme 5.45

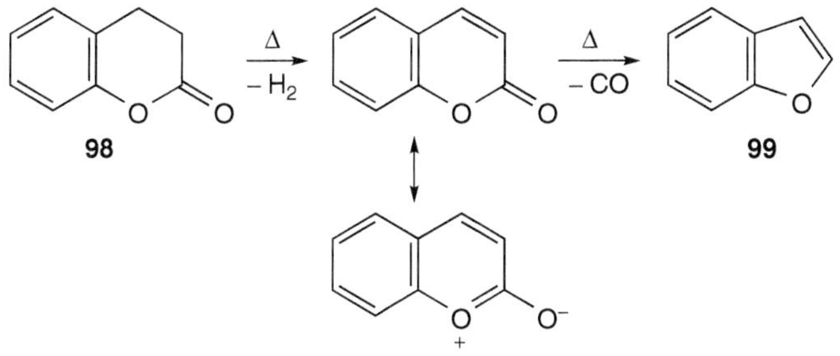

Scheme 5.46

produces intermediate o-quinodimethane **101** with a log A value of
$14.3\,s^{-1}$, which is typical for six-membered electrocyclic elimination.
Electrocyclic ring closure of the intermediate **101** produced benzocyclobutene (Scheme 5.47). On the other hand, phenylacetate **102**, the
acyclic analogue of **98**, eliminates ketene via the evidently preferred
four-center process, which might be due to the nucleophilicity of
the OAr group and retention of aromaticity in the transition state
(Scheme 5.48).

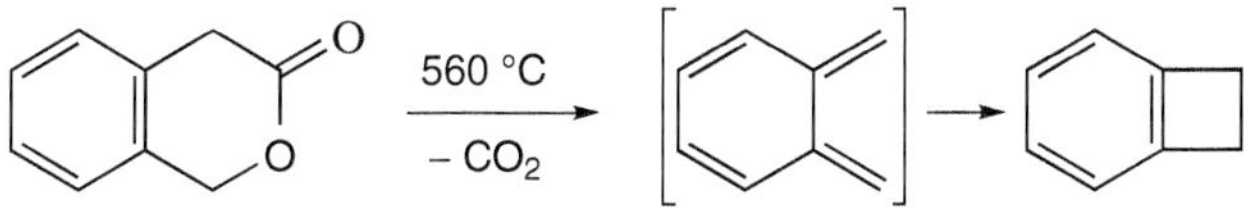

Scheme 5.47

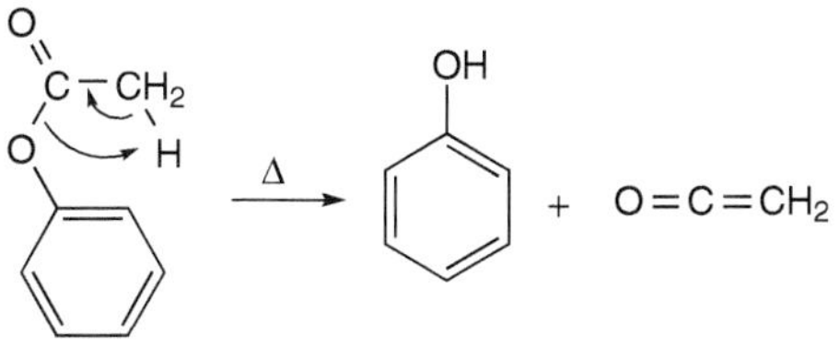

Scheme 5.48

5.2.1 Alkyl Heterocycles

2-Ethoxypyridine **103** eliminates rapidly in a gas-phase pyrolysis reaction at $400\,°C$, into 2-pyridone **104** and ethene (Scheme 5.49), which is

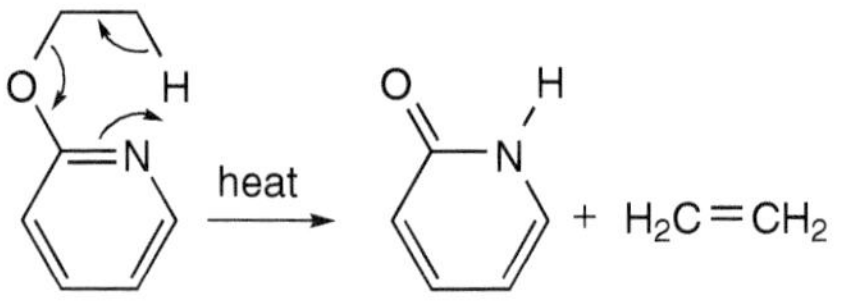

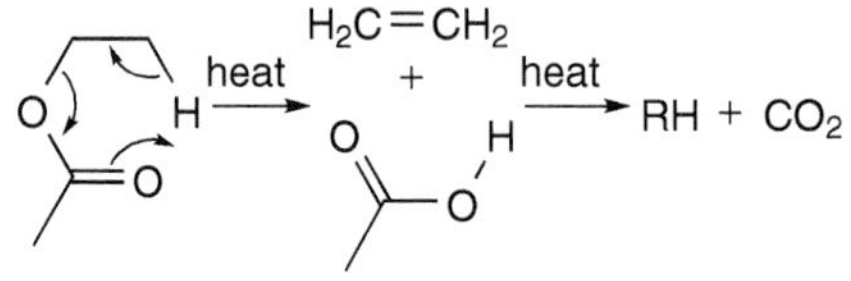

Scheme 5.49

remarkable evidence for the participation of the aromatic π electrons of C=N of the heterocyclic system. For this type of reaction, it was the first kinetic study in which heteroaromatic π electrons [57] participated in a six-membered cyclic transition state. This is a nitrogen analogue of ester pyrolysis (Scheme 5.49).

To access the polarity of the transition state, 2-ethoxy, 2-isopropoxy, and 2-*tert*-butoxypyridines were examined [58a]. The relative rates of elimination at 600 K were found to be $1.0 : 18.0 : 1675$, which are lower than those obtained for pyrolysis of the corresponding acetates, $1 : 28.8 : 3315$ [58b]. This suggests the transition state is less polar for the pyridines than for ester pyrolysis. An alternative decomposition pathway is the reaction of N-alkylated 2-pyridones **105** to ethene and 2-hydroxypyridine **106** [59] (Scheme 5.50). This pathway was ruled out because of the effect of methyl substituents at each carbon atom of the pyridine ring (Scheme 5.50).

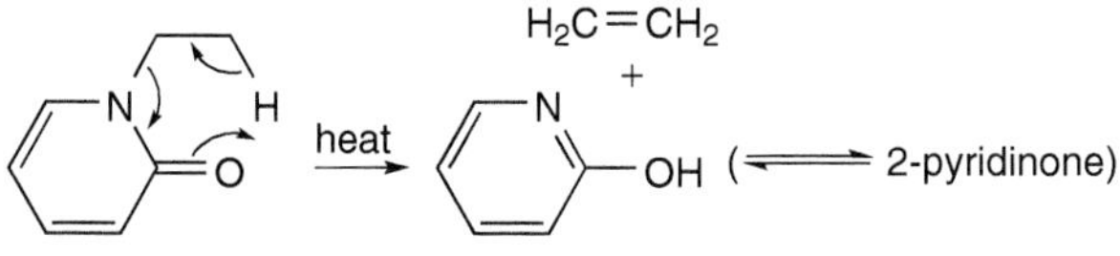

Scheme 5.50

We have further examined the effect of aza-substituent and π-bond order on the molecular reactivity of ethoxy, pyrazine **107**, pyridazine **108**, pyrimidines **109** and **110**, quinoline **111**, and isoquinolines

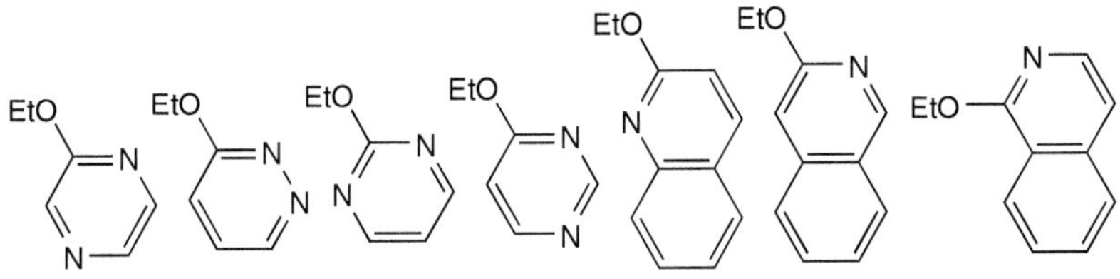

	107	108	109	110	111	112	113
$10^3\ k\ s^{-1}$ (377 °C)	0.139	2.56	0.262	0.285	0.798	0.054	1.65

Scheme 5.51

Table 5.7 Relationship between relative rates of pyrolysis of ethoxyheterocycles at 377 °C and C—N bond lengths.

Compound	C—N bond length (Å)	k_{rel}
2-Ethoxypyridine	1.340	1.0
2-Ethoxypyridazine	1.281	9.8
1-Ethoxyisoquinoline	1.300	6.47
2-Ethoxyquinoline	1.330	3.13
3-Ethoxyisoquinoline	1.366	0.21

112 and **113** [59]. The rate of pyrolysis reaction of these structurally related heterocycles is given in Scheme 5.51. The highest rate factor of 3-ethoxypyridazine confirms the importance of the nucleophilicity of the C=N and, accordingly, its bond order. This parallelism between the relative rates of pyrolysis and C=N bond order is shown in Table 5.7. The bond order is not directly available experimentally, but the bond length can be taken as a measure.

References

1 Hoffmann, H.M.R. (1969) Angew. Chem., 81, 597–618.
2 Ripoll, J.L. and Vallee, Y. (1993). *Synthesis* (7): 659–677.
3 Jenneskens, L.W., Hoefs, C.A.M., and Wiersum, U.E. (1989). *J. Org. Chem.* 54: 5811–5814.

4 Viola, A. and Iorio, E.J. (1970). *J. Org. Chem.* 35: 856–858.

5 Mutterer, F., Morgen, J.M., Biedermann, J.M. et al. (1970). *Tetrahedron* 26: 477–495.

6 Bailey, W.J. and Di Pietro, J. (1977). *J. Org. Chem.* 42: 3899–3902.

7 Viola, A. and Locke, J.S. (1984). *J. Chem. Soc. Chem. Commun.*: 1429–1431.

8 Rodler, M., Brown, R.D., Godfrey, P.D., and Tack, L.M. (1984). *Chem. Phys. Lett.* 110: 447–451.

9 Wentrup, C. (1988). *J. Am. Chem. Soc.* 110: 1337–1343.

10 Kroto, H.W., Matti, G.Y., Suffolk, R.J. et al. (1990). *J. Am. Chem. Soc.* 112: 3779–3784.

11 Giles, H.G., Marty, R.A., and de Mayo, P. (1974). *J. Chem. Soc. Chem. Commun.*: 409–410.

12 Giles, H.G., Marty, R.A., and de Mayo, P. (1976). *Can. J. Chem.* 54: 537–542.

13 Bogey, M., Demuynck, C., Destombes, J.L. et al. (1989). *J. Am. Chem. Soc.* 111: 7399–7402.

14 Martin, G., Martinez, H., Suhr, H., and Suhr, U. (1986). *Int. J. Chem. Kinet.* 18: 355–362.

15 Lasne, M.C. and Ripoll, J.L. (1985). *Synthesis* (2): 121–140.

16 Vallée, Y., Ripoll, J.L., and Masume, D.J. (1988). *Anal. Pyrolysis*: 14–171.

17 (a) Bourdon, F., Ripoll, J.L., and Vallee, Y. (1990). *Tetrahedron Lett.* 31: 6183–6184. (b) Bourdon, F., Ripoll, J.L., and Vallee, Y. (1990). *J. Org. Chem.* 55: 2596–2600.

18 McNab, H., Morel, G., and Stevenson, E. (1997). *J. Chem. Res. (s)* (6): 207–207.

19 Banert, K., Huckstadt, H., and Vrobel, K. (1992). *Angew. Chem. Int. Ed. Engl.* 31: 90–92.

20 Banert, K. and Toth, C. (1995). *Angew. Chem. Int. Ed. Engl.* 34: 1627–1629.

21 Ruchardt, C., Meier, M., and Haff, K. (1991). *Angew. Chem. Int. Ed. Engl.* 30: 893–901.

22 Chuchani, G. and Rotinov, A. (1989). *Int. J. Chem. Kinet.* 21: 367–371.

23 Chuchani, G., Dominguez, R.M., and Rotinov, A. (1991). *Int. J. Chem.* 23: 779–783.

24 Chuchani, G., Martin, I., and Dominguez, A. (1993). *J. Phys. Org. Chem.* 6: 54–58.

25 Chuchani, G., Martin, I., and Rotinov, A. (1995). *Int. J. Chem. Kinet.* 27: 849–853.

26 Lowry, T.H. and Richardson, K.S. (1976). *Mechanism and theory in organic chemistry*, Ch. 3, 162. New York: Harper and Row.

27 Chuchani, G., Rotinov, A., and Dominguez, R.M. (1996). *J. Phys. Org. Chem.* 9: 787–794.

28 Rotinov, A., Chuchani, G., Andrews, J. et al. (1999). *Chem. Phys.* 246: 1–12.

29 Chuchani, G., Dominguez, R.M., Rotinov, A., and Martin, I. (1999). *J. Phys. Org. Chem.* 12: 612–618.

30 Safont, V.S., Moliner, V., Andres, J., and Domingues, L.R. (1997). *J. Phys. Chem.* 119: 1859–1865.

31 Domingues, L.R., Andres, J., Moliner, V., and Safont, V.S. (1997). *J. Am. Chem. Soc.* 119: 6415–6422.

32 Domingues, L.R., Picher, M.T., Andres, J. et al. (1997). *Chem. Phys. Lett.* 274: 422–428.

33 Domingues, L.R., Picher, M.T., Safont, V.S. et al. (1999). *J. Phys. Chem. A* 103: 3935–3943.

34 Xue, Y., Kand, C.H., Kim, C.K., and Lee, I. (2003). *J. Comput. Chem.* 24: 963–972.

35 Al-Awadi, N.A., Kaul, K., and El-Dusouqui, O.M.E. (2000). *J. Phys. Org. Chem.* 13: 499.

36 Al-Awadi, S., Abdullah, M., Hasan, M., and Al-Awadi, N.A. (2004). *Tetrahedron* 60 (13): 3045–3049.

37 Tosta, M., Oliver, J.C., Mora, J.R. et al. (2010). *J. Phys. Chem. A* 114: 2483–2488.

38 Ensuncho, A., Lafont, J., Rotinov, A. et al. (2001). *Int. J. Chem. Kinet.* 33: 465–471.

39 Lafont, J., Ensuncho, A., Rotinov, A. et al. (2003). *J. Phys. Org. Chem.* 16: 84–88.

40 Dominguez, R.M., Tosta, M., and Chuchani, G. (2003). *J. Phys. Org. Chem.* 16: 869–874.

41 Rosas, F., Monsalve, A., Tosta, M. et al. (2005). *Int. J. Chem. Kinet.* 37: 384–389.

42 Chuchani, G., Dominguez, R.M., Herize, A., and Romero, R. (2000). *J. Phys. Org. Chem.* 13: 757–764.

43 Lapidus, G., Baron, D., and Yankwich, P. (1964). *J. Phys. Chem.* 68: 1863–1865.

44 Taylor, R. (1987). *Int. J. Chem. Kinet.* 19: 709–713.

45 (a) Taylor, R. (1991). *Int. J. Chem. Kinet.* 23: 247–250. (b) Cordova, T., Rotinov, A., and Chuchani, G. (2004). *J. Phys. Org. Chem.* 17: 148–151.

46 Al-Awadi, N.A. and El-Dusouqui, O.M.E. (1997). *Int. J. Chem. Kinet.* 29: 295–298.

47 Reyes, A., Dominguez, R., Tosta, M. et al. (2011). *J. Phys. Org. Chem.* 24: 74–82.

48 Safont, V.S., Andres, J., Castillo, R. et al. (2004). *J. Phys. Chem. A* 108: 996–1007.

49 Taylor, R. (1978). *J. Chem. Soc. Chem. Commun.*: 732–733.

50 Al-Awadi, S.A., Abdullah, M.R., Dib, H.H. et al. (2005). *Tetrahedron* 61: 5769–5777.

51 Al-Juwaiser, I.A., Al-Awadi, N.A., and El-Dusouqui, O.M.E. (2002). *Can. J. Chem.* 80: 499–503.

52 Black, M., Cadogan, J.I.G., McNab, H. et al. (1997). *J. Chem. Soc. Perkin Trans.* 1: 2483–2494.

53 Brown, A., Cadogan, J.I.G., McPherson, A.D., and McNab, H. (2007). *ARKIVOV* XI: 64–72.

54 Taylor, R. and Thorne, M.P. (1976). *J. Chem. Soc. Perkin Trans.* 2: 799–802.

55 Brent, D.H., Hribar, J.D., and deJongh, D.C. (1970). *J. Org. Chem.* 35: 135–137.

56 Spangler, J., Beckmann, B.G., and Kim, J.H. (1977). *J. Org. Chem.* 42: 2989–2996.

57 Taylor, R. (1978). *Chem. Commun.*: 732–733.

58 (a) Al-Awadi, N.A., Ballam, J., Hemblade, P., and Taylor, R. (1982). *J. Chem. Soc. Perkin Trans. II*: 1175–1177. (b) Taylor, R. (1975). *J. Chem. Soc. Perkin Trans.* II: 1025–1029.

59 Al-Awadi, N.A. and Taylor, R. (1986). *J. Chem. Soc. Perkin Trans.* II: 1255–1258.

6

Gas-Phase Pyrolysis of Hydrazones

The hydrazone moiety plays a key role in heterocyclic chemistry. It is formally derived from ketones or aldehydes by replacement of the oxygen of the carbonyl group with a hydrazine group [1] by elimination of water (Scheme 6.1).

The alpha-carbanion of hydrazones is estimated to be 10 times more nucleophilic than its ketone or aldehyde counterpart [1]. This is considered the most important aspect of the reactivity of hydrazones. Thus reactions such as Mannich, coupling, and halogenation take place readily at such carbon atoms, whereas the hydrazone nitrogen atom remains the main site for attack by acylating and alkylating agents. It has been noted that the nitrogen atom (NHR) is preferentially attacked by hard nucleophiles, whereas soft nucleophiles attack the carbanion carbon atom [2, 3].

In Chapter 2, we have shown how flash vacuum pyrolysis (FVP) was used as a valuable alternative tool for the Synthesis of novel heterocyclic and condensed heterocyclic compounds for potential biological and pharmaceutical applicability, from N-substituted cyclic amides and thioamides via a six-membered transition state and elimination of aryl-nitriles. Homolytic cleavage of the N—N bonds is a common feature in the gas-phase pyrolysis of hydrazines and their derivatives [4].

In this chapter, we will discuss the structural differences of various hydrazones by dividing them into three main groups based on the

Gas-Phase Pyrolytic Reactions: Synthesis, Mechanisms, and Kinetics,
First Edition. Nouria A. Al-Awadi.
© 2020 John Wiley & Sons, Inc. Published 2020 by John Wiley & Sons, Inc.

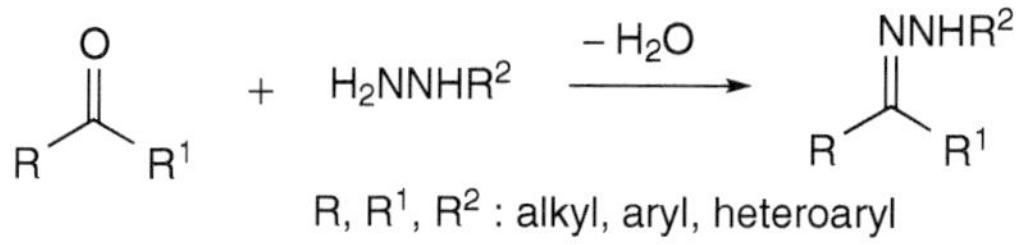

$$\text{R}\overset{\text{O}}{\underset{}{\math13C}}\text{R}^1 \quad + \quad \text{H}_2\text{NNHR}^2 \quad \xrightarrow{-\text{H}_2\text{O}} \quad \text{R}\overset{\text{NNHR}^2}{\underset{}{\mathbb{C}}}\text{R}^1$$

R, R^1, R^2 : alkyl, aryl, heteroaryl

Scheme 6.1

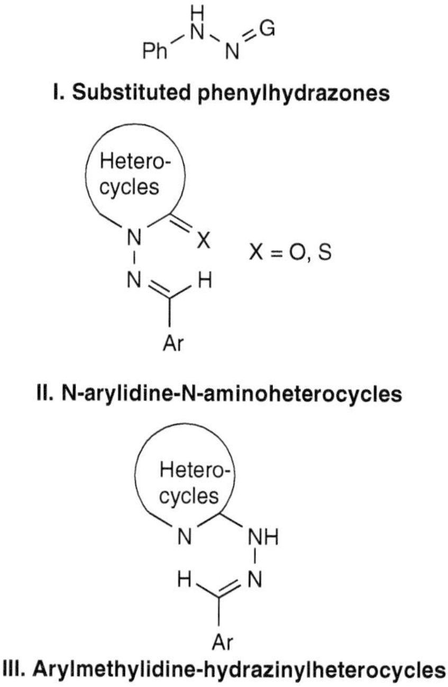

$$\text{Ph}\overset{\text{H}}{\underset{}{\diagdown}}\text{N}\diagdown\text{N}\diagup^{\text{G}}$$

I. Substituted phenylhydrazones

Heterocycles

$$\underset{\text{N}}{\overset{\text{N}}{\mid}}\diagup^{\text{X}} \qquad \text{X = O, S}$$

Ar

II. N-arylidine-N-aminoheterocycles

Heterocycles

N   NH

H   N

Ar

III. Arylmethylidine-hydrazinylheterocycles

Scheme 6.2

positions of the hydrazine moiety relative to the heteroarenes and other substituents (Scheme 6.2).

The only structural difference between the N-arylideneamino-N-heterocycles of group II and the arylideneheteroaryl-hydrazines of group III is that in group II, the nitrogen of an N-arylideneamino group is attached to the nitrogen atom of a heteroaromatic, whereas in group III, the hydrazine moiety is a substituent on the nitrogen heterocycle. Although in compounds of group I substituents at the phenyl hydrazone vary among non-aromatic, aromatic, and heterocycles, this structural feature together with the nature of the heteroaryl ring has resulted in interesting products. The behavior of these hydrazines has been investigated by gas-phase pyrolysis, and the kinetics were evaluated in terms of reaction rates and entropies of activation to correlate molecular reactivities and to provide evidence in support of the suggested pyrolysis mechanisms.

6.1 Substituted Phenylhydrazones

Various heterocyclic compounds of synthetic importance have been prepared by gas-phase pyrolysis of arylhydrazones. We have described efficient routes to cinnolines [5, 6], pyridines [7–9], pyridazines [10–12], pyrazoles, and isoxazoles [13–15] by flash vacuum and static pyrolysis techniques. Phenylhydrazones **1** have been pyrolyzed using a silica tube packed with silica tubing under reduced pressure. The pyrolysates were analyzed quantitatively by gas-chromatography and qualitatively by IR, UV, and NMR spectra [16]. The pyrolytic reactions involve elimination of benzonitriles and aniline, which were explained by two mechanistic pathways: (A) involving rupture of the N—N bond and leading to radicals that ultimately gave the major products, aniline and benzonitriles; and (B) involving a concerted four-centered reaction (Scheme 6.3).

Pyrolysis of 1,5-diaryl-1,2,5-triazapentadienes **2** in a flow system at 600 °C was initiated by cleavage of the N—N bond to give the conjugated iminyl radical **3**, which either fragments by loss of HCN to give minor products or cyclizes to quinoxaline **4** [17] as shown in Scheme 6.4.

Further investigation on the cyclization step of the iminyl radical **3**, leading to quinoxaline **4** (Scheme 6.4), was performed by pyrolyzing the hydrazone **5** at 600 °C and 10^{-2} mmHg, which was expected to lead exclusively to quinoxaline **6** by direct cyclization [18]. Instead, two isomeric quinoxalines **6** and **7** were formed. Similarly, hydrazone **8** leads to the same products (Scheme 6.5); this was explained in terms of interconversion of the two iminyls **9** and **10** via the spirodienyl radical **11**. Control experiments established the intramolecularity of the reaction and demonstrated that neither the products nor the starting materials interconvert significantly under the pyrolysis conditions used.

This investigation was extended to prove that the cyclization reaction is independent of the electronic nature of the substituents in the aryl rings [19]. Also, a ^{15}N-labeling study to probe the degenerate rearrangement of the aryliminoiminyl radicals suggested complete involvement of the spirodienyl radical [20].

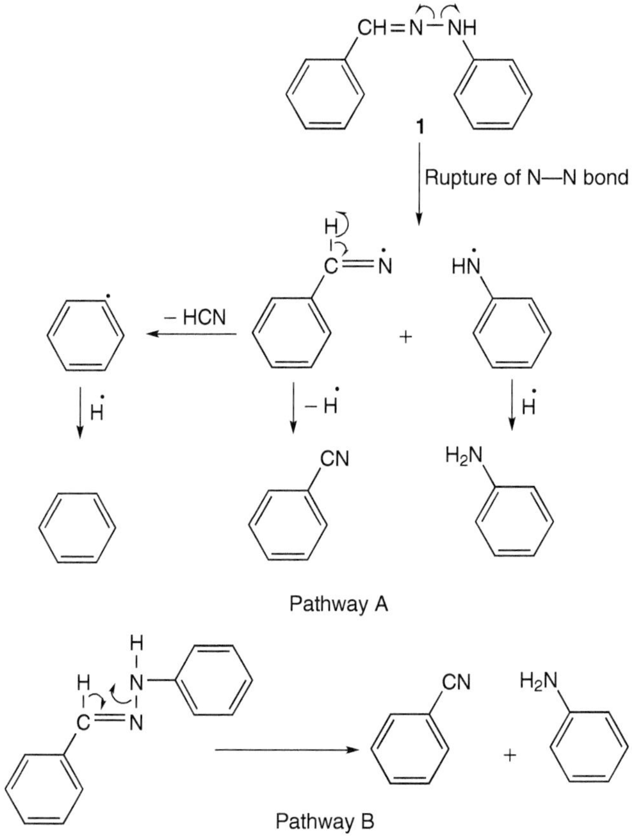

Scheme 6.3

Flash pyrolysis of 4-arylhydrazono-3-methyl-isooxazol-5-ones **12a–e** at 400–500 °C resulted in clean fragmentation into acetonitrile, carbon dioxide, phenylcyanamide **13**, and indazole **14** by a suggested mechanism [21] shown in Scheme 6.6.

A deuterium labeling experiment suggests that only the isocyanoamines **15**, and not the isomeric nitrile imines, were the intermediates in the formation of indazoles by an aromatic C—H insertion, leading to indazole **14** or rearrangement of **15** to cyanamides **13**.

It is worth mentioning here that until 1981, no aromatic isocyanoamines had been prepared, and the entire class of secondary amine derivatives RNH-N≡C was unknown. The FVP technique, coupled with low-temperature spectroscopy, allowed the identification of isocyanoamines **16a–d** [22]. Chemical evidence that the absorptions

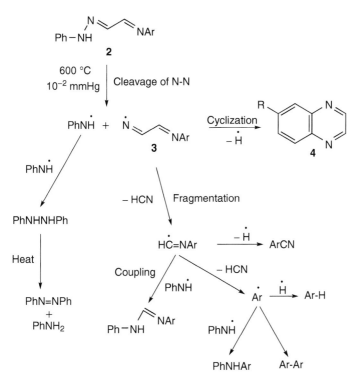

Scheme 6.4

are due to **16** is provided by thermal isomerization to stable, isolable cyanamide **17** (Scheme 6.7).

The hydrazone derivatives of Meldrum's acid **18a–c**, pyrolyzed at 400–500 °C, were found to also give isocyanide **19** and phenylcyanamide **20** [23]. Interestingly, the IR spectra were identical to those previously observed for the compounds generated by pyrolysis of isoxazolone **21**. A preparative pyrolysis of **18b** at 500 °C and 10^{-3} torr resulted in an 85% yield of phenylcyanamide **20b** together with 15% of indazole **22** as a result of gas-phase cyclization of the isocyanide **19b** (Scheme 6.8).

A series of arylhydrazono derivatives of α-benzotriazolyl ketones **23a–e**, **24f–g** were pyrolyzed (Scheme 6.9). They undergo pyrolytic transformation, leading to benzimidazoles; a detailed kinetic and mechanistic study provided data to evaluate the effect of the arylhydrazono substituent on reactivity and its effect on the thermal gas-phase elimination reaction of these ketones [24a].

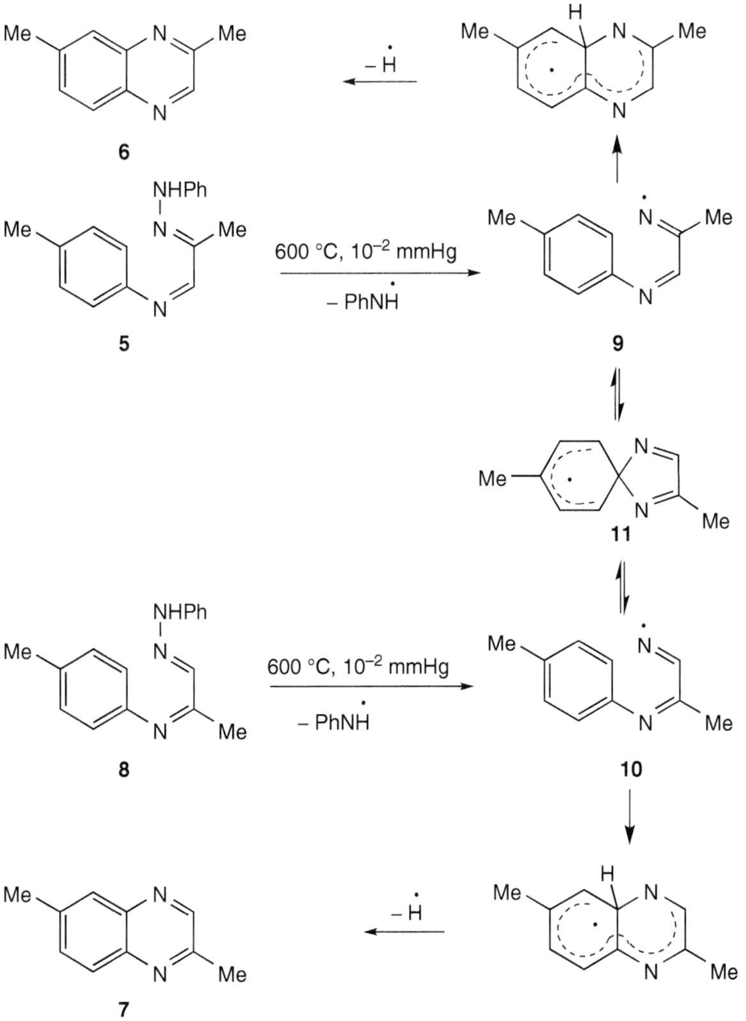

Scheme 6.5

The pyrolysates from arylhydrazonophenyl α-benzotriazol-1-yl ketones **23a–e** and **24f–g** were analyzed and found to include N-phenylaminobenzimidazoles **25a–e**, 2-benzoylbenzimidazoles **26**, and aniline and p-substituted anilines **27a–e**, together with benzimidazo [1,2-b]-cinnoline **28a–e** and p-substituted benzanilide **29a–e** (Scheme 6.10).

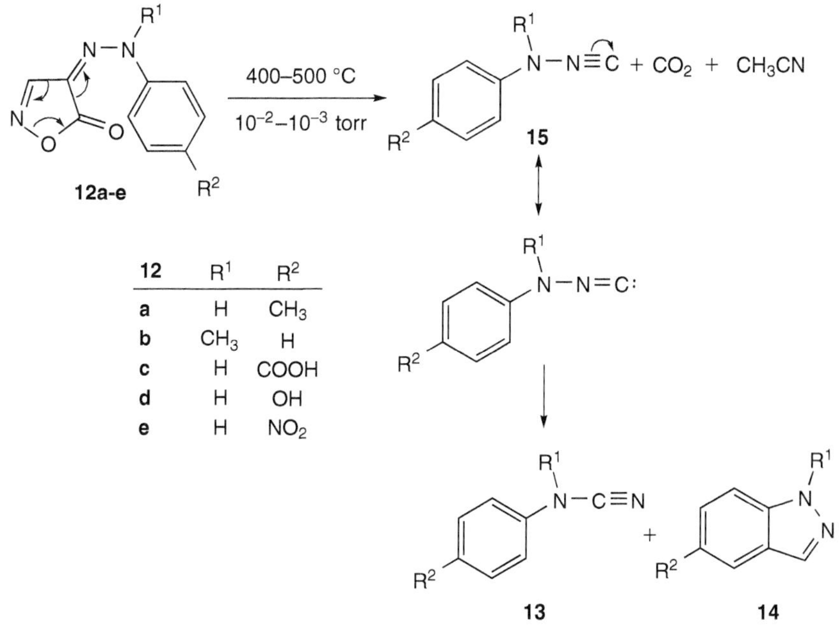

Scheme 6.6

12	R¹	R²
a	H	CH₃
b	CH₃	H
c	H	COOH
d	H	OH
e	H	NO₂

Scheme 6.7

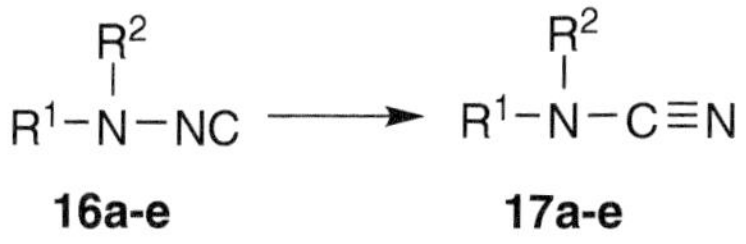

16	R¹	R²
a	phenyl	H
b	3-pyridyl	H
c	2,6-dimethylphenyl	H
d	phenyl	methyl
e	phenyl	phenyl

The suggested mechanism shown in Scheme 6.11 accounts for the formation of **25a–e** and **27a–e**. The mechanism includes elimination of an N_2 molecule, leading to a biradical or carbene intermediate [24b] followed by cyclization and then fragmentation processes.

Isolation and further pyrolysis of **25a–e** confirmed that it is the likely precursor of **26** and **27a–e**. It is worth noting that no reaction products corresponding to **25f–j** or **27f–j** were detected in the pyrolysis of the hydrazones **24f–j** when R=CH₃ (Scheme 6.12).

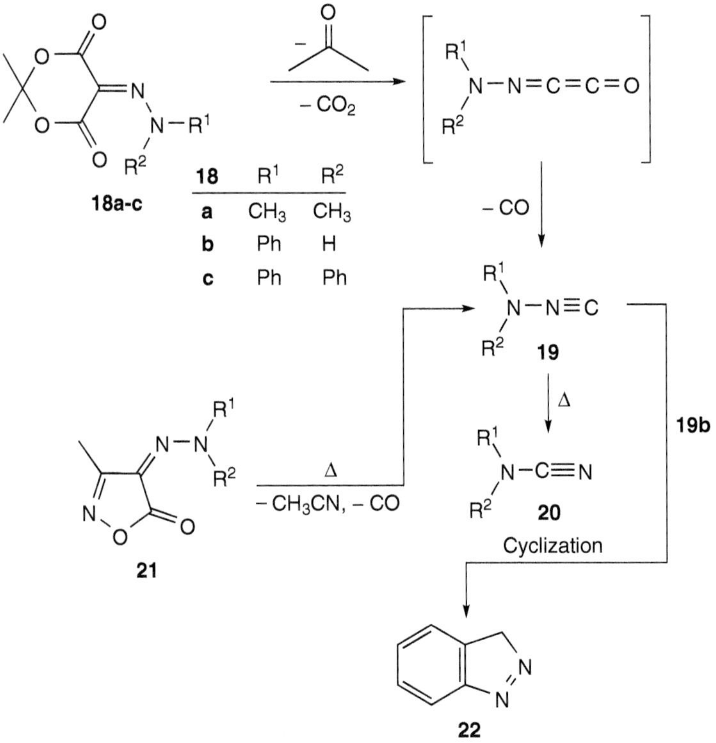

18	R^1	R^2
18a-c		
a	CH_3	CH_3
b	Ph	H
c	Ph	Ph

Scheme 6.8

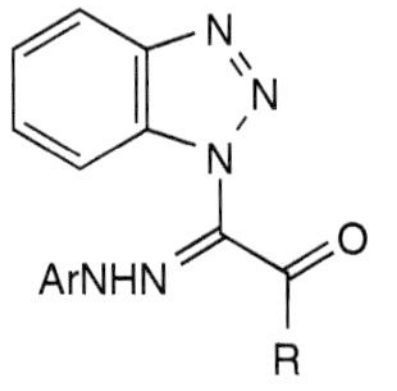

23	Ar	R		24	Ar	R
a	C_6H_5	C_6H_5		f	C_6H_5	CH_3
b	p-$CH_3OC_6H_4$	C_6H_5		g	p-$CH_3OC_6H_4$	CH_3
c	p-$CH_3C_6H_4$	C_6H_5		h	p-$CH_3C_6H_4$	CH_3
d	p-ClC_6H_4	C_6H_5		i	p-ClC_6H_4	CH_3
e	p-$NO_2C_6H_4$	C_6H_5		j	p-$NO_2C_6H_4$	CH_3

23a–e & 24 f–j

Scheme 6.9

The structure of the novel ring system of benzimidazocinnoline **28a–j** was confirmed by full proton and carbon signal assignment of compound **28g** as a representative example, based on ^1H NMR, ^{13}C NMR, NOE-difference spectra, 2D H,H-COSY, HMQC, and HMBC. From coupling constants in the ^1H NMR spectra and the cross peak in the H,H-COSY experiment, all the protons were assigned. From ^{13}C

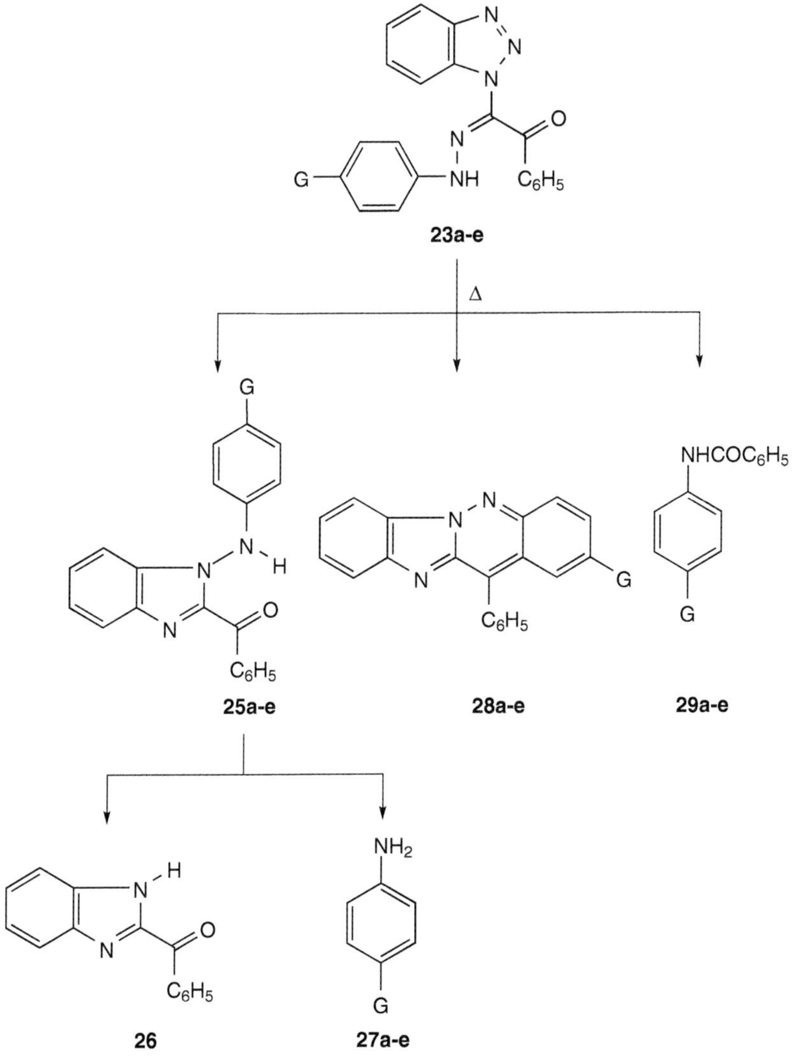

Scheme 6.10

NMR, HMBC, and HMQC experiments, the different carbon signals were also assigned [24a]. And the formation of **28g** was explained in terms of intramolecular nucleophilic addition of the arylhydrazono group to the ketone carbonyl followed by cyclization and fragmentation with loss of H_2O followed by loss of an N_2 molecule (Scheme 6.13).

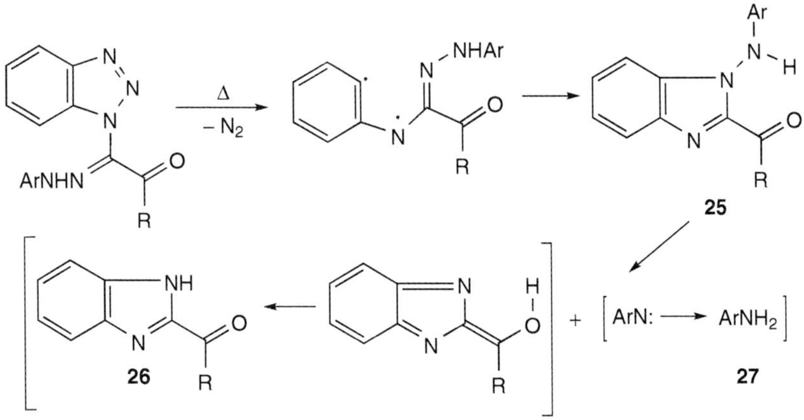

Scheme 6.11

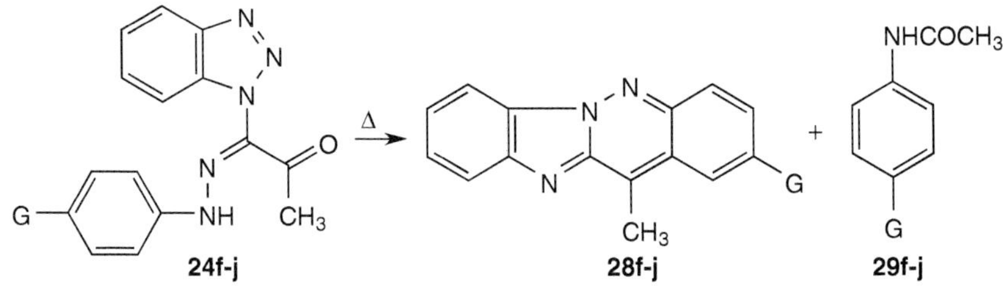

Scheme 6.12

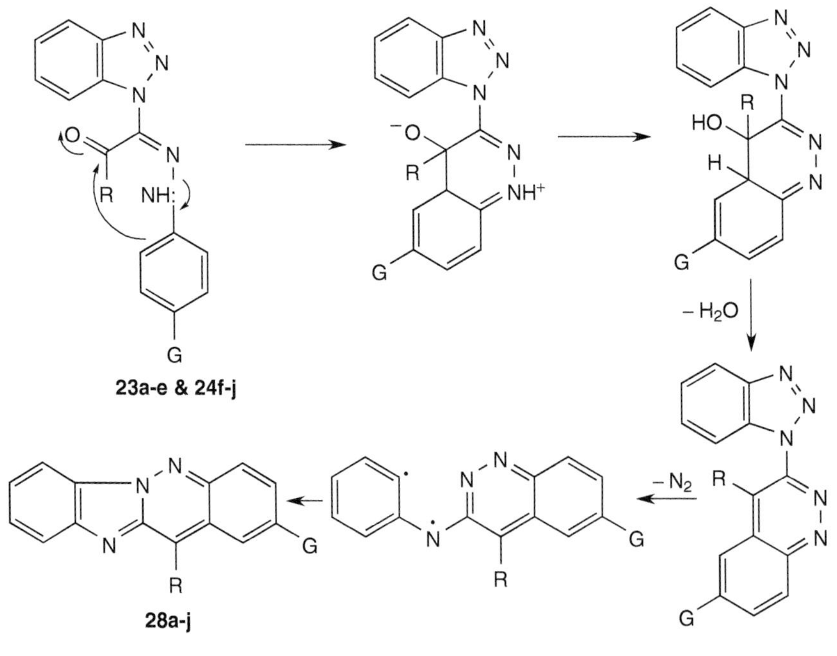

Scheme 6.13

Table 6.1 Rate coefficients and Arrhenius parameters for pyrolysis of **23a–e** and **24f–j** at 227 °C.

Substrate	Log A s^{-1}	E_a kJ mol^{-1}	$10^4\ k$ s^{-1}
23a	11.49 ± 0.13	139.5 ± 1.13	90.68
23b	13.71 ± 0.43	141.2 ± 3.64	915.4
23c	12.23 ± 0.18	134.4 ± 1.72	153.9
23d	12.12 ± 0.09	134.4 ± 0.753	121.1
23e	10.83 ± 0.42	124.4 ± 3.64	68.75
24f	11.837 ± 0.15	146.2 ± 1.55	3.947
24 g	7.842 ± 0.14	100.2 ± 1.21	23.84
24 h	8.110 ± 0.33	107.3 ± 3.10	7.958
24i	9.351 ± 0.09	118.5 ± 0.962	9.307
24j	8.203 ± 0.04	103.3 ± 0.377	26.07

The rate coefficient at 227 °C and the Arrhenius parameters for the reactions of **23a–e** and **24f–j** are shown in Table 6.1.

An analysis of the kinetic data reveals the following:

i) The arylhydrazono group leads to a dramatic jump in the rate of the pyrolysis reaction: **23a** is 10^3-fold more reactive than the unsubstituted ketone **30** [25]. The increase in the rate factor for the acetone counterpart **24f** is 8.6×10^2 (Scheme 6.14).

ii) The arylhydrazophenyl ethanone benzotriazole ketones **23a–e** are more reactive than the acetone counterpart **24f–j** (Table 6.1).

iii) Interestingly, there seems to be no opposing substituent effect from the para position of the aryl group of the arylhydrazono group.

Gas-phase cyclization of oximes **31a–d** as a route to 1,2,3-triazoles **32a–d** was investigated [26]. The suggested mechanism shown in Scheme 6.15 involves nucleophilic attack by the hydrazone nitrogen on the oxime nitrogen followed by H_2O elimination. Interestingly, compounds **31e,f** behaved differently. The products from **31e,f** were characterized as β-keto nitrile derivatives **33e,f**, which was attributed to the electron-withdrawing effect of the nitrile group reducing the nucleophilicity of the hydrazone nitrogen; so, **32e** and **32f** eliminate water from the oxime moiety via a four-membered transition state (Scheme 6.16).

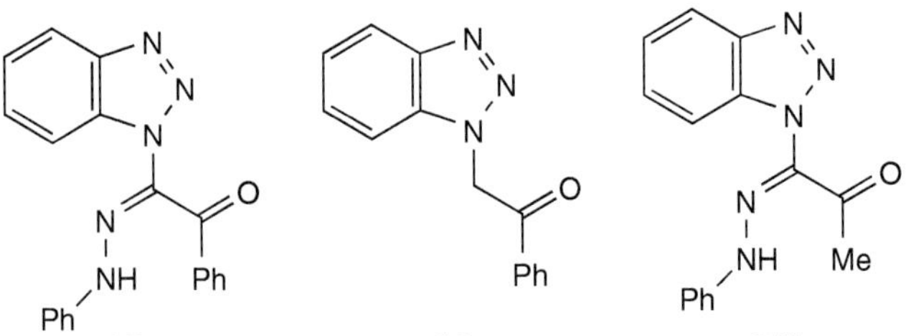

23a   **30**   **24f**

90.68×10^{-4} 8.59×10^{-6} 3.947×10^{-4}

Scheme 6.14

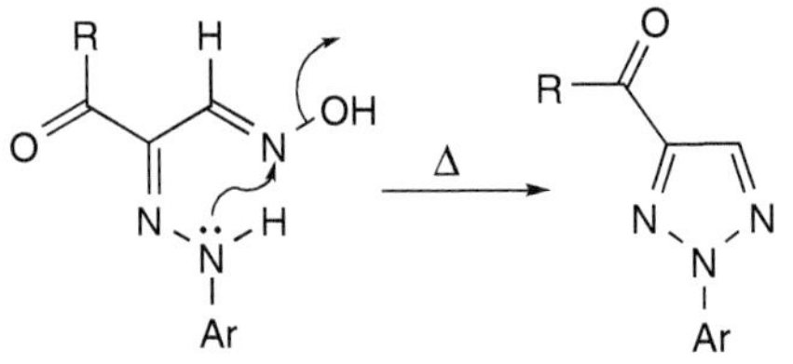

31	R	Ar
a	C_6H_5	C_6H_5
b	C_6H_5	$4\text{-}CH_3C_6H_4$
c	2-thienyl	C_6H_5
d	2-thienyl	$4\text{-}CH_3C_6H_4$

31a-d   **32a-d**

Scheme 6.15

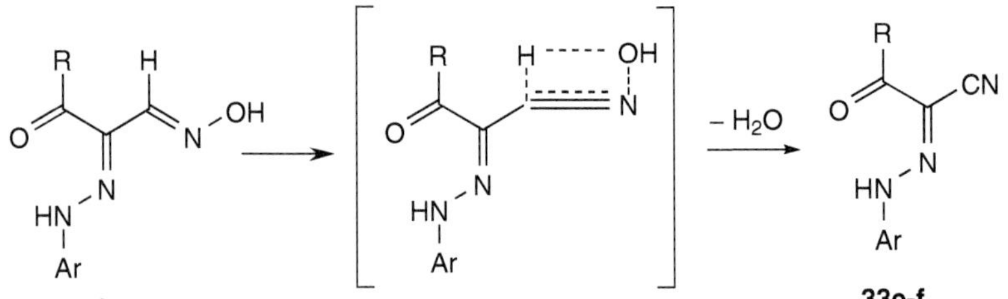

31e-f   **33e-f**

31	R	Ar
e	C_6H_5	$4\text{-}NCC_6H_4$
f	2-thienyl	$4\text{-}NCC_6H_4$

Scheme 6.16

A remarkably straightforward synthetic approach to cinnolines by gas-phase pyrolysis of 2-arylhydrazonopropanoals was investigated [5]. Cyclization of 3-oxo-3-aryl-2-arylhydrazonopropanals **34a–k** was expected to result in the formation of 3-aroyl derivatives **35** or the isomeric 3-carboxaldehyde **36** (Scheme 6.17). Spectral analysis proved the formation of the 3-aroyl derivatives **35** as the sole reaction products.

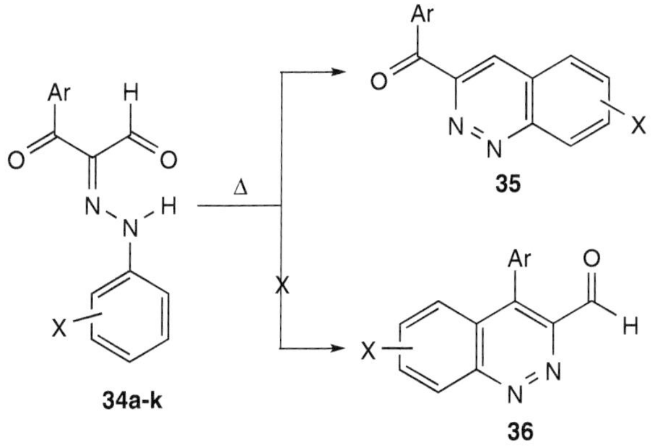

34	R	Ar
a	C$_6$H$_5$	H
b	4-ClC$_6$H$_4$	H
c	4-MeC$_6$H$_4$	H
d	4-MeOC$_6$H$_4$	H
e	2-Furyl	H
f	2-Thienyl	H
g	C$_6$H$_5$	4-OMe
h	C$_6$H$_5$	4-NO$_2$
i	C$_6$H$_5$	4-Cl
j	C$_6$H$_5$	3-OMe
k	C$_6$H$_5$	3-Cl

Scheme 6.17

Formation of **35** could be explained by either route A (Scheme 6.18), which involves electrophilic aromatic substitution, or route B, which involves thermal isomerization followed by 6-π-electro cyclization and finally loss of water to form **35**. The kinetic data in Table 6.2 show no significant electronic effect of the substituents on the rate of pyrolysis, which might well rule out route A.

6.2 N-Arylidineamino Heterocycles

Pyrolysis of the hydrazones **37a–i** at 200–250 °C for 45–60 minutes gave the corresponding nitriles via six-membered transition states [27] in yields of 71–91%. This was regarded as a practical procedure for conversion of amines used in the synthesis of hydrazones to nitriles (Scheme 6.19).

Hydrazones **38** could have been expected to yield the alkyl cyanates, but the major products were the corresponding ether or alcohol and 1-cyano-4,6-diphenyl-2-pyridone **39** formed by α-elimination and rearrangement of the presumed isonitrile **40** to cyano-2-pyridone **39**, as shown in Scheme 6.20. However, it was not certain whether the alkyl cyanates were formed or not in this reaction [27].

Upon heating at 180 °C, the hydrazono 1-amino-4,6-diphenyl-2-pyridone derivative **41** gave isocyanates **42** and the corresponding 4,6-diphenylpyridone [27], as shown in Scheme 6.21.

N-Arylidineamino heterocycles **43a–e** undergo elimination to cyanoarenes and 2-hydroxy-4,6-dimethylpyridine-3-carbonitrile **44** [28] (Scheme 6.22) following the same mechanism as described for

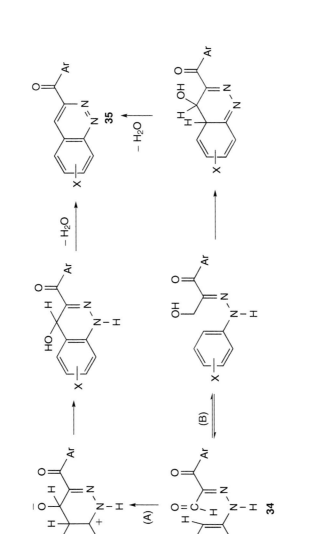

Scheme 6.18

Table 6.2 Rate coefficients and relative rates at 277 °C for gas-phase cyclization of 2-arylhydrazonopropanals **34a–k**.

Compound (34)	$10^4\ k\ s^{-1}$	k_{rel}
a	9.37	1
b	7.48	0.8
c	5.59	0.6
d	6.83	0.7
e	5.99	0.6
f	5.90	0.6
g	14.45	1.5
h	23.03	2.4
i	5.22	0.5
j	3.57	0.4
k	6.49	0.7

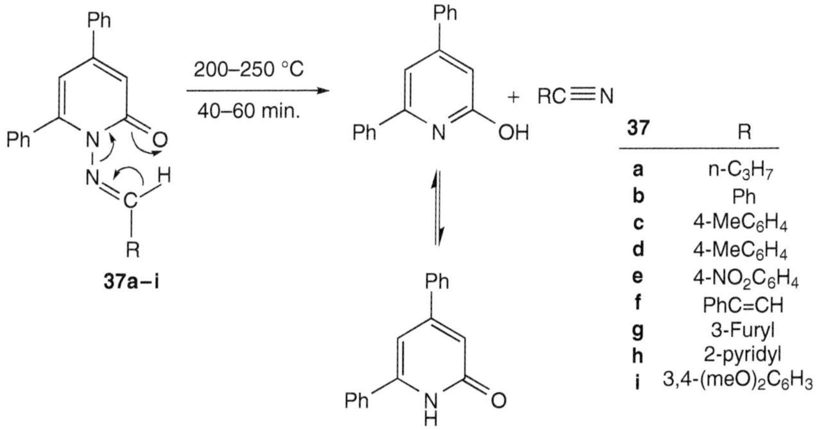
Ph
Ph
N
O
N
N
H
C
R
37a–i
200–250 °C
40–60 min.
Ph
Ph
N
OH
+ RC≡N
Ph
Ph
N
H
O
37
R
a n-C3H7
b Ph
c 4-MeC6H4
d 4-MeC6H4
e 4-NO2C6H4
f PhC=CH
g 3-Furyl
h 2-pyridyl
i 3,4-(meO)2C6H3

Scheme 6.19

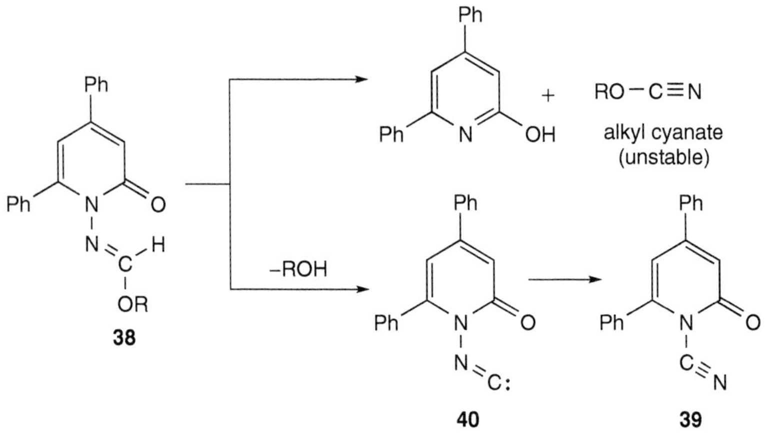

Scheme 6.20

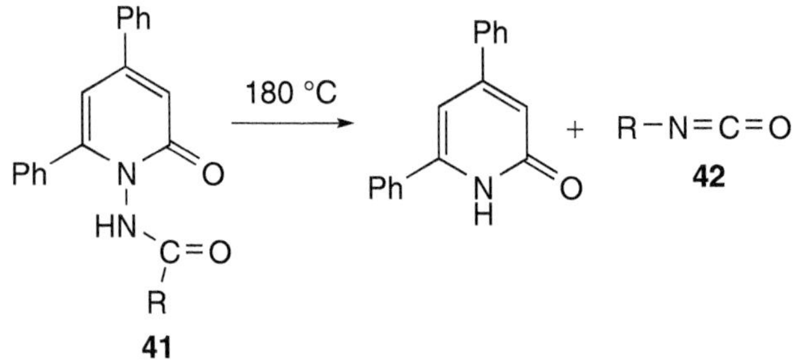

Scheme 6.21

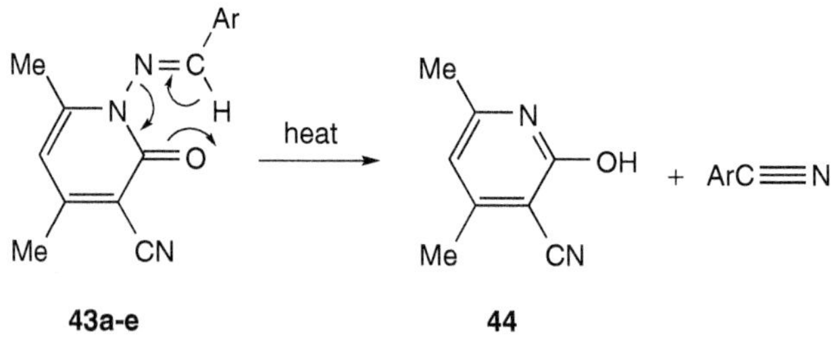

43	R
a	4-NO$_2$C$_6$H$_4$
b	4-ClC$_6$H$_4$
c	C$_6$H$_5$
d	4-MeC$_6$H$_4$
e	4-MeOC$_6$H$_4$

Scheme 6.22

hydrazones **37**. The molecular reactivity of these arylhydrazones was investigated by relating the structure to rate constants in the pyrolysis reactions of **43a–e**, which were found to be unimolecular and first-order. The rate constants correlate well with Hammett σ^0 values of the aryl substituents. This correlation resulted in a ρ-factor of 0.83 at 247 °C, which indicates that elimination of nitriles **44** from hydrazones **43** in Scheme 6.22 is aided by electron-withdrawing substituents at the aryl group. The reactivity of such hydrazones is due to the acidity of the hydrogen atom involved in a six-membered transition state. This gas-phase pyrolytic reaction may be considered an efficient approach to the synthesis of cyanoarenes with functional substituents that are unstable in a basic and/or acidic medium.

Gas-phase pyrolysis of 4-arylidenimino-3-thioxo-1,2,4-triazines-5-ones **45** revealed an interesting mechanistic feature. Here, elimination of a cyanoarene could be explained by participation of either the C=O or C=S group in the six-membered cyclic transition state (Scheme 6.23). Theoretical calculations [29] indicated that the transition state involves the C=O moiety rather than the C=S group. This

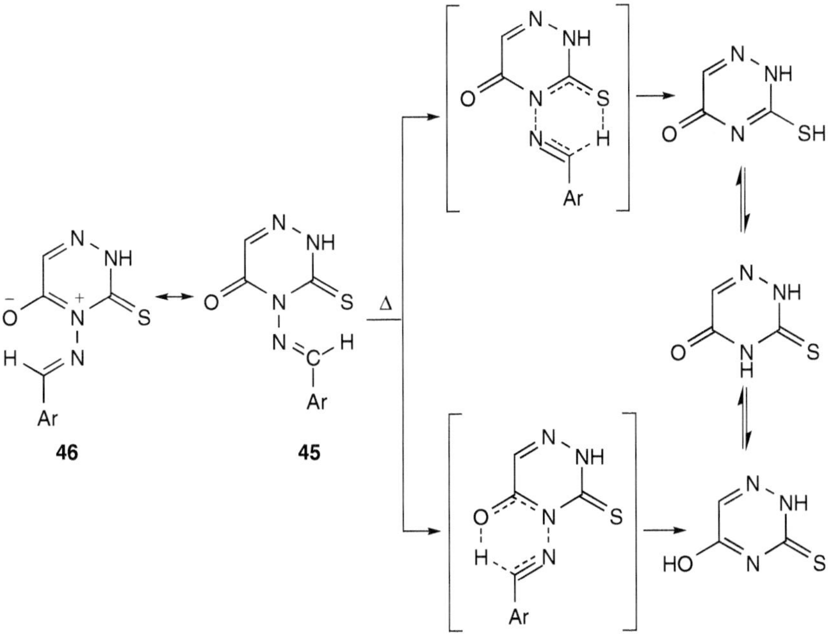

Scheme 6.23

is explained by the fact that the electron pair of the nitrogen atom is more effectively delocalized over the C=O than the C=S group. Conversely, the aromatic, zwitterionic betaine resonance structure **46** with a negative charge on O rather than S is preferred.

In further investigations, the gas-phase pyrolysis of 4-arylideneamino-1,2,4-triazole-3(2H)-ones **47** and their thione analogues **48** was found to yield benzonitriles and 1,2,4-triazolones **49** and thiones **50**, respectively [30] (Scheme 6.24).

Kinetic investigations show that the triazolethiones **48** are 10^3-fold more reactive than **47** (Table 6.3), which is ascribed to the greater protophilicity and the thermodynamic lability of the C=S group.

Computational studies at the DFT level [31] of the pyrolysis of **47** and **48** supported the one-step reaction mechanism resulting in **49** and **50** with elimination of benzonitrile via a six-membered transition state. The data indicate that the transition states are intermediate between reactants and products; however, the calculated synchronicities showed

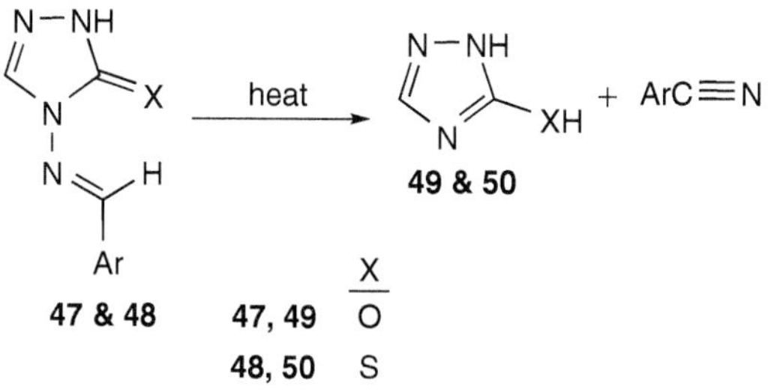

Scheme 6.24

Table 6.3 Absolute and relative rate constants at 227 °C for pyrolysis of **47** and **48**.

Substituent on aryl ring	47	48	k_{rel}
	$10^6\ k\ s^{-1}$	$10^3\ k\ s^{-1}$	
H	2.35	4.83	2.1×10^3
4-Cl	2.19	5.20	2.4×10^3
4-Me	2.80	7.05	2.5×10^3
4-OMe	8.15	4.75	0.6×10^3

that the pyrolysis of triazolthiones **48** is slightly asynchronous with bond-breaking slightly more advanced than bond-forming.

A similar reactivity difference was observed between 3-arylideneamino pyrimidine-2-one **51** and the thio-analogue (Scheme 6.25); thione **52** is 114 times more reactive than **51** [32].

This can be attributed to the enhanced nucleophilicity (protophilicity) of the carbonyl oxygen atom of **51** due to the fact that the electron pairs of N1 and N2 are delocalized over the carbonyl oxygen atom involved in the transition state, whereas in **52**, not only is the pair of electrons on N1 less effectively delocalized onto thione sulfur, but also the electron pair of N3 is delocalized preferentially onto the carbonyl oxygen more than on the thione sulfur. This will decrease the protophilic character of the thione sulfur. The same argument explains the 192-fold increased reactivity of compound **53** relative to **54**. This

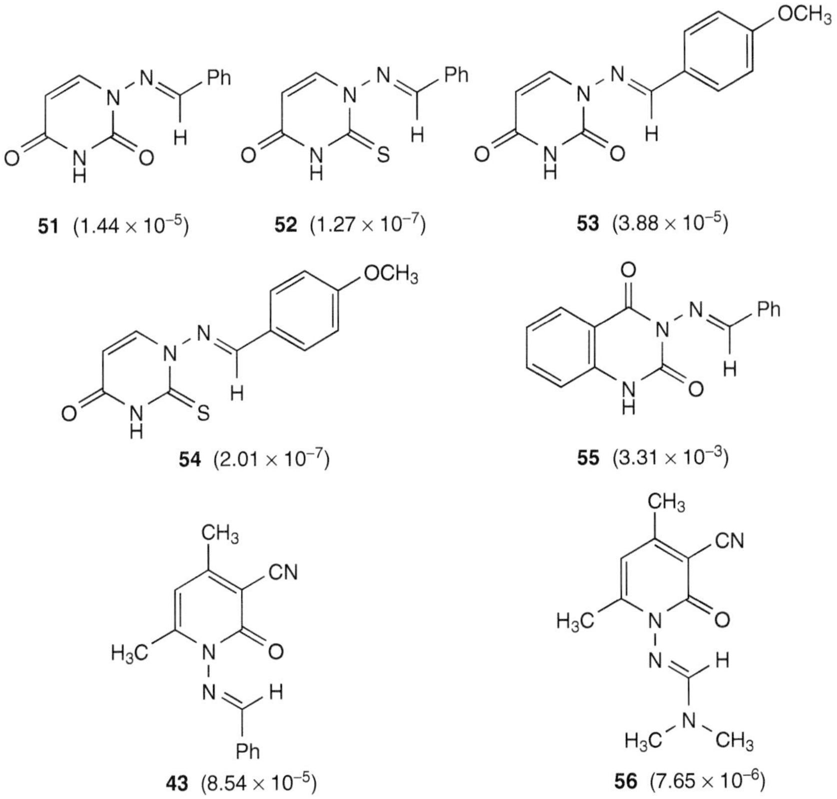

51 (1.44×10^{-5}) **52** (1.27×10^{-7}) **53** (3.88×10^{-5})

54 (2.01×10^{-7}) **55** (3.31×10^{-3})

43 (8.54×10^{-5}) **56** (7.65×10^{-6})

Scheme 6.25

enhanced protophilicity of C=O relative to the C=S group was also observed in the thermal elimination reactions of N-acetylthiourea [33] and N-acetyl-N-phenyl-thiourea [34], where a phenyl group in the latter compound affected the protophilicity of the C=S moiety so much that it restored the H-bonding ability of the acetyl C=O group and, further, increased the acidity of the other hydrogen atom of the amino group. Thus, the interaction with the C=O group becomes preferred.

Compound **55** is found to be 230 times more reactive than **51**. This is explained in terms of the conjugation of the electron pair of the nitrogen atom with the two carbonyl groups bonded to this nitrogen atom. This leads to an effective delocalization by cross-conjugation (Scheme 6.26), which will increase the polarity of the N—N bond involved in the transition state, enhance its breaking, and thereby increase the reaction rate relative to compound **51**.

Compound **43** is about 10 times more reactive than compound **56**. This is because of the delocalization of the electron pair of the nitrogen atom of the NMe_2 moiety, which will give the N—N involved in the transition state some double-bond character (Scheme 6.27), thus rendering its cleavage more difficult.

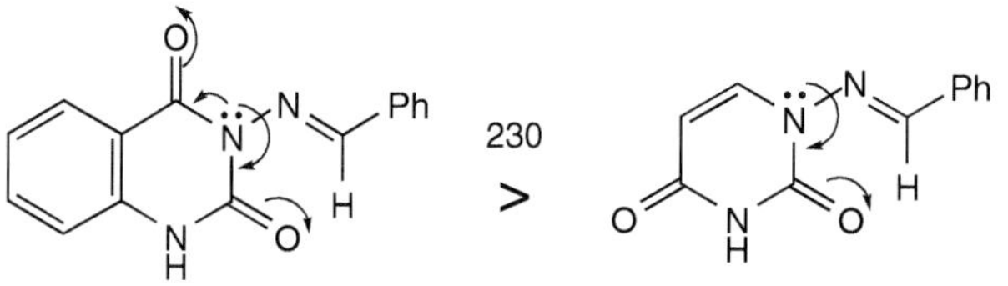

55   230   **51**

Scheme 6.26

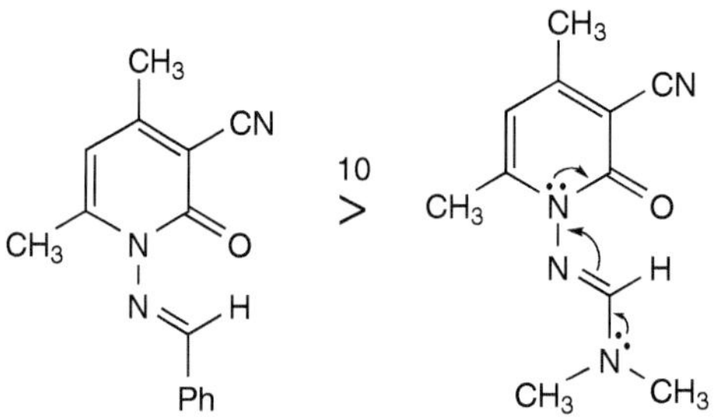

43   10   **56**

Scheme 6.27

6.3 Arylidene Hydrazine Heterocycles

In these compounds, the hydrazine moiety is coupled to an arylidene group at one nitrogen and a heteroaryl residue at the other (Scheme 6.2; III). These arylidene hydrazines not only offer interesting new routes toward various heterocycles and condensed heterocycles but also are considered a novel kinetic and mechanistic probe in gas-phase pyrolysis reactions of diketones, ketoanilides, and other related compounds [28, 35–42], because the effects of substituents on the heteroaryl and aryl groups provide a rationale for the mechanistic and kinetic behavior of such compounds.

Heating neat hydrazone **57** at 200 °C for about 40 minutes led to the formation of *s*-triazoloquinoxaline **58** and **59** via intermediate **60**, which in several cases was isolated and characterized [43] (Scheme 6.28).

The liberation of a hydrocarbon from the intermediate **60** is thought to give the s-triazolo ring. The hydrazone **61**, however, gave only a poor yield of triazine ring **62** upon pyrolysis (Scheme 6.29).

When hydrazone **63** was pyrolyzed, it failed to give a product analogous to **62**, but led instead to **64** by a reaction similar to that described for hydrazones **57** (Scheme 6.30).

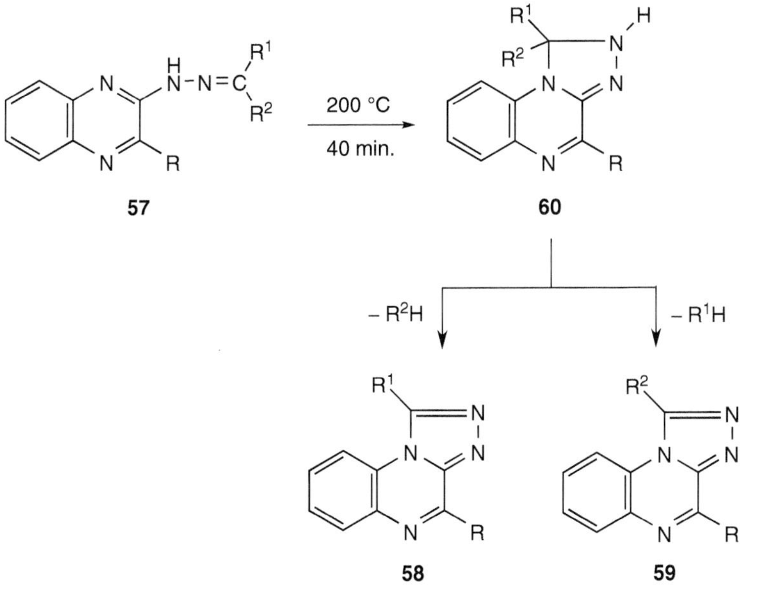

Scheme 6.28

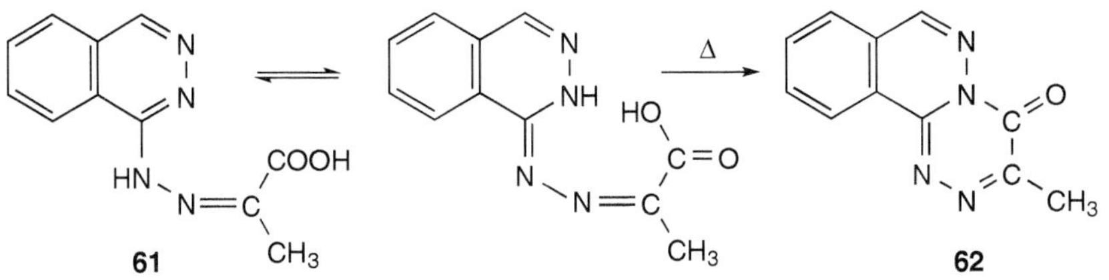

Scheme 6.29

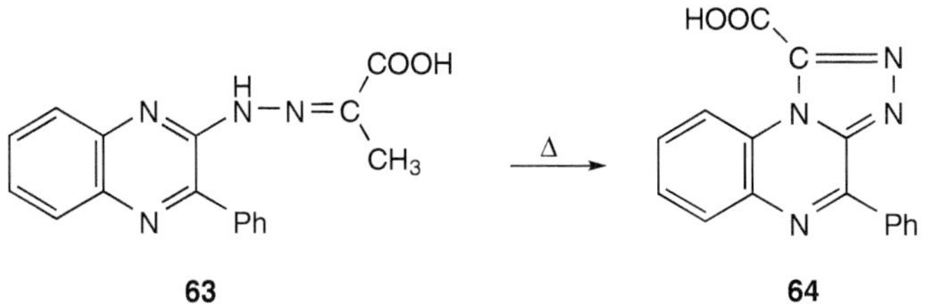

Scheme 6.30

The hydrazines **65a–e** were pyrolyzed by both FVP and static pyrolysis (STP) techniques, and the products were found to be affected by the mode of the pyrolysis [35a]. It is noteworthy that the substrate residence time under STP reaction conditions was 280 °C/900 seconds in a closed sealed tube, while under FVP it is 700 °C/10 minutes; both processes were conducted at low pressure.

The products of FVP of **65a–e** were identified as benzonitriles **66a–e** and 2-aminobenzimidazole **67**. The mechanism postulated in Scheme 6.31 has two alternative pathways via tautomers **A** and **B**, the latter involving radical intermediates that finally led to **66a–e** and **67** [35a].

A detailed study of the STP of hydrazine **65a** was performed at 280 °C, 0.045 torr with a contact time of 900 seconds in a sealed pyrolysis tube. The products of complete pyrolysis were collected, separated by column chromatography, and then analyzed qualitatively and quantitatively by different spectroscopic techniques. The products were identified as benzonitrile **66a**, 2-aminobenzimidazole **67**, 2,4,5-triphenylimidazole **68**, 1,3-diphenylimidazo[3,4-a]benzimidazole **69**, and 5,11-diphenyldibenzimidazolo pyrazine **70**, as shown in Scheme 6.32.

It was postulated that elimination of H_2 from **65a** led to intermediate **71**, which eliminates an N_2 molecule to form the diradical **72**. This

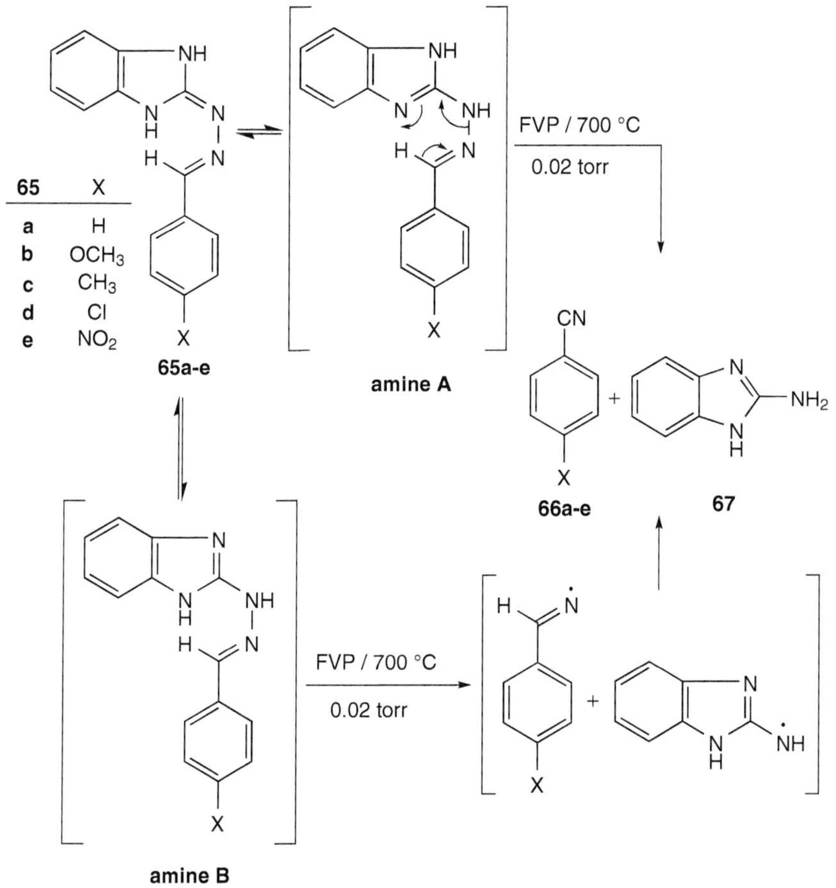

Scheme 6.31

diradical may then react with the benzonitrile **66a** already present in the reaction mixture to form the condensed heterocycle **69**. Two units of the diradical **72** led to the formation of the heterocycle **70** (Scheme 6.33).

Three moles of the radical intermediate **73** resulting from the pyrolysis of tautomer **B** of **65a** gave one mole of intermediate **74**, which cyclizes to **75** followed by dehydrogenation to give 2,4,5-triphenylimidazole **68**. An alternative route to **68** was suggested from further reaction of product **69** with H_2 to give intermediate **76**: it undergoes consecutive 1,3-H shifts to yield **77**, which upon 1,3-shift and loss of an NH_3 molecule yields 2,4,5-triphenylimidazole **68**, as shown in Scheme 6.34. It is noteworthy that reaction of the radical intermediate **73** with

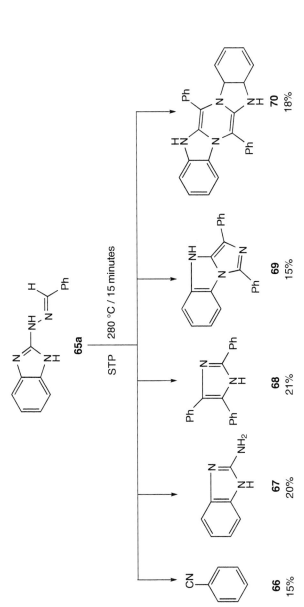

Scheme 6.32

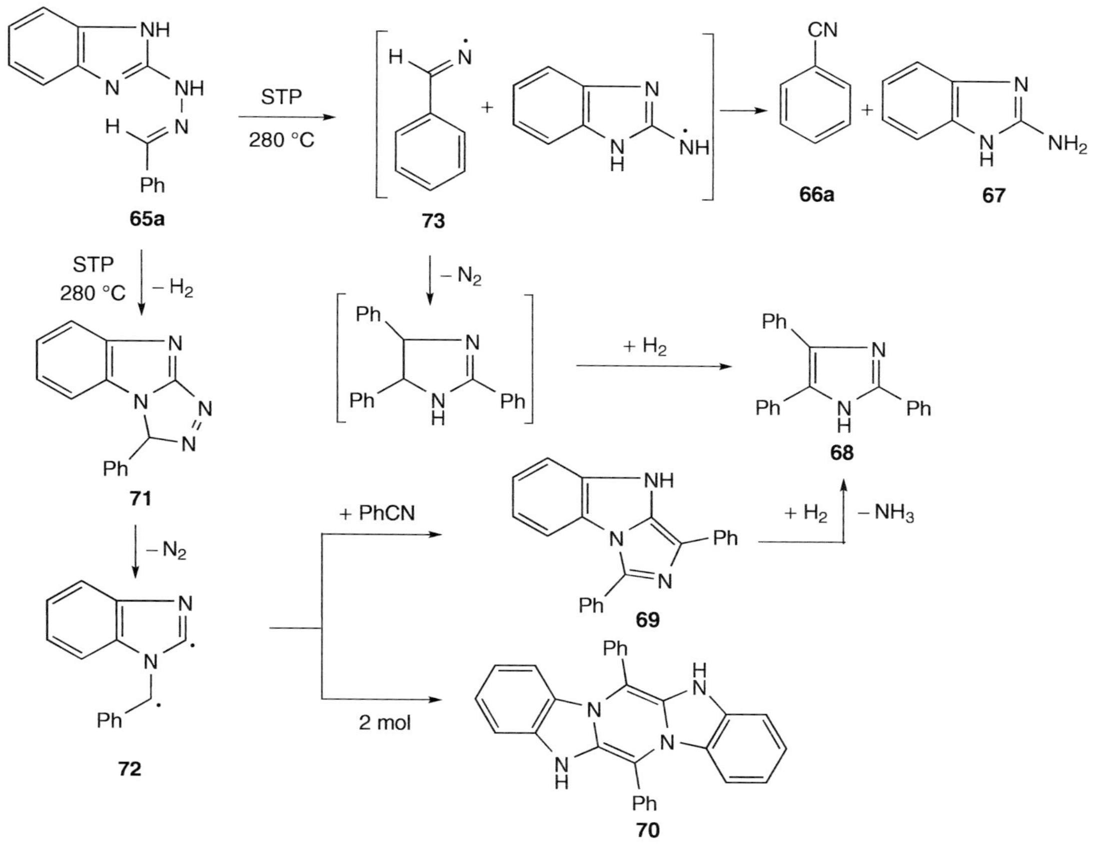

Scheme 6.33

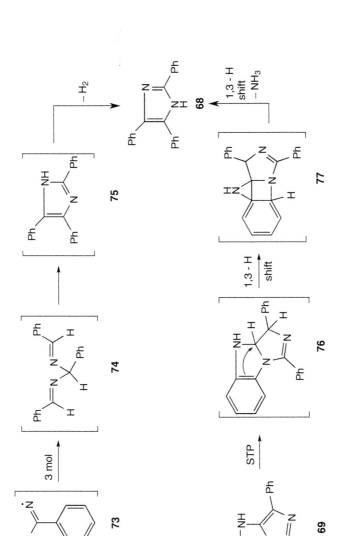

Scheme 6.34

hydrogen leads to the formation of the corresponding arylimine, which has been reported to yield **74** by elimination of ammonia [35b].

Arylidenepyridyl-2-hydrazines **78a–e** were subjected to flash vacuum and static pyrolysis [44]. Rates of their pyrolytic reactions were measured. The composition of the pyrolysates from the two pyrolysis modes showed remarkable similarities as well as some differences that are attributed mainly to differences in temperatures and residence times.

The products were identified as cyanoarenes **79**, 2-aminopyridine **80**, and 1,2,4-triazolopyridines **81**, and explained by a mechanism analogous to the one suggested for the imidazolehydrazines **65a–e** and other hydrazines [28, 35–42] (Scheme 6.35). It was noted that at higher temperatures, the yields of the cyanoarenes **79** and 2-aminopyridine **80** increased, whereas the 1,2,4-triazole **81** almost disappeared (Table 6.4). This may be attributed to an isomerization of the triazole intermediate **81** to its counterpart **81*** followed by fragmentation into two radical intermediates **82** and **83**; whereas **82** can produce cyanoarenes **79**, capture of H_2 by **83** gives 2-aminopyridine **80** [44] (Scheme 6.35).

It is noteworthy that both modes of pyrolysis, FVP and STP, of hydrazines **78a–e** involve oxidative loss of H_2 coupled with intramolecular cyclization to give 1,2,4-triazole **81**, which undergoes valence bond isomerization at higher temperatures to give 1,3,4-isomers **81*** as shown in Scheme 6.36. This process is suggested to proceed by the formation of a nitrile imine intermediate, which through N/N insertion gives a nitrene intermediate that cyclizes to give the isomer **81***. This isomerization represents a Dimroth-like rearrangement [45].

The two isomeric intermediates (**81**, **81***) formed in both modes of pyrolysis were identified and characterized by GC–MS analysis. The independent synthesis of the two triazolo isomers **81** and **81*** has provided a basis for the authentication [44] of the products of pyrolysis and conformational details resulting from their gas-phase pyrolysis reactions. An example of the MS pattern of analysis is that obtained for 1-(4-chlorobenzylidene)-2-pyridine-2-yl-hydrazine **78d**. The two triazolo isomers (**81** and **81***) from both modes of pyrolysis reactions have characteristic GC retention times of 14.84 and 17.67, respectively; the electron impact MS spectra gave the same molecular ion with the two isotopic chlorine-atom tags shown in their correct ratios. Isomer **81** gave a major fragment ion at m/z = 202, due to elimination of N2, while isomer **81*** gave a major fragment ion at m/z = 92 for the elimination of a cyanoarene fragment.

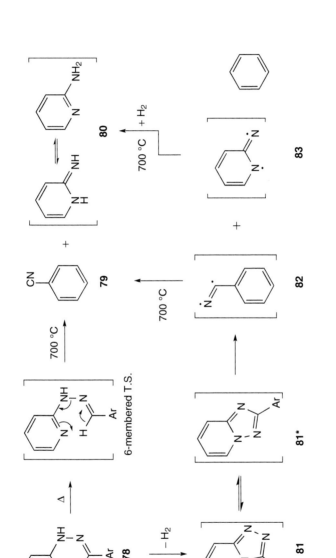

Scheme 6.35

Table 6.4 Percentage yields of products of STP (900 seconds) of hydrazines **78a–e**.

78	Substituent on aryl ring	340 °C 79	80	81	375 °C 79	80	81	Unreacted substrate 340 °C	375 °C
a	H	28	23	20	62	58	—	12	44
b	p-CH$_3$	38	40	15	58	62	11	14	37
c	p-OCH$_3$	34	33	18	85	82	8	6	41
d	p-Cl	40	41	10	76	67	6	18	32
e	p-NO$_2$	26	24	17	21	24	—	22	38

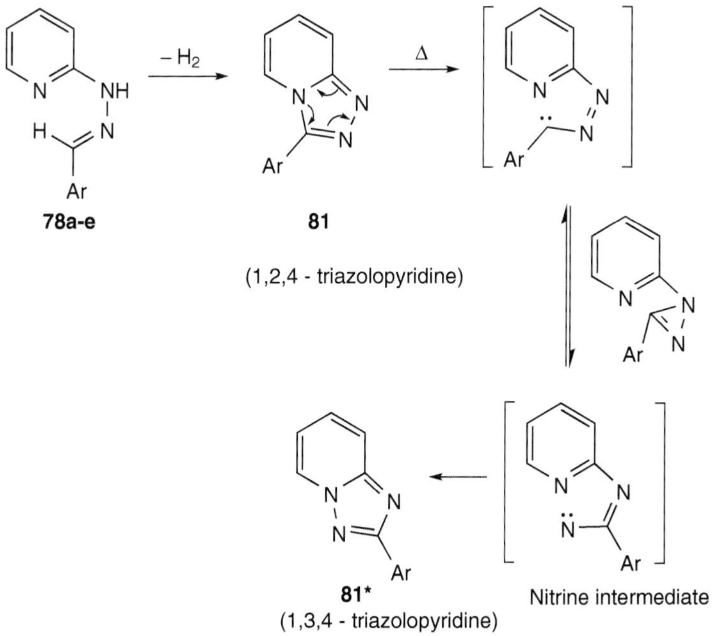

Scheme 6.36

Extensive work on structurally related hydrazines was carried out in our laboratories [46]; each compound features an aromatic azine ring joined to an arylhydrazone group (Scheme 6.37). This is considered a valuable study of the interplay of structural and substituent effects on the gas-phase pyrolysis of heterocycles and the modifications resulting from fused benzene rings. The main structural feature of these

pyrazines pyrimidines quinolines isoquinolines

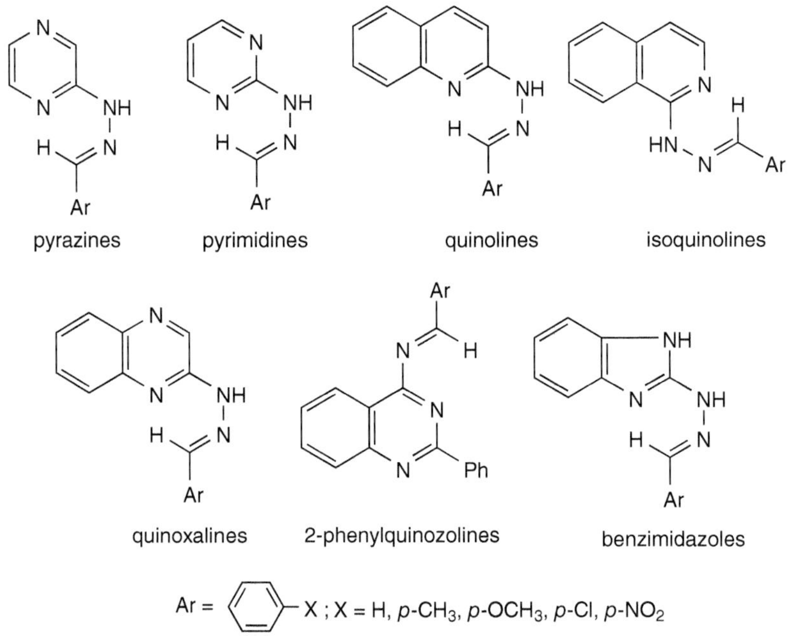

quinoxalines 2-phenylquinozolines benzimidazoles

$Ar = $ ⟨benzene ring⟩ $- X ; X = H, p\text{-}CH_3, p\text{-}OCH_3, p\text{-}Cl, p\text{-}NO_2$

Scheme 6.37

compounds is the common hydrazone substituent (HN—N=), bridging a heterocycle with an aryl group.

Analysis of the pyrolysates indicates that all the hydrazones shown in Scheme 6.37 behave in the same way. A lower temperature (400 °C) yielded triazole derivatives of condensed heterocycles, but at a higher temperature (700 °C) the pyrolysates were limited to amines and arenes. The formation of triazoles at lower pyrolysis temperatures and their absence at higher temperature are common features in these pyrolysis reactions [47]. The reaction mechanisms shown in Schemes 6.35 and 6.36 explain the products formed from the hydrazines shown in Scheme 6.37. The mechanism was further supported by kinetic and structure/reactivity correlations. Imine/amine tautomerism in aromatic N-heterocycles favors the amine tautomer in the gas phase.

The cyclization process shown in Scheme 6.38 is similar to the oxidative intramolecular cyclization reaction to remove H_2 by oxidizing agents and is considered an interesting example of gas-phase oxidative cyclization pyrolysis [48–50].

The large negative values of the entropy of activation ($\Delta S^{\#}$) shown in Table 6.5 suggest a favorable structural preorganization and provide

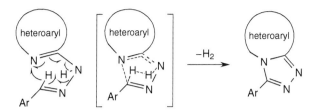

Scheme 6.38

Table 6.5 Entropy of activation ($\Delta S^{\#}$) of N-heteroaryl-hydrazones.

Heteroaryl	$\Delta S^{\#}$ J K^{-1} mol^{-1} (247 °C)	Heteroaryl	$\Delta S^{\#}$ J K^{-1} mol^{-1} (227 °C)
Pyrazine	-118 ± 48	Pyridine	-118 ± 45
Pyrimidine	-169 ± 41	Quinoline	-215 ± 18
Quinoxaline	-153 ± 66	Isoquinoline	-173 ± 38
Thienopyrimidine	-202 ± 20	Phenylquinazoline	-194 ± 38
—	—	Phenylthienopyrimidine	-150 ± 16

support for the reaction mechanism for the pyrolysis of hydrazones in Scheme 6.37, in support of the formation of condensed 1,2,4-triazole intermediates and their isomerization to 1,2,3-triazoles. It is noteworthy that in reports of a pyrolysis reaction that involves regular fused triazolotriazines, they were found to isomerize into their linearly fused triazolotriazines associated with significantly high and negative entropies of activation [51].

References

1 Singh, N., Ranjana, R., Kumari, M., and Kumar, B. (2016). *Int. J. Pharm. Clin. Res.* 8 (3): 162–166.
2 Elassar, A.A., Dib, H.H., Al-Awadi, N.A. et al. (2007). *Arkivoc* 2007 (2): 272–315.
3 Ali, M.R., Ali, M.R., Marella, A. et al. (2012). *Indian J. Pharm.* 23 (4): 193–202.

4 Gaber, A.M. and Wentrup, C. (2017). *J. Anal. Appl. Pyrolysis* 125: 258–278.

5 Kumar, A., Al-Awadi, N.A., Elnagdi, M.H. et al. (2001). *Int. J. Chem. Kinet.* 33: 402–406.

6 Al-Awadi, N.A., Elnagdi, M.H., Ibrahim, Y.A. et al. (2001). *Tetrahedron* 57: 1609–1614.

7 Agamy, S.M., Abdelkhalik, M.M., Mohamed, M.H., and Elnagdi, M.H. (2001). *Z. Naturforsch.* 56b: 1074–1078.

8 Abdel-Khalik, M.M. and Elnagdi, M.H. (2002). *Synth. Commun.* 32: 159–164.

9 Abdel-Khalik, M.M., Eltoukhy, A.M., Agamy, S.M., and Elnagdi, M.H. (2004). *J. Heterocyclic Chem.* 41: 431–434.

10 Al-Zaydi, K.M., Hafez, E.A., and Elnagdi, M.H. (2000). *J. Chem. Res. (S)* 154 (M): 510–527.

11 Abdel-Khalik, M.M., Agamy, S.M., and Elnagdi, M.H. (2001). *Synthesis* (12): 1861–1865.

12 Al-Omran, F., Abdel-Khalik, M.M., Abou El-Khair, A., and Elnagdi, M.H. (1997). *Synthesis* (1): 91–94.

13 Abdelkhalik, M.M., Agamy, S.M., and Elnagdi, M.H. (2000). *Z. Naturforsch.* 55b: 1211–1215.

14 Al-Saleh, B., El-Anasery, M.A., and Elnagdi, M.H. (2004). *J. Chem. Res.*: 578–580.

15 Al-Shiekh, M.A., Salah El-Din, A.M., Hafez, E.A., and Elnagdi, M.H. (2004). *J. Chem. Res.*: 174–179.

16 Crow, W.D. and Solly, R.K. (1966). *Aust. J. Chem.* 19: 2119–2126.

17 McNab, H. (1980). *J. Chem. Soc. Perkin Trans.* 1: 2200–2204.

18 McNab, H. (1980). *J. Chem. Soc. Chem. Commun.*: 422–423.

19 McNab, H. (1982). *J. Chem. Soc. Perkin Trans.* 1: 1941–1945.

20 McNab, H. and Smith, G.S. (1984). *J. Chem. Soc. Perkin Trans.* 1: 381–384.

21 Reichen, W. and Wentrup, C. (1976). *Helvetia Chemica Acta* 59 (7): 282, 2618–282, 3620.

22 Wentrup, C. and Winter, H.W. (1981). *J. Org. Chem.* 46: 1046–1078.

23 Briehl, H., Lukosch, A., and Wentrup, C. (1984). *J. Org. Chem.* 49: 2772–2779.

24 (a) Dib, H.H., Al-Awadi, N.A., Ibrahim, Y.A., and El-Dusouqui, O.M.E. (2003). *Tetrahedron* 59: 9455–9464. (b) Bégué, D., Santos-Silva, H., Dargelos, A., and Wentrup, C. (2017). *J. Phys. Chem. A* 121: 5998–6003.

25 Dib, H.H., Al-Awadi, N.A., Ibrahim, Y.A., and El-Dusouqui, O.M.E. (2004). *J. Phys. Org. Chem.* 17: 267–272.

26 El-Dusouqui, O.M.E., Abdelkhalik, M.M., Al-Awadi, N.A. et al. (2006). *J. Chem. Res.*: 291–298.

27 Katritzky, A.R. and Molina-Buendia, P. (1979). *J. Chem. Soc. Perkin Trans. 1*: 1957–1960.

28 Al-Awadi, N.A., Elnagdi, M.H., Mathew, T. et al. (1996). *Int. J. Chem. Kinet.* 28: 741–748.

29 Al-Etibi, A., Abdullah, M., Al-Awadi, N.A. et al. (2004). *J. Phys. Org. Chem.* 17: 49–55.

30 Al-Awadi, N.A., Ibrahim, Y.A., Kaul, K., and Dib, H.H. (2001). *J. Phys. Org. Chem.* 14: 521–525.

31 Leon, L., Notario, R., Quijano, J. et al. (2003). *Theor. Chem. Accounts* 110: 387–394.

32 George, B.J., Dib, H.H., Abdallah, M.R. et al. (2006). *Tetrahedron* 62: 1182–1192.

33 Al-Awadi, N.A., Elnagdi, M.H., Mathew, T., and El-Ghamry, I. (1996). *Heteroat. Chem.* 7: 417–420.

34 Al-Awadi, N.A., Elnagdi, M.H., Mathew, T., and El-Ghamry, I. (1997). *Heteroat. Chem.* 8: 63–66.

35 (a) Ibrahim, M.R., Al-Azemi, T.F., Al-Etaibi, A. et al. (2010). *Tetrahedron* 66: 4243–4250. (b) Chou, C.H., Chu, L.-S., Chiu, S.-J. et al. (2004). *Tetrahedron* 60: 6581–6584.

36 El-Dusouqui, O.M.E., Abdelkhalik, M.M., Al-Awadi, N.A., and Elnagdi, M.H. (2008). *J. Heterocyclic Chem.* 45: 1751–1753.

37 Al-Awadi, N.A., Elnagdi, M.H., Al-Awadhi, H.A., and El-Dusouqui, O.M.E. (1998). *Int. J. Chem. Kinet.* 30: 457–462.

38 El-Dusouqui, O.M.E., Abdelkhalik, M.M., Al-Awadi, N.A. et al. (2006). *J. Chem. Res.*: 295–302.

39 Malhas, R.N., Al-Awadi, N.A., and El-Dusouqui, O.M.E. (2007). *Int. J. Chem. Kinet.* 39: 82–91.

40 Al-Awadi, N.A., Elnagdi, M.H., Mathew, T., and El-Gamry, I. (1996). *Int. J. Chem. Kinet.* 28: 749–754.

41 Al-Awadi, N.A., Elnagdi, M.H., Kaul, K. et al. (1998). *Tetrahedron* 54: 4633–4640.

42 Al-Awadi, N.A., Elnagdi, M.H., Kaul, K. et al. (1999). *J. Phys. Org. Chem.* 12: 654–658.

43 Shiho, D. and Tagami, S. (1960). *J. Am. Chem. Soc.* 82: 4044–4054.

44 Al-Qallaf, M.A., Dib, H.H., Al-Awadi, N.A., and El-Dusouqui, O.M.E. (2017). *J. Anal. Appl. Pyrolysis* 124: 446–453.

45 Brown, D.J. and Nagamatsu, T. (1977). *Aust. J. Chem.* 30: 2515–2525.

46 Al-Awadi, S. (2018). Synthesis, kinetics and mechanism of gas-phase pyrolysis of arylidenehydrazine heterocycles. PhD thesis. Kuwait University.

47 Duffy, E.F., Foot, J.S., McNab, H., and Milligan, A.A. (2004). *J. Org. Biomol. Chem.* 2: 2677–2683.

48 Ciesielski, M., Pufky, D., and Döring, M. (2005). *Tetrahedron* 61: 5942–5947.

49 El-Sherief, H.A., Mahmoud, A.M., and Esmaiel, A.A. (1984). *Bull. Chem. Soc. Jpn.* 57: 1138–1142.

50 Gibson, M.S. (1963). *Tetrahedron* 19: 1587–1589.

51 Peláez, W.J., Yranzo, G.I., Gróf, C. et al. (2005). *Tetrahedron* 61: 7489–7498.

7

Gas-Phase Pyrolysis of Phosphorus Ylides

This chapter represents an extensive discussion of the valuable synthetic routes to alkynes, dialkynes, enynes, and dienes using flash vacuum pyrolysis (FVP) of phosphorus ylides. Several examples are provided to illustrate how changing the molecular structure of ylides with thermal extrusion of triphenylphosphine oxide leads to useful synthesis of functionalized alkynes. Kinetics, thermal reactivities, and mechanisms of the gas-phase pyrolysis reactions are specifically addressed. This chapter further reviews the chemistry of sulfonyl- and sulfinyl-stabilized phosphorus ylides.

7.1 Synthetic Applications

The chemistry of ylides is rich and diverse, covering all aspects of synthesis, structure, and reactivity [1]. The structure of phosphorus ylides is characterized by the polar (P—C) bond. The barrier to rotation of this bond is only $4.5\,kJ\,mol^{-1}$, which makes the bond order >1 and <2 [2]. The electron pair on the carbanion forms a back-bond

Gas-Phase Pyrolytic Reactions: Synthesis, Mechanisms, and Kinetics,
First Edition. Nouria A. Al-Awadi.
© 2020 John Wiley & Sons, Inc. Published 2020 by John Wiley & Sons, Inc.

by overlap with an unoccupied orbital at the phosphorus moiety, resulting in an approximately planar geometry around the carbon. These considerations allow the structure of the unsubstituted triphenylphosphonium meth-ylide to be represented as either dipolar ylide **1** or ylene **2** (Scheme 7.1); form **1** is more appropriate for the discussion and rationalization of ylide chemical reactivity [2, 3].

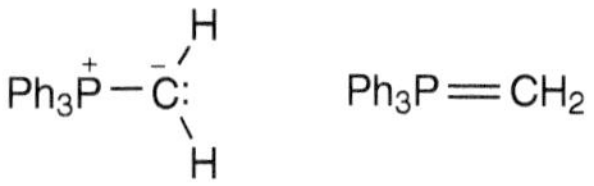

$$Ph_3\overset{+}{P}-\overset{-}{\underset{H}{\overset{H}{\text{C}}}}{:} \qquad Ph_3P = CH_2$$

    1   **2**

Scheme 7.1

Replacement of one of the (H) atoms by an electron-withdrawing group leads to molecular stability, and the planar conformation at the (C) atom provides the ideal geometry for conjugation and hyper-conjugation effects to dominate the molecular reactivity of stabilized triphenylphosphonium meth-ylides.

In 1959, Trippett and Walker reported that stabilized phosphorus ylides of structure **3** react with phenyl isocyanate/isothiocyanate to form carbamoyl-stabilized ylides **4** [4] (Scheme 7.2).

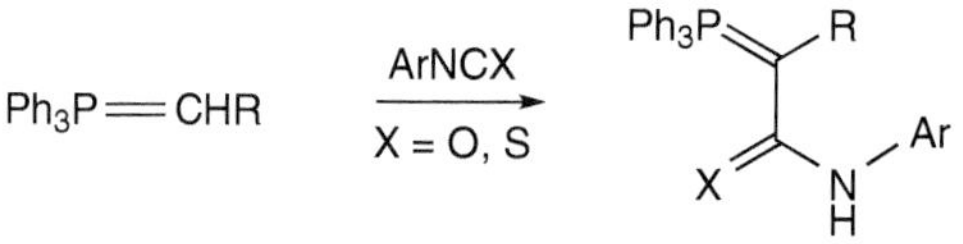

$$Ph_3P = CHR \quad \xrightarrow[X=O,\,S]{ArNCX} \quad \begin{array}{c} Ph_3P \diagdown \diagup R \\ X \diagdown \underset{H}{N} \diagup Ar \end{array}$$

    3   **4**

Scheme 7.2

There have since been several reports on the formation of such adducts between a variety of ylides and both isocyanates [5, 6] and isothiocyanates [7–10]. Where R contains a carbonyl group, these are known to exhibit intramolecular hydrogen bonding as shown in **5** [5–7, 9]. Compound **6** reverts to its starting compounds upon heating in boiling toluene [11] (Scheme 7.3).

Trippett et al. pyrolyzed [4] a series of β-ketoalkylidenetriphenylphosphorane **7** in the hope of obtaining acetylene. The phosphoranes **7a–d** gave no acetylene, but phosphorane **7e** at 300 °C gave almost quantitative yields of triphenylphosphine oxide and diphenylacetylene (Scheme 7.4).

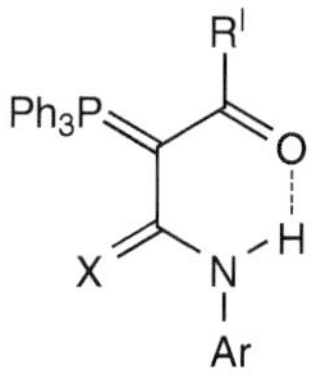

5

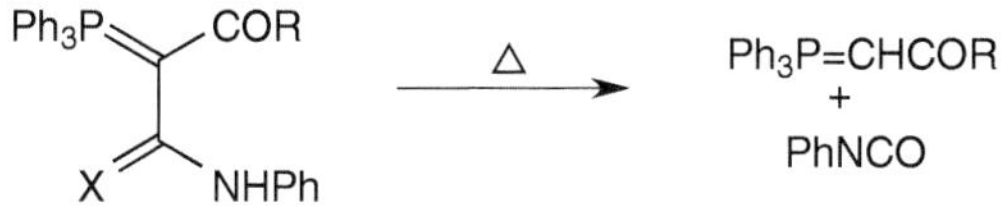

6

Scheme 7.3

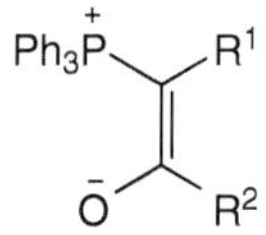

7a-h

7	R^1	R^2
a	H	H
b	H	me
c	H	Ph
d	H	OEt
e	Ph	Ph

7	R^1	R^2
f	R	Ph
g	COOEt	Ph
h	CHN	Ph

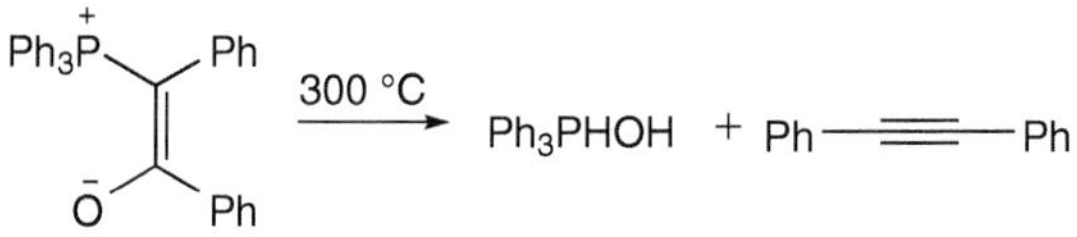

7e

Scheme 7.4

They then reported that the pyrolysis of phosphorane 7 constitutes a general synthesis of acetylene [12], provided that:

a) Neither R^1 nor R^2 is hydrogen.
b) R^1 or R^2 is a phenyl or carbonyl group or the equivalent.

In this way, phenylalkylacetylenes were obtained from **7f**; α,β-acetylenic esters were obtained from phosphoranes **7 g** and α,β-acetylenic nitriles from **7 h**. However, later it was reported that FVP of stabilized phosphoranes where R and R1 are H or alkyl groups provides good yields of acetylenes [13, 14].

An interesting series of polyaromatic acetylenes (Scheme 7.5) has been reported: dianthrylacetylenes **8a–c** and diphenanthrylacetylenes

8a

8b

8c

9a

9b

9c

Scheme 7.5

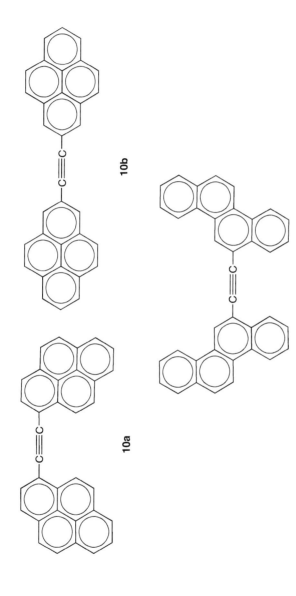

10a

10b

11

Scheme 7.5 (*Continued*)

9a–c, dipyrenylacetylenes **10a,b,** and 6,6′-dichrysenylacetylenes **11** were successfully prepared according to the general Scheme 7.6 [15].

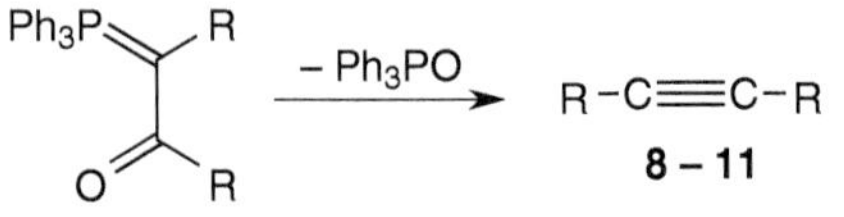

Scheme 7.6

Gottfried Märkl in 1961 reported on the behavior of various phosphorus ylides of structure **12** that were found to be useful compounds for the synthesis of various substituted acetylenes [16] (Scheme 7.7).

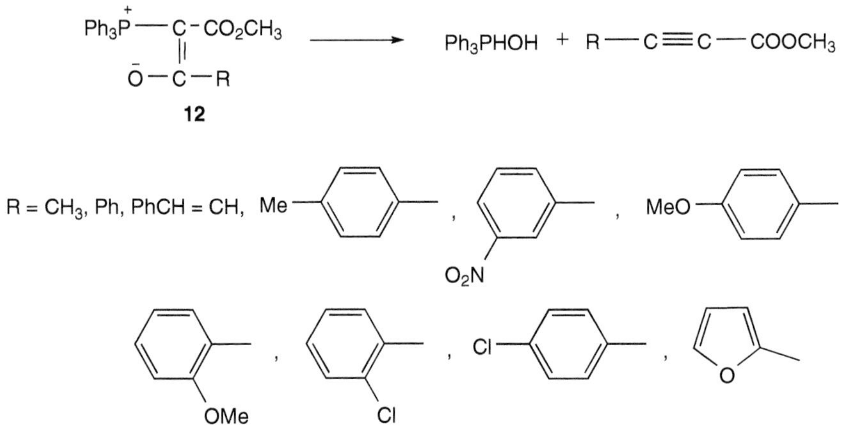

Scheme 7.7

Thioacetylenes **13** were prepared through the pyrolysis of α-acyl-α-thiomethylene triphenylphosphoranes **14,** taking advantage of the well-known ability of sulfur to stabilize carbanions [17]. This thiophosphorane route is of general applicability, as shown in Scheme 7.8 and Table 7.1.

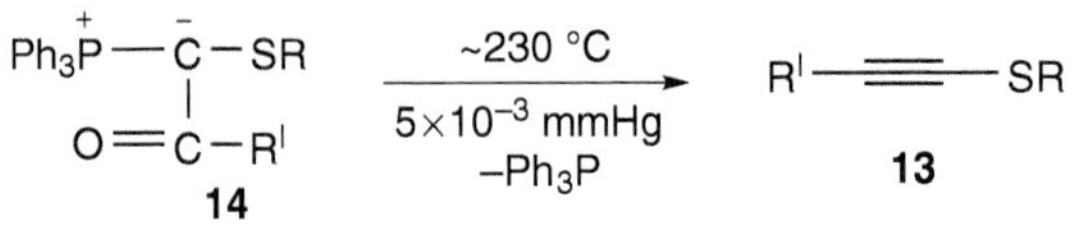

Scheme 7.8

Table 7.1 Percentage yields of thioacetylenes **13**.

R	R^I	Yield (%)
CH$_3$	Ph	80
CH$_3$	4-MeC$_6$H$_4$	68
CH$_3$	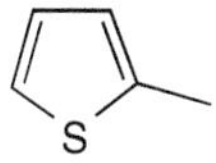 (2-methylthiophene)	69
Ph	Ph	71
Ph	4-MeC$_6$H$_4$	69
Ph	CF$_3$	74
Ph	CH$_3$	41
Ph	(CH$_3$)$_3$C	43
Ph	n-C$_5$H$_{11}$	45

7.2 Haloalkynes

Shen et al. have successfully applied the intramolecular Wittig reaction to the synthesis of perfluoro- and polyfluoroacetylenic thiolesters **15** in excellent yields [18] from perfluoro- and polyfluoroacetyl phosphoranes **16** under vacuum at 190–220 °C (Scheme 7.9).

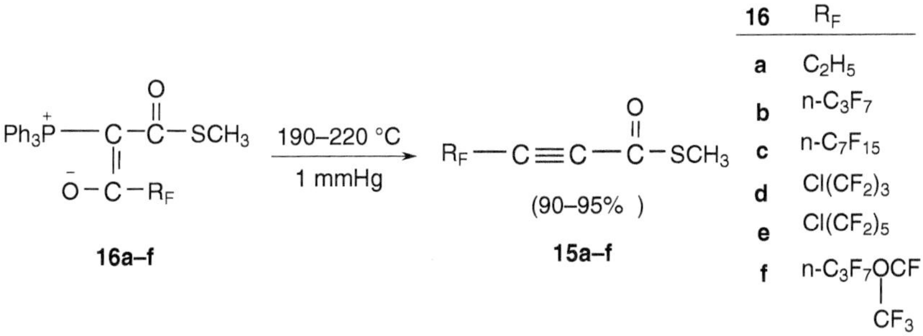

16	R$_F$
a	C$_2$H$_5$
b	n-C$_3$F$_7$
c	n-C$_7$F$_{15}$
d	Cl(CF$_2$)$_3$
e	Cl(CF$_2$)$_5$
f	n-C$_3$F$_7$OCF(CF$_3$)

Scheme 7.9

The phosphorane compounds **17** were pyrolyzed under vacuum to give phenyltrifluoromethylacetylene derivatives **18** in moderate to good yields [19] according to Scheme 7.10 and Table 7.2.

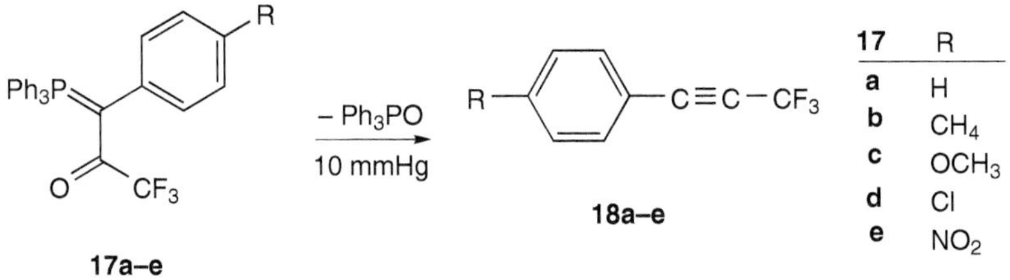
R
Ph_3P
O
CF_3
$- Ph_3PO$
10 mmHg
R
$C\equiv C-CF_3$
18a–e
17 R
a H
b CH_4
c OCH_3
d Cl
e NO_2

17a–e

Scheme 7.10

Table 7.2 Reaction temperatures and yields of **18a–e.**

18	Temperature (°C)	Yield (%)
a	280	85
b	280	54
c	410	54
d	280	40
e	280	85

During further investigation by Shen et al. [20] of the synthesis of substituted fluoroacetylenes by the intramolecular Wittig reaction of phosphorane ylides **19a–e**, pyrolysis under reduced pressure produced acetylenes **20a–d**. However, no acetylene was obtained from **20e** (Scheme 7.11).

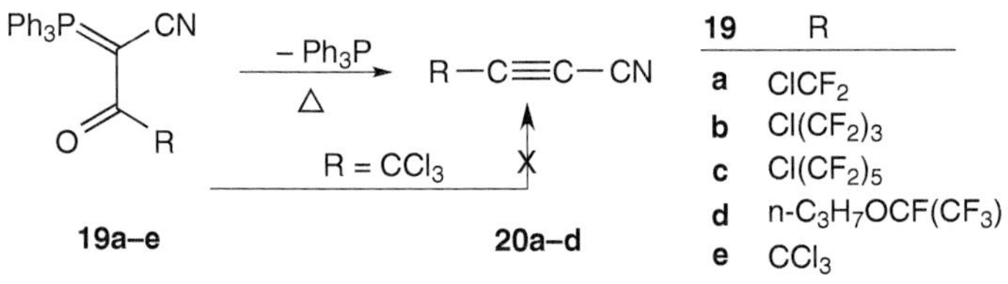
Ph_3P
CN
O
R
$- Ph_3P$
Δ
$R-C\equiv C-CN$
$R = CCl_3$
19 R
a $ClCF_2$
b $Cl(CF_2)_3$
c $Cl(CF_2)_5$
d $n\text{-}C_3H_7OCF(CF_3)$
e CCl_3

19a–e

20a–d

Scheme 7.11

1-Aryloxyperfluoro-alkynes **21a–f** were obtained from the pyrolysis of phosphoranes **22a–f** in 28–40% yields [21] (Scheme 7.12).

Following the same approach, 1-perfluoroalkynyl phosphonates **23a–c** were obtained from the pyrolysis of phosphoranes **24a–c** in very good yields [22] (Scheme 7.13).

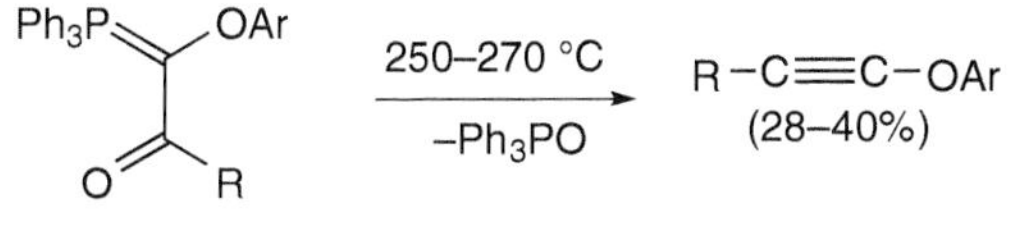

$$\text{22a–f} \xrightarrow[-\text{Ph}_3\text{PO}]{250\text{–}270\ °\text{C}} \text{21a–f}$$

Ph$_3$P=C(OAr)–C(O)–R → R–C≡C–OAr (28–40%)

22	a	b	c	d	e	f
R	C$_2$F$_5$	n-C$_3$F$_7$	n-C$_3$H$_7$OCF(CF$_3$) —	C$_2$F$_5$	n-C$_3$F$_7$	n-C$_3$H$_7$OCF(CF$_3$) —
Ar	C$_6$H$_5$	C$_6$H$_5$	C$_6$H$_5$	4-ClC$_6$H$_4$	4-ClC$_6$H$_4$	4-ClC$_6$H$_4$

Scheme 7.12

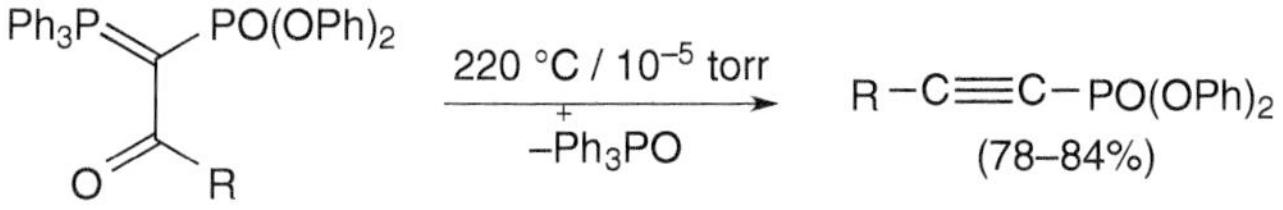

Ph$_3$P=C(PO(OPh)$_2$)–C(O)–R

$$\text{24a–c} \xrightarrow[-\text{Ph}_3\text{PO}]{220\ °\text{C}\ /\ 10^{-5}\ \text{torr}} \text{R–C≡C–PO(OPh)}_2\ (78\text{–}84\%)\ \ \text{23a–c}$$

	R
a	CF$_3$
b	CF$_3$CF$_2$
c	n-C$_3$F$_7$

Scheme 7.13

Shen has extended the work to the synthesis of haloalkynes by synthesizing α-(perfluoroalkanoyl)-pentafluorobenzylidenetriphenyl-phosphoranes **25a–d** and α-(ω-chloroperfluoroalkanoyl)-perfluoroben-zylidenetriphenylphosphoranes **25e–f**. The six phosphorane ylides were pyrolyzed at temperatures of 230–260 °C and a reduced pressure of 2 torr to produce fluoro and chloro alkynes **26a–f** in 85–96% yield [23], as shown in Scheme 7.14.

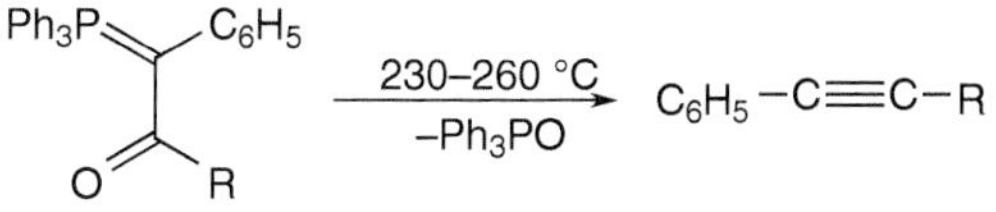

Ph$_3$P=C(C$_6$H$_5$)–C(O)–R

$$\text{25a–f} \xrightarrow[-\text{Ph}_3\text{PO}]{230\text{–}260\ °\text{C}} \text{C}_6\text{H}_5\text{–C≡C–R} \quad \text{26a–f}$$

25	R
a	CF$_3$
b	C$_2$F$_5$
c	n-C$_3$F$_7$
d	n-C$_7$F$_{15}$
e	(CF$_2$)$_3$Cl
f	(CF$_2$)$_5$Cl

Scheme 7.14

Pyrolysis of the halophosphoranes **27** resulted in α-haloacetylene **28** according to Scheme 7.15 in moderate yields [12] (Table 7.3).

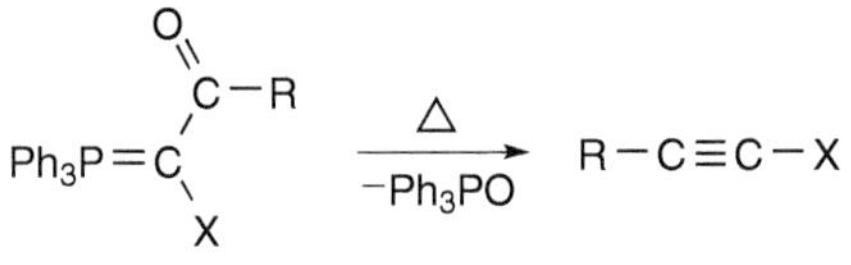

27a–h → **28a–h**

Scheme 7.15

Table 7.3 Percentage yields of α-haloacetylene **28**.

28	R	Yield (%)	28	R	Yield (%)
a	$PhC\equiv CCl$	50	e	$p\text{-}ClPhC\equiv CCl$	55
b	$PhC\equiv CBr$	48	f	$p\text{-}ClPhC\equiv CBr$	53
c	$p\text{-}CH_3PhC\equiv CCl$	50	g	$(CH_3)_3CC\equiv CCl$	38
d	$p\text{-}CH_3PhC\equiv CBr$	48	h	$(CH_3)_3CC\equiv CBr$	37

The moderate yields obtained in the pyrolysis of the halophospho-ranes **27** were attributed to the degradation of the halophosphoranes prior to the elimination of triphenylphosphene oxide to give acetylene. This instability of the halophosphorane is probably associated with the low energy of the C—X bonds (C—Cl = 95 kcal mol^{-1}; C—Br = 67 kcal mol^{-1}) [12].

7.3 Terminal Alkynes

Trippett and Walker were the first to describe the thermal extrusion of triphenylphosphine oxide from α-oxophosphorus ylides **7** to give alkynes **29** (Scheme 7.16) in quantitative yields [4].

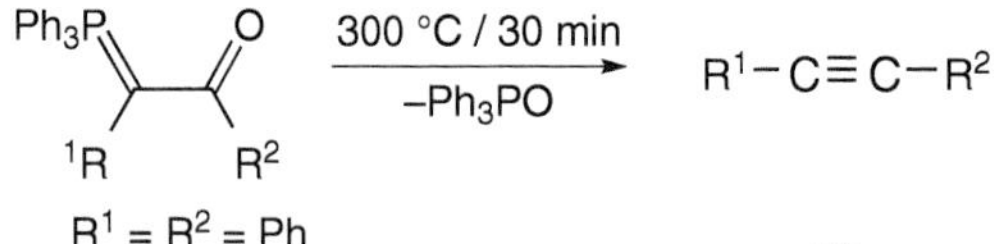

$R^1 = R^2 = Ph$

7

29

Scheme 7.16

Serious limitations became apparent in the early stages [12, 24] in compounds where R^1 = H or alkyl did not provide the expected alkynes. Under conventional pyrolysis conditions, side reactions occurred, including partial extrusion of Ph_3P and isomerization of the desired alkynes to allenes, making the method useless for formation of simple aliphatic and terminal alkynes [25]. In order to overcome these problems, Bestmann described two separate ways in which compound 7 could be converted into **29** indirectly [26, 27], but these involved additional steps that result in reducing the overall yield. Later, Aitken et al. described a successful application of FVP, a technique that gives excellent results in a wide variety of thermal extrusion reactions [25]. They have demonstrated the use of FVP to overcome the longstanding limitation on the thermolysis of ylide 7 to give the alkyne **29**. In so doing, they have provided a convenient new route for construction of aliphatic and terminal alkynes, and this has opened the door for the pyrolysis of a wide variety of other substituted phosphorus ylides.

Ylides **7a–q** were prepared in acceptable yield and subjected to FVP in a conventional flow system at 10^{-2} mmHg, contact time ≈ 10 ms. They sublimed unchanged at temperatures up to 600 °C but at 750 °C underwent complete reaction to give Ph_3PO and alkyne **29** in moderate to excellent yield (Table 7.4).

Pyrolysis was generally performed using 0.5 g of ylide. The feasibility of using this methods to prepare multigram quantities of alkynes was demonstrated by FVP of compound **7d** (18 g) to give compound **29d** (3.4 g) and compound **7n** (13.2 g) to give compound **29n** (3.7 g).

Only when R^2 is cyclobutyl in compounds **7f** and **7 m** does an additional fragmentation occur by loss of ethene to produce vinylalkynes. FVP at 750 °C of **7f** gave the expected cyclobutylalkynes **29** accompanied by a comparable amount of vinylalkenes **30** (Scheme 7.17).

Table 7.4 Percentage yields of alkynes **29**.

29	R[1]	R[2]	Yield of 29 (%)	29	R[1]	R[2]	Yield of 29 (%)
a	H	Et	78	j	Pr	Pr^i	93
b	H	Pr	59	k	Pr	Bu^i	81
c	H	Pr^i	72	l	Pr	MeCH—CH	82
d	H	Bu^i	82	m	Pr	n-C_4H_7	81^a
e	H	MeCH—CH	59	n	Pr	n-C_6H_{11}	81
f	H	n-C_4H_7	67^a	o	Bu	Bu	80
g	H	n-C_6H_{11}	64	p	Bu	Bu	69
h	H	Ph	82	q	Me	2-Thienyl	64
i	Me	Me	74				

^{a}Yield of vinylalkyne **30** formed at 800–850 °C.

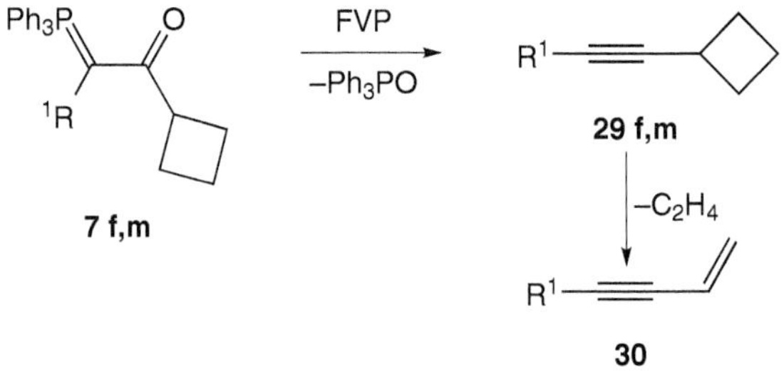

Scheme 7.17

Aitken described another method for the synthesis of terminal alkynes by subjecting α-acyl-α-ethoxycarbonyl ylides **31** to FVP at 750 °C; Ph_3PO was again eliminated, but this was accompanied by complete loss of the COOEt ester group to give terminal alkynes **32** in moderate yield [28]. When the ylides **31** were subjected to 500 °C, the expected extrusion of Ph_3PO was observed, and acetylene esters **33** were recovered in pure form and excellent yield (Table 7.5, Scheme 7.18).

Table 7.5 Yields of **32** and **33**.

31	R	Yield of 33 (%)	Yield of 32 (%)
a	Ph	—	48
b	4-MeC$_6$H$_4$	91	58
c	4-MeOC$_6$H$_4$	92	16
d	4-ClC$_6$H$_4$	92	46
e	4-NO$_2$C$_6$H$_4$	88	—
f	2-MeC$_6$H$_4$	90	34
g	2-MeOC$_6$H$_4$	88	–
h	2-MeSC$_6$H$_4$	82	38
i	2-Furyl	90	40
j	2-Thienyl	83	34
k	3-Me-2-thienyl	88	28
l	3-Thienyl	86	66
m	Cyclohexyl	66	40

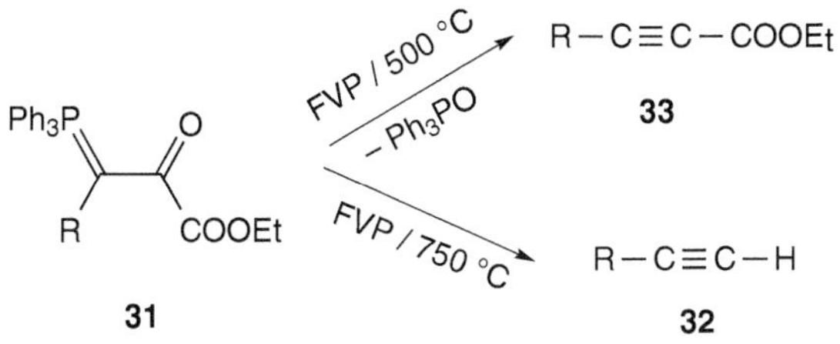

Scheme 7.18

This method is suitable for preparation of multigram quantities of acetylenic esters **33**. The identity and purity of the acetylenic esters were checked and characterized by ^{1}H and ^{13}C NMR. The loss of the ester group took place at higher temperatures, which was sufficient to eliminate CO_2 and ethane from the COOEt group and was explained by a secondary reaction via a six-membered transition state as shown in Scheme 7.19.

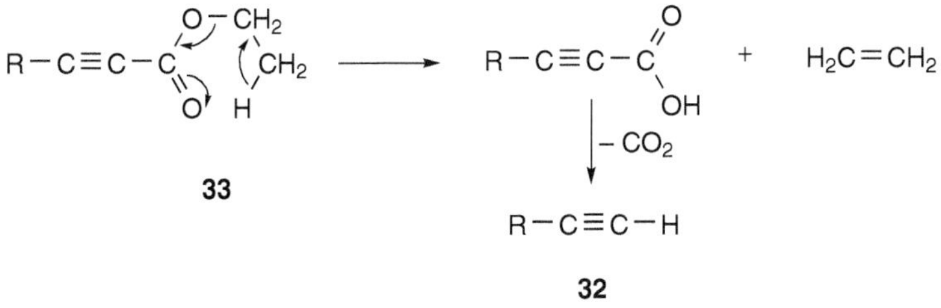

Scheme 7.19

7.4 Diynes

When the stable crystalline ylides **34** were subjected to FVP at 500 °C, extrusion of Ph_3PO took place as expected to give diacetylene esters **35** in moderate yield [29] (Table 7.6). The identity and purity of the diacetylenic ester were readily confirmed by their ^{13}C NMR. When **34** was subjected to FVP at 750 °C, extrusion of Ph_3PO was accompanied by loss of ester grouping, as described in Scheme 7.20, to produce the terminal 1,3-diynes **36** again in moderate yields (Table 7.6).

It is noteworthy that the heterocyclic substituted examples **34i–m** all gave poor yields; in the case of **34i**, for example, this was attributed to a further reaction of **35i**, which involves a radical intermediate formed by loss of Me• followed by cyclization and hydrogen abstraction to produce benzothiophene derivatives **37** (Scheme 7.21).

Table 7.6 FVP of ylides **34** to give **35** and **36**.

34	R	Yield of 35 (%)	Yield of 36 (%)
a	Pr	37	47
b	Bu	—	33
c	Ph	53	18
d	4-MeC$_6$H$_4$	68	33
e	4-MeOC$_6$H$_4$	35	6
f	4-ClC$_6$H$_4$	24	21
g	2-MeC$_6$H$_4$	28	53
h	2-MeOC$_6$H$_4$	48	∼ 0
i	2-MeSC$_6$H$_4$	6	5
j	2-Furyl	12	10
k	2-Thienyl	20	—
l	3-Me-2-thienyl	6	2
m	3-Thienyl	23	15

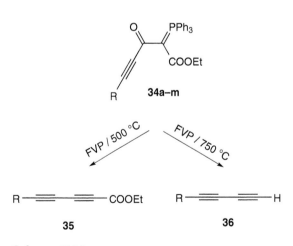

Scheme 7.20

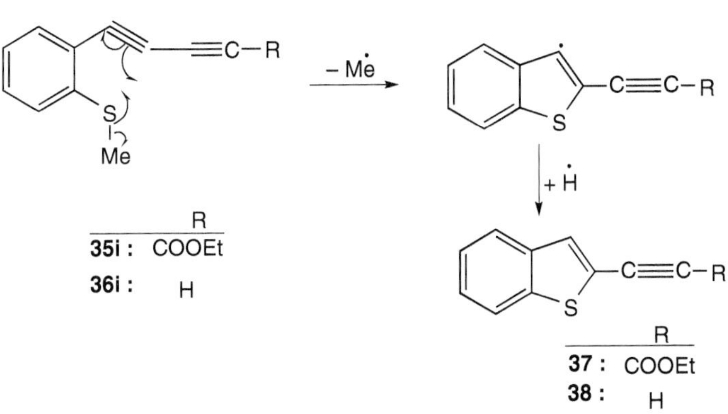

Scheme 7.21

During the course of this work, Aitken [29] managed to prepare and characterize an interesting triyne **39** from the diacetylene ester **35d** that resulted from FVP of **34d**. **35d** was hydrolyzed and chlorinated to give diacetylenic acid chloride, which was then treated with Ph$_3$P=CHCOOEt to produce **40** in 68% yield. The latter was pyrolyzed at 500 °C, and triacetylenic ester **39** was formed in 30% yield (Scheme 7.22).

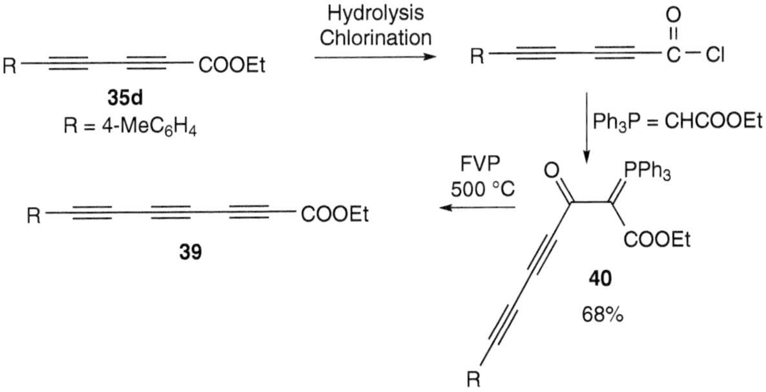

Scheme 7.22

Diylides **41a,b** were found to be highly stable up to 850 °C, reflected in an extraordinary resistance toward extrusion of Ph$_3$PO [30]. Only at 900 °C did the required diynes **42a,b** (Scheme 7.23) result, at 64% and 10%, respectively.

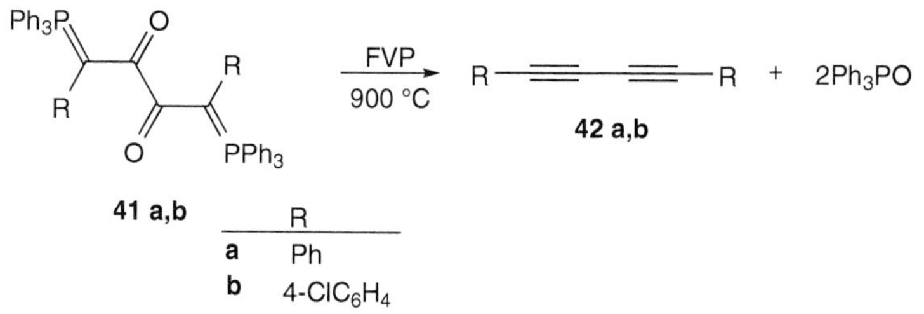

Scheme 7.23

7.5 Enynes and Dienes

A series of cinnamoylalkidene-triphenylphosphenyl phosphoranes **43**, **44** were subjected to FVP in order to offer an attractive alternative to the synthesis of conjugated enynes of considerable interest as synthetic intermediates [31]. The FVP of these ylides was accompanied by double-bond isomerization of the enynes. At 500 °C, there is little double-bond isomerization; but at 700 °C this does occur, to give an almost 1 : 1 mixture of E and Z isomers, which was followed by ^{1}H and ^{13}C NMR spectroscopy. To minimize the double-bond isomerization, FVP was applied at 500 °C and resulted in extrusion of Ph$_3$PO to give enynes in >95% as the E isomers (Scheme 7.24). When this condition was applied to ylides **43a–l**, in most cases they produced the required enynes in moderate yield with double-bond isomerization (Table 7.7).

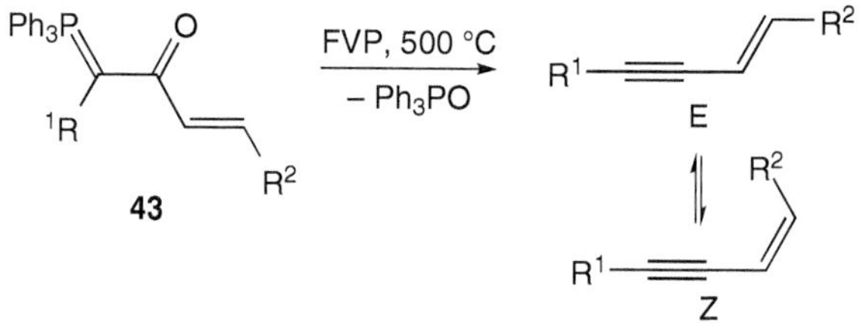

Scheme 7.24

Pyrolysis of two examples **44a,b** gave enynes with little double-bond isomerization (Scheme 7.25, Table 7.7).

This work was followed by FVP of a series of 2-methylcinnamoyl phosphorus ylides **45**. Upon their FVP at 500 °C, Ph$_3$PO was eliminated to give the corresponding styrylalkynes **46** almost exclusively as the E isomers [32] (Scheme 7.26).

Table 7.7 Formation of E and Z enyne isomers at 500 °C and 700 °C.

43/44	R^1	R^2	500 °C		700 °C	
			Yield (%)	E/Z ratio	Yield (%)	E/Z ratio
43a	Pr	Ph	42	95/5	21	53/47
43b	Pr^i	Ph	—	—	67	56/44
43c	Bu	Ph	58	96/4	25	50/50
43d	Ph	Ph	88	97/3	99	57/43
43e	Ph	$4\text{-MeC}_6\text{H}_4$	82	>95/5	—	—
43f	Ph	$4\text{-ClC}_6\text{H}_4$	39	97/3	—	—
43g	Ph	$2\text{-ClC}_6\text{H}_4$	58	95/5	—	—
43h	Ph	$2\text{-NO}_2\text{C}_6\text{H}_4$	19	77/23	—	—
43i	Ph	$3,4\text{-OCH}_2\text{OC}_6\text{H}_3$	58	85/15	—	—
43j	Ph	2-Furyl	55	90/10	—	—
43k	Ph	2-Thienyl	47	85/15	—	—
43l	Ph	5-Me-2-thienyl	34	79/21	—	—
44a	Me	–	67	95/5	—	—
44b	Ph	–	55	95/5	—	—

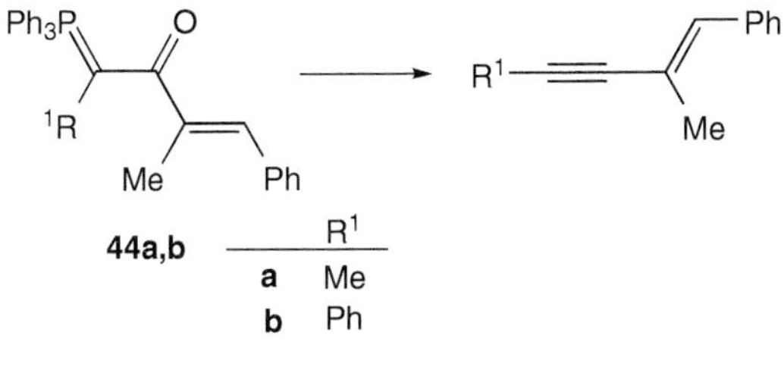

Scheme 7.25

In 1999, Aitken and his group elegantly described a method that allows potential access to a wide range of significantly useful 1,3-dienes from ylide **47**, which were readily prepared by conjugate addition of ester-stabilized ylides to α,β-unsaturated ketones [33] (Scheme 7.27). When ylides **47** were pyrolyzed at 650 °C and 10^{-2} torr, there was complete elimination of Ph_3PO, and the functionalized 1,3-dienes **48** formed in moderate yields (Scheme 7.27). The dienes' formation

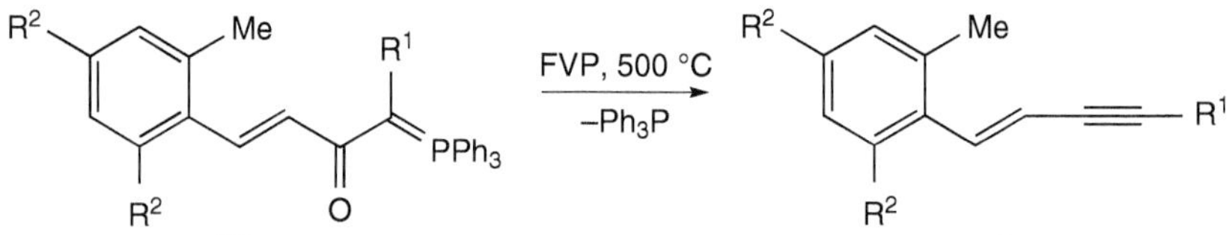

45 → **46**

R^1	Me	Et	Pr^1	Ph	COOMe	Me	Et	Pr^1	Ph
R^2	H	H	H	H	H	Me	Me	Me	Me

Scheme 7.26

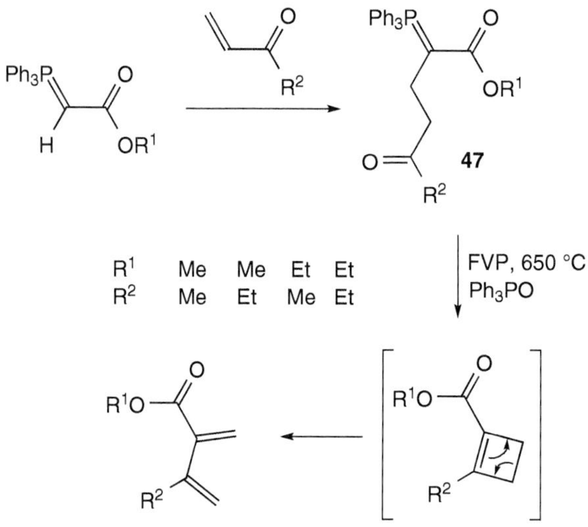

R^1	Me	Me	Et	Et
R^2	Me	Et	Me	Et

47 → **48**

Scheme 7.27

was explained by cycloreversion of the initially formed cyclobutenes under the FVP reaction conditions. Some spectroscopic evidence was obtained for the formation of the cyclobutenes [δH 2.6–2.8 (m)], but this was accompanied by a high yield of **48** and unreacted **47**.

7.6 Selective Elimination of Ph$_3$PO from Di- and Tri-oxo-stabilized Phosphorus Ylides

Thermal extrusion of triphenylphosphine oxide from α-oxoalkylidene triphenylphosphoranes **7** is considered to be a well-established method for the synthesis of alkynes $R^1C\equiv CR^2$, which proceeds particularly well using the technique of FVP. This has been successfully demonstrated for R=H or alkyl, aryl, CN, Cl or Br, SR, SeAr, OAr, and other groups [34].

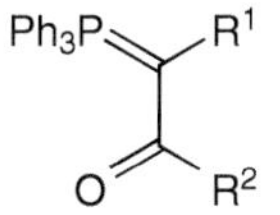

7

In the case of ester and keto carbonyl groups present in the phosphoranes, as in **49**, it was observed that oxygen is eliminated exclusively from the keto carbonyl group to give acetylenic esters **50** in good yields [12, 16, 23, 35] (Scheme 7.28).

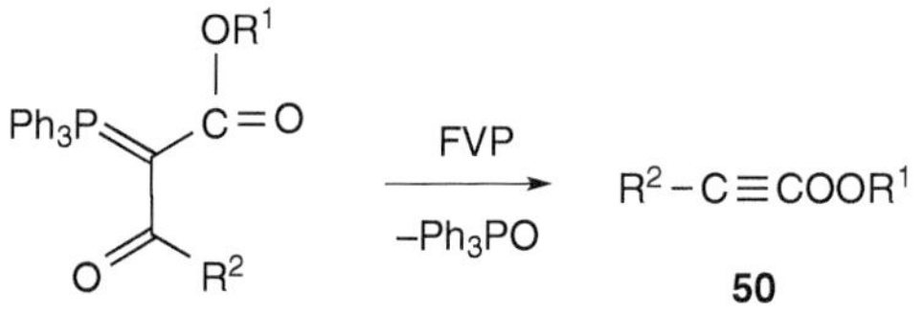

Scheme 7.28

The same applies to thioester groups, as shown in Scheme 7.8. The choice between two different carbonyl groups present in phosphoranes as **51** showed poor selectivity, as an equal mixture of acetylenic products formed (Scheme 7.29).

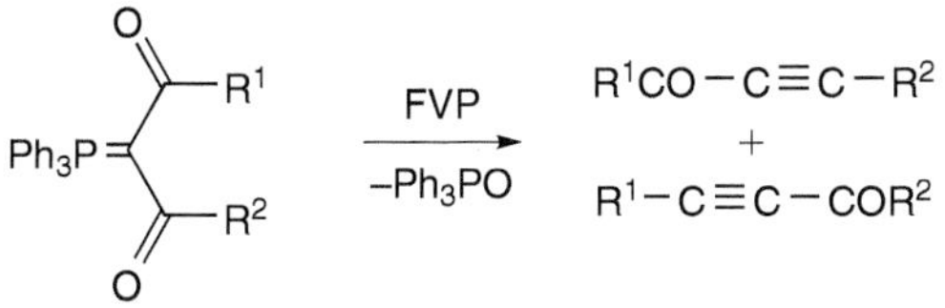

Scheme 7.29

1-Ethoxycarbonyl-2-oxoalkylidene triphenylphosphoranes **52a–e**, **53a,b**, and **54** were prepared and efficiently pyrolyzed at 500–540 °C by the FVP technique to give dialkylhex-2-yne-1,6-dioates **55a–e**, dialkylhex-2-yne-1,6-dioates **56a,b**, and dialkyl-5-oxahept-2-yne-1,7-dioates **57**, respectively [35] (Schemes 7.30–7.32).

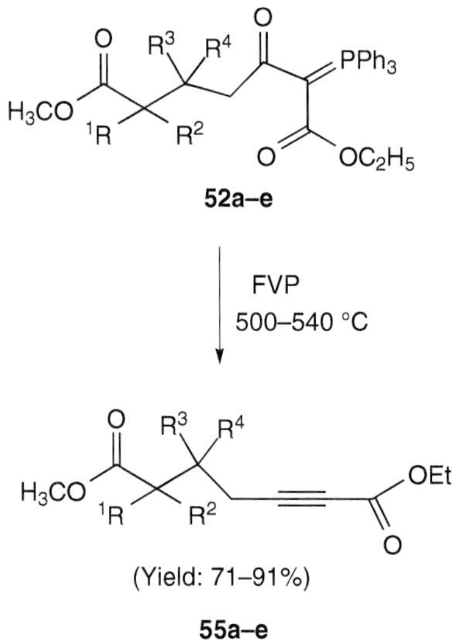

52	R¹	R²	R³	R⁴
a	H	H	H	H
b	CH₃	CH₃	H	H
c	H	H	CH₃	H
d	H	H	CH₃	CH₃
e	H	H	C₆H₅	H

(Yield: 71–91%)

55a–e

Scheme 7.30

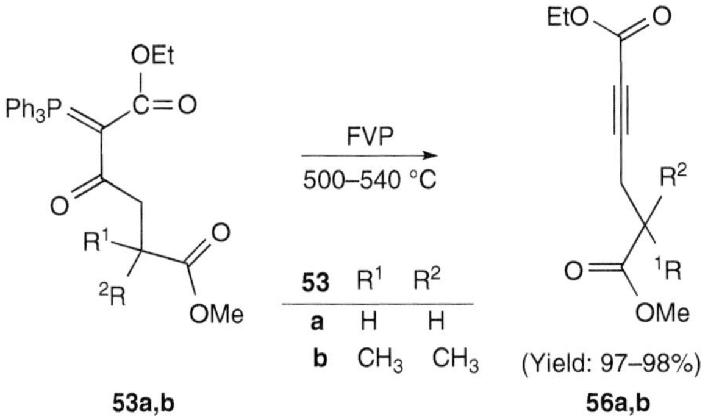

53	R¹	R²
a	H	H
b	CH₃	CH₃

(Yield: 97–98%)

53a,b   **56a,b**

Scheme 7.31

The formation of acetylenic esters **55–57** by FVP of ylides **52–54** shows that the oxygens of the ester carbonyl groups were not involved in the extrusion of Ph₃PO; they were maintained in the products, while the oxygen of the oxo group was completely eliminated.

In a further investigation by Shen [36] on the applications of the intramolecular Wittig reaction in the synthesis of functionalized

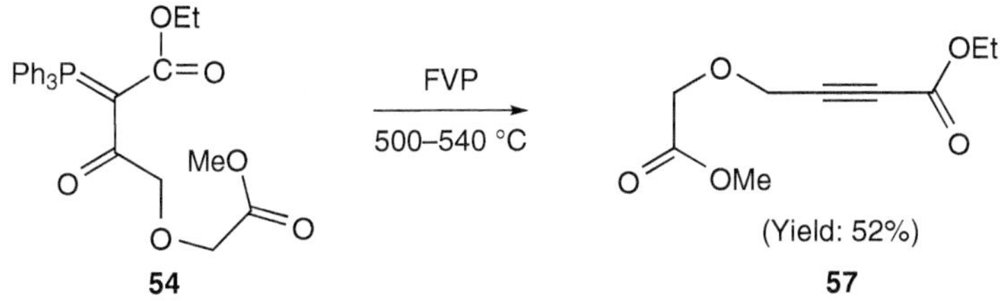

Scheme 7.32

fluoroacetylenes to synthesize both terminal perfluoroacylacetylenes and perfluoroalkynals in one reaction, phosphoranes **58a–c** were pyrolyzed under vacuum to give perfluoroacylacetylenes **59** and perfluoroalkynals **60** (Scheme 7.33).

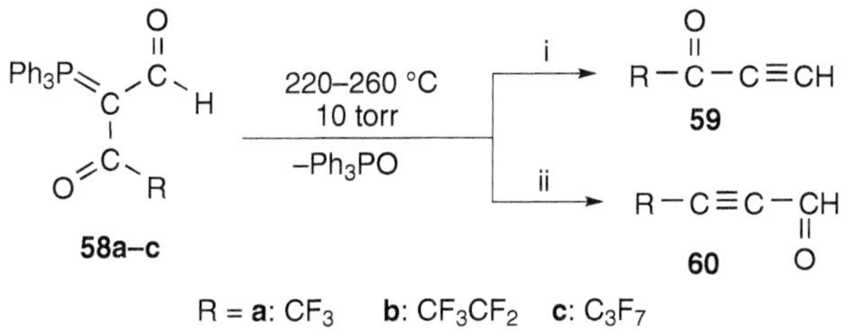

Scheme 7.33

There are two choices for the Ph$_3$P: to eliminate with either (i) oxygen of the aldehyde carbonyl group or (ii) acyl oxygen, as shown in Scheme 7.33. The ratio of the two products **59** : **60** was found to be 3 : 2, 6 : 5, and 3 : 4 for phosphoranes **58a**, **58b**, and **58c**, respectively. This result assumes that the larger perfluoroalkyl group in the phosphorane ylides facilitates the pyrolysis via route (ii), forming **60** in higher yield.

Higher homologs – phosphorus ylides **61a–j** stabilized by an ester or keto group on one side of the phosphorus and an α-diketo or α-ketoester group on the other – were synthesized by Aitken [34, 37]. The diacylalkynes **62** obtained in excellent yield by FVP of these ylides suggest that Ph$_3$PO is lost exclusively across the central position, as shown in Scheme 7.34.

This selectivity does allow convenient preparation of multigram quantities of diacylalkynes.

Blitzk described the first synthesis of diethyl-2-oxopent-3-ynedioate trioxoalkyne **63** by FVP of the tetraoxo-bis-ylide **64** and made use of it in the synthesis of various heterocyclic molecules [38a] (Scheme 7.35).

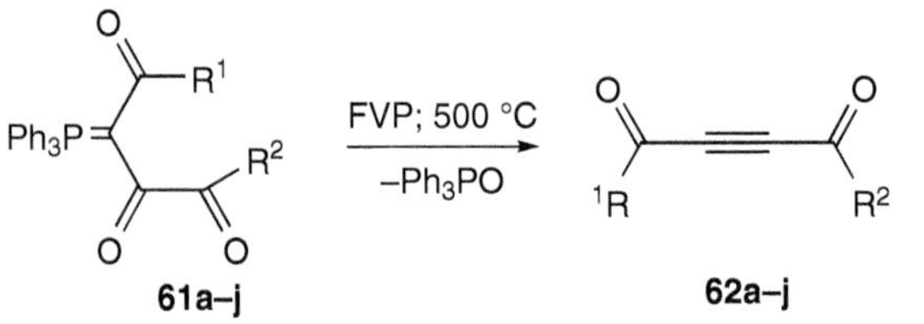

61	R^1	R^2
a	Ph	Ph
b	Ph	OMe
c	Ph	OEt
d	OMe	Ph
e	OMe	OMe
f	OMe	OEt
g	OEt	Ph
h	OEt	OMe
i	OEt	OEt
j	OEt	Me

Scheme 7.34

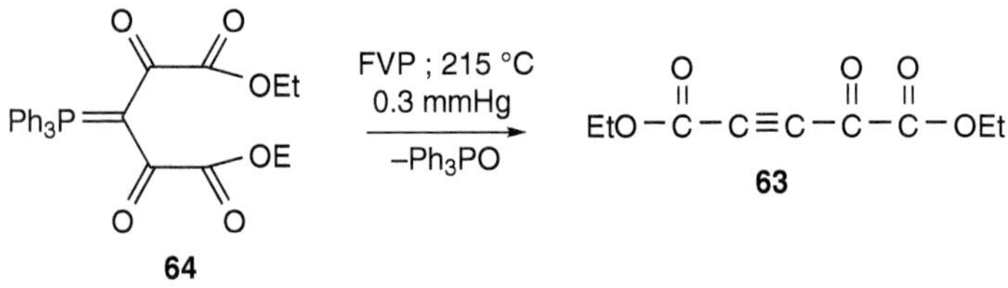

Scheme 7.35

Later, Aitken prepared higher homologs, the tetraoxo-ylides **65** and the hexaoxobis-ylides **66** (Scheme 7.36). With these ylides, the FVP technique failed to yield acetylenic products, and this was attributed possibly to ready fragmentation of polyoxoalkynes under high vacuum conditions [38b].

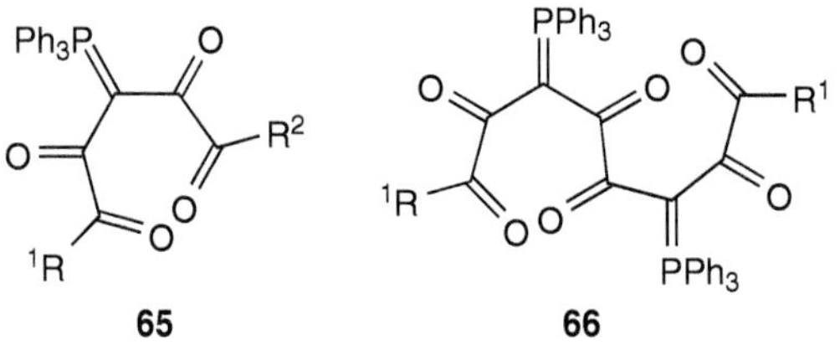

Scheme 7.36

Selected examples of the bis (oxoylides) **67a,b** and **68a–d** were pyrolyzed by the FVP technique at 500 °C and 10^{-2} torr. They behaved well and produced diynes **69 a,b** and the bis (acetylenic esters) **70a–d**, respectively, in moderate yields [39] (Scheme 7.37).

Complete FVP of compound **71** produced triphenylphosphene, and the expected 1,4-bis (phenylbutadiynyl)benzene **72** did not form [39] (Scheme 7.38).

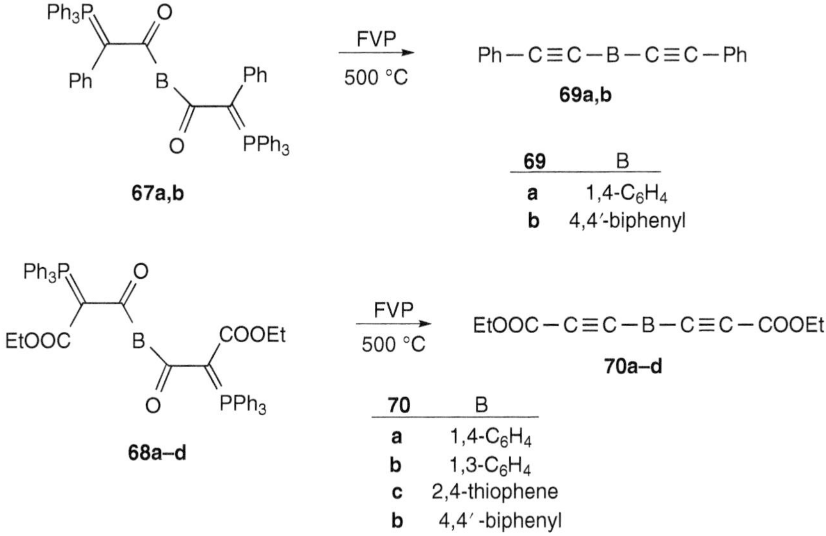
Ph3P
O
Ph
B
Ph
O
PPh3
67a,b
FVP
500 °C
Ph—C≡C—B—C≡C—Ph
69a,b
69 B
a 1,4-C6H4
b 4,4′-biphenyl
Ph3P
O
EtOOC
B
COOEt
O
PPh3
68a–d
FVP
500 °C
EtOOC—C≡C—B—C≡C—COOEt
70a–d
70 B
a 1,4-C6H4
b 1,3-C6H4
c 2,4-thiophene
b 4,4′-biphenyl

Scheme 7.37

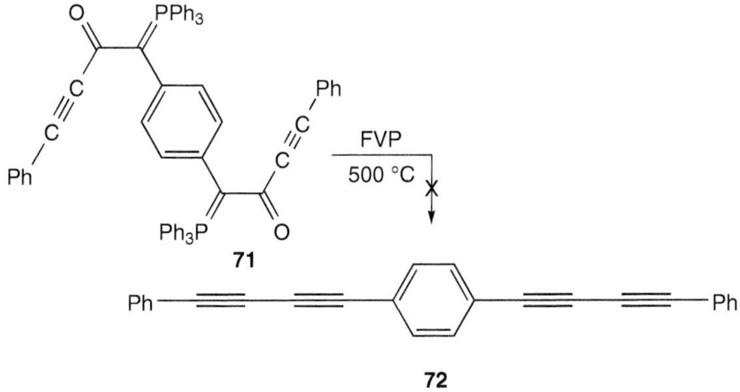
O
PPh3
C
C
Ph
Ph
C
C
FVP
500 °C
Ph3P
O
71
Ph
Ph
72

Scheme 7.38

7.7 Sulfonyl-Stabilized Phosphorus Ylides

Although the corresponding sulfonyl-stabilized phosphorus ylides are well known, their pyrolysis was not studied until 1991 by Aitken et al. [40, 41]. They first prepared a range of sulfonyl-stabilized phosphorus ylides of general structure **73** (Scheme 7.39), which on FVP at 500 °C underwent mainly extrusion of Ph_3P to give products derived from sulfonyl carbene intermediates [40]. 1,3-Insertion of sulfonyl carbene **74**

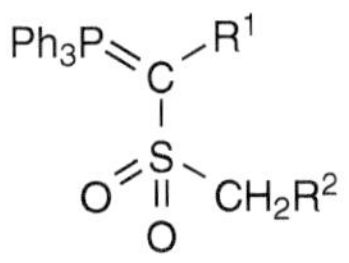

73	R^1	R^2
a	Ph	Ph
b	4-MeC$_6$H$_4$	Ph
c	2-MeOC$_6$H$_4$	Ph
d	2-MeSC$_6$H$_4$	Ph
e	Me	Ph
f	Me	2-MeC$_6$H$_4$
g	Me	2-MeOC$_6$H$_4$

Ph$_3$P=C(R^1) attached to S(=O)(=O)CH$_2$R^2

73a–g

Scheme 7.39

into the C—H bond of the OMe or SMe groups for ylides **73c** and **73d** gave **75**, which then lose SO$_2$ to produce **76**.

Sulfonyl-stabilized phosphorus ylides **73a–f** were pyrolyzed at 600 °C and 0.01 torr to give a mixture of Ph$_3$P, Ph$_3$PO, and Ph$_3$PS together with the other products shown in Scheme 7.40.

Alkenes **76** were formed in all cases, accompanied sometimes by other products depending on the nature of R^1. The formation of alkenes **76** was explained as shown in Scheme 7.40 by extrusion of Ph$_3$P to give sulfonylcarbenes **74**, which undergo intramolecular 1,3-insertion in the case of ylides **73a,b,e–g**, and 1,3-insertion in ylides **73c,d**, where the carbenes insert into the benzylic C—H bond to give the thiirane dioxide **75**. The latter readily loses SO$_2$ to give the main product alkene **76**. The formation of Ph$_3$PO and Ph$_3$PS can be explained in terms of the reaction of Ph$_3$P with SO$_2$, and the validity of this was confirmed by control experiments.

For ylides **73e–g**, the major product is **79**, formed from the sulfonyl-carbenes **74** followed by a 1,2-shift to give the vinyl sulfone. Loss of SO$_2$ would give vinyl and benzyl radicals, which will then give **79**.

We have performed a study on the pyrolysis of a series of substituted sulfonyl-stabilized phosphorus ylides **80a–i** using conventional sealed-tube static pyrolysis. The products of reaction were compared with those obtained from FVP at 327 °C [42]. The static method allows for ample residence time to ensure maximum pyrolysis. The constituents of the pyrolysates were isolated using preparative liquid chromatography (PLC) and analyzed using GC–MS, FT–IR, and 1H NMR techniques. Both modes of pyrolysis, static and FVP, gave identical products of Ph$_3$P, Ph$_3$PO, Ph$_3$PS and symmetrical and unsymmetrical alkenes. However, static pyrolysis of ylides followed by PLC gave better separation and isolation profiles of alkenes, and allowed the identification of more reaction products including p-anisaldehyde

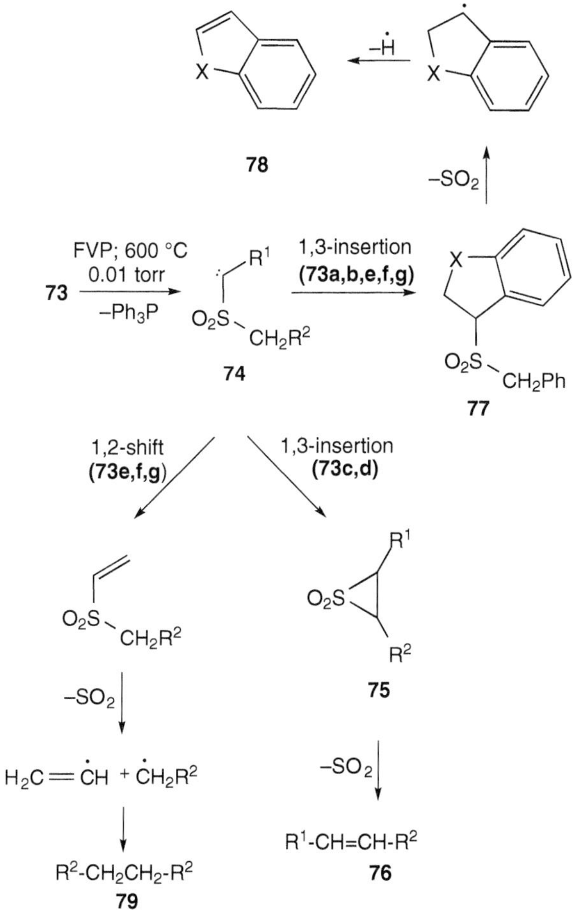

Scheme 7.40

and mixed sulfones, which were not detected in the pyrolysates from FVP. Based on the overall pyrolysate composition, a feasible reaction mechanism was proposed to account for the products obtained from ylides **80a–i**, as shown in Scheme 7.41; the mechanism involves the formation and further reaction of sulfonylcarbene intermediates **80a–i**. In the case of ylides **80a–h**, the carbene obtained from the extrusion of Ph_3O reacts by an intramolecular CH insertion process to give thiirane dioxide followed by the loss of SO_2, alkene metathesis, and partial reaction of Ph_3P with SO_2 to give Ph_3PO and Ph_3PS. It was observed that the carbene intermediate abstracts hydrogen in a novel reaction only in the static pyrolysis of ylides **80a–e** (all with electron-withdrawing

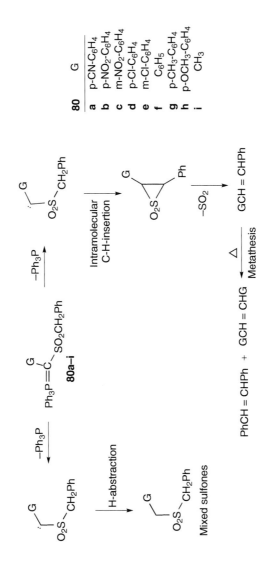

80	G
a	p-CN-C$_6$H$_4$
b	p-NO$_2$-C$_6$H$_4$
c	m-NO$_2$-C$_6$H$_4$
d	p-Cl-C$_6$H$_4$
e	m-Cl-C$_6$H$_4$
f	C$_6$H$_5$
g	p-CH$_3$-C$_6$H$_4$
h	p-OCH$_3$-C$_6$H$_4$
i	CH$_3$

Scheme 7.41

groups in the substituent G) to give mixed sulfones (Scheme 7.41); it is to be noted that for ylides **80,g,h** (electron-donating group in G), no such sulfone products were detected (Scheme 7.42).

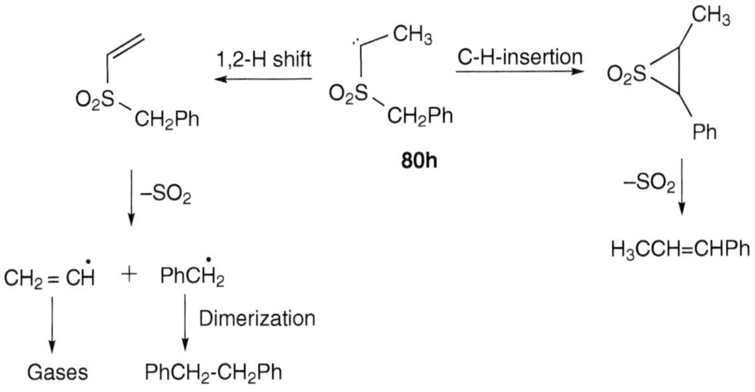

Scheme 7.42

The first-order rate constants of the gas-phase pyrolytic reaction of **80a–i** at 227 °C and the relative rates of these ylides are shown in Table 7.8. The rate of pyrolysis of ylide **80i,** where G=CH$_3$, is ca. 3 times faster than that of ylide **80f,** where G=C$_6$H$_5$; this represents the highest increment in molecular reactivity in this series of ylides and was attributed to the fact that the reactive carbene intermediate of

Table 7.8 Rate constants k(s^{-1}) at 227 °C and relative rate (k_{rel}) for the sulfonyl-stabilized phosphorus ylides **80a–i.**

80	G	$10^4\,k\,s^{-1}$	k_{rel}
a	p-CNC$_6$H$_4$	3.926	0.738
b	p-NO$_2$C$_6$H$_4$	6.406	1.20
c	m-NO$_2$C$_6$H$_4$	6.109	1.15
d	p-ClC$_6$H$_4$	3.782	0.711
e	m-ClC$_6$H$_4$	2.881	0.542
f	C$_6$H$_5$	5.317	1.00
g	p-CH$_3$C$_6$H$_4$	12.00	2.26
h	p-CH$_3$OC$_6$H$_4$	12.48	2.35
i	CH$_3$	16.17	3.04

ylide **80i** can react by either 1,3-H insertion or 1,2-H insertion prior to elimination of SO_2, whereas when G is an aromatic phenyl group, the carbene can react only by 1,3-H insertion. Introducing p-methyl and p-methoxy groups into the aromatic ring **80g,h**, respectively, increased the rate of pyrolysis by almost 2.3 times for both ylides, and this could be explained in terms of the electron-donating nature of these substituents and their suggested influence on the carbene. Electron-withdrawing substituents *p*-CN, *p*-Cl, and *m*-Cl, as in **80a,d,e**, reduced the molecular reactivity due to their destabilizing effect on the carbene intermediate. The nitro substituent in **80b,c** resulted in no change in reactivity compared to G=Ph in **80 f**. This unique behavior of the nitro group in the gas-phase elimination reactions of organic compounds has been reported in other studies [43].

It is important to note that these sulfonylcarbene intermediates are already stabilized by conjugative interaction with the sulfonyl group, as in Scheme 7.43, and this might be the reason for the relatively small substituent effect obtained.

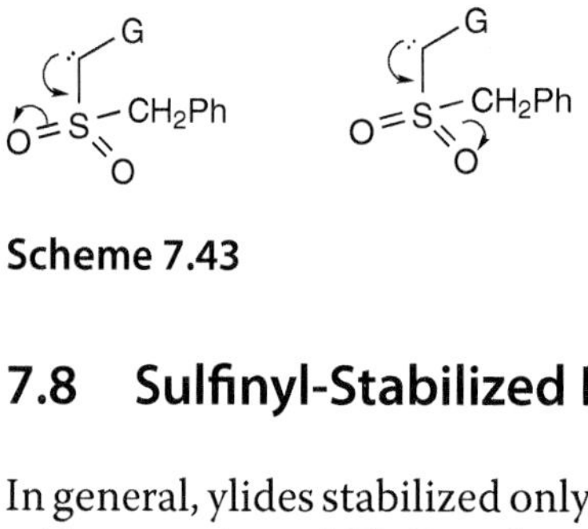

Scheme 7.43

7.8 Sulfinyl-Stabilized Phosphorus Ylides

In general, ylides stabilized only by alkanesulfinyl group **81** undergo FVP with extrusion of Ph_3P to form a sulfinylcarbene intermediate that rearranges to give a major product of thioester [44] (Scheme 7.44).

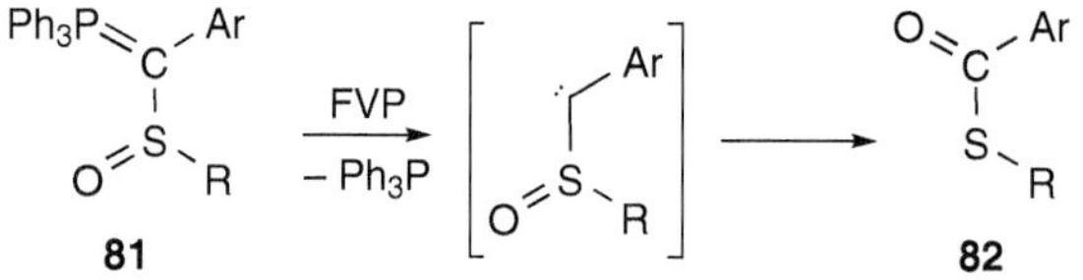

Scheme 7.44

A detailed study of the FVP of a series of sulfinyl-stabilized phosphorus ylides was performed by Aitken et al. [45, 46]. Alkanesulfinyl ylides **81a–f** (Scheme 7.45) were prepared and subjected to FVP at 500 °C and

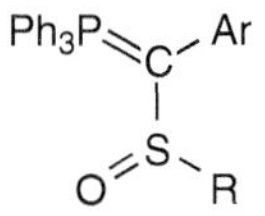

81a–f

81	R
a	Et
b	Pr^i
c	CH_2Ph
d	Ph
e	4-Me-C_6H_4
f	4-Cl-C_6H_4

Scheme 7.45

10^{-2} torr. Compounds **82a–c** were formed by extrusion of Ph_3P to form sulfinylcarbenes followed by (1,2-O) insertion to form oxathiirane intermediate **83** and to finally produce product **82**, as shown in Scheme 7.46.

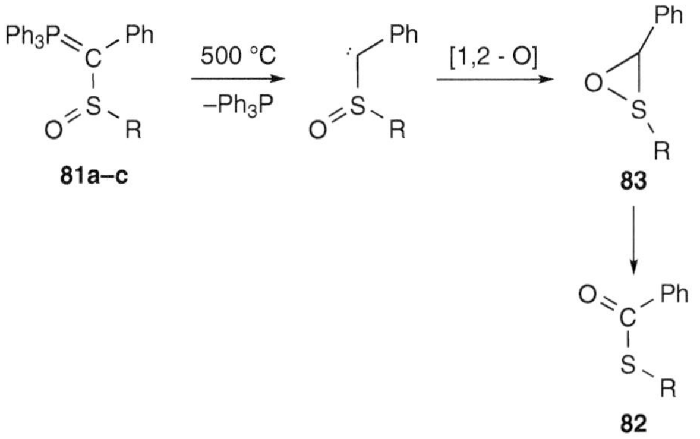

Scheme 7.46

FVP of ylides **81d–f** produced an intense dark blue color in the cold trap, which faded rapidly upon warming. This was attributed to the formation of trace quantities of a thioketone. The formation of compounds **84** is the result of extrusion of Ph_3PO from the ylides **81** to form sulfinylcarbenes, as shown in Scheme 7.47.

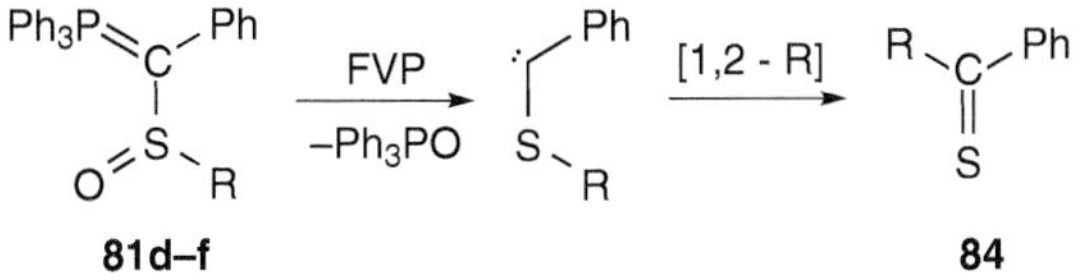

Scheme 7.47

Ketone **85** was a major product detected from FVP of **81e,f**, which most likely resulted from rearrangement of the sulfinylcarbene to a sulfine followed by loss of sulfur, as shown in Scheme 7.48.

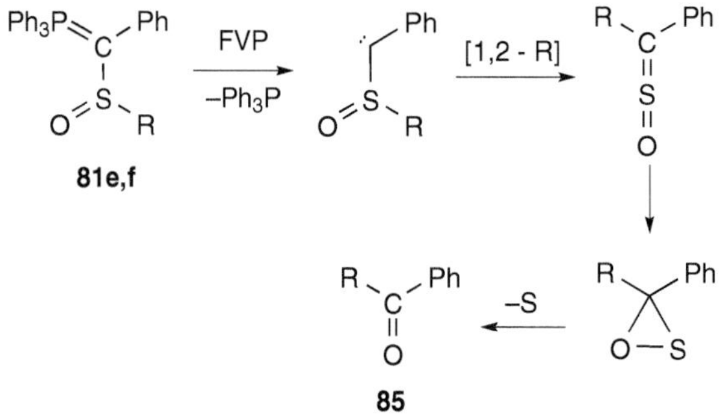

81e,f

85

Scheme 7.48

Phosphorus ylides **86** stabilized by both sulfinyl and ester groups undergo FVP with elimination of Ph_3PO resulting in a carbene intermediate, which will finally produce vinylsulfide (Scheme 7.49).

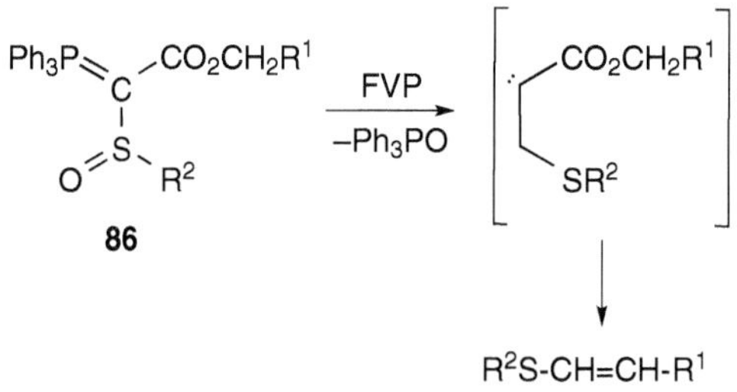

86

$$R^2S\text{-}CH{=}CH\text{-}R^1$$

Scheme 7.49

The FVP at $600\,^\circ C$ of sulfinyl phosphorus ylides stabilized by both alkoxy carbonyl and sulfinyl groups with various substituents on each group **86a–l** was studied by Aitken [47, 48] (Scheme 7.50).

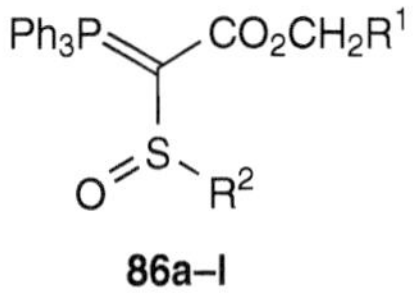

86a–l

86	R^1	R^2
a	H	Me
b	H	Et
c	Me	Et
d	H	Ph
e	Me	Ph
f	H	$4\text{-MeC}_6\text{H}_4$
g	Me	$4\text{-MeC}_6\text{H}_4$
h	Et	$4\text{-MeC}_6\text{H}_4$
i	Me	$4\text{-ClC}_6\text{H}_4$
j	H	$4\text{-BrC}_6\text{H}_4$
k	Me	$4\text{-BrC}_6\text{H}_4$
l	Ph	Et

Scheme 7.50

All of these sulfinyl ylides produced Ph_3PO accompanied by Ph_3P and small amounts of Ph_3PS in the case of **86a–c**. Alkenyl sulfides **87** were the major products identified by independent synthesis for compounds **86a** and **86i**, and their formation was suggested as shown in Scheme 7.51.

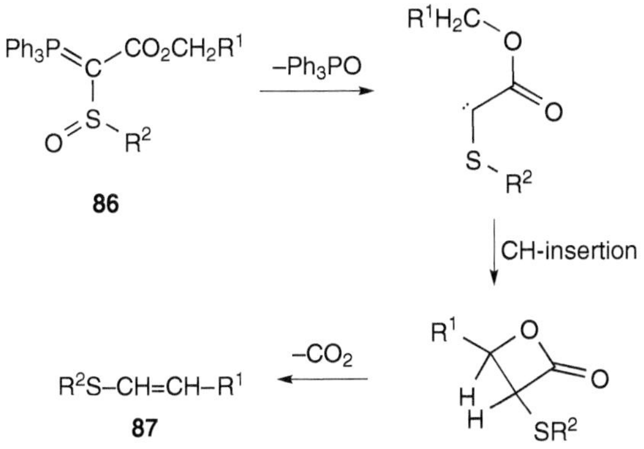

Scheme 7.51

Loss of Ph_3PO gives carbene intermediates that undergo intramolecular CH insertion to give β-lactones, which then lose CO_2 to produce alkenyl sulfides **87**. An elegant experiment in support of this mechanism was carried out [47] by preparing ^{13}C-labeled ylides, labeled **88** on the ylide carbon and **89** on the carbonyl carbon, and subjecting them to FVP under the same conditions. The alkenyl sulfide produced from **88** was labeled, whereas ylide **89** produced only unlabeled products (Scheme 7.52).

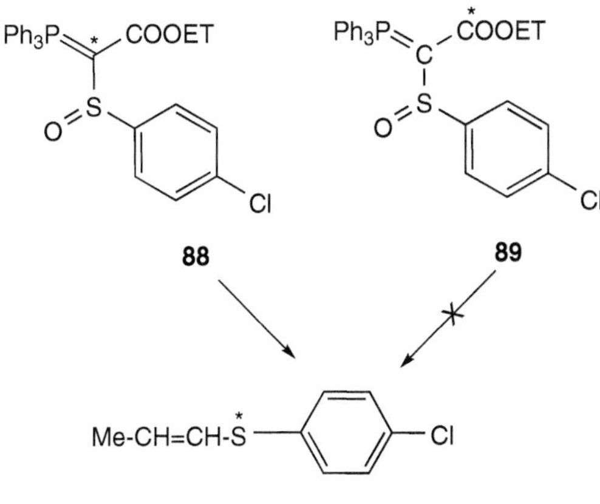

Scheme 7.52

Sulfide **90** was identified as another major product; its formation is shown in Scheme 7.53.

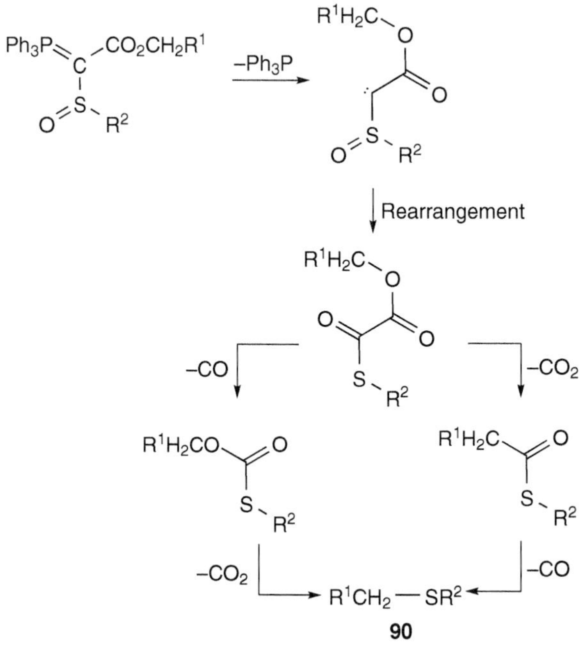

Scheme 7.53

Monothiooxalates resulting from the rearrangement of the carbene intermediates from ylides **86e** and **86g** were pyrolyzed at 600 °C and gave the corresponding sulfides **90e,g** (Scheme 7.53).

7.9 Kinetic and Thermal Reactivity of Carbonyl-Stabilized Phosphonium Ylides

Apart from a brief note mentioning that the compound reverts to the starting components upon heating in boiling toluene [11], there has been no systematic study of the thermal reactivity of phosphorus ylides until 2005, when the pyrolysis of a series of substituted [(benzoyl) (phenylcarbonyl) methylene] (triphenylphosphoranes) **91** and their thiocarbonyl analogues were examined, and detailed kinetic measurements were performed [49] in the first collaborative work between the Kuwait and St. Andrew's Chemical Laboratories (Scheme 7.54).

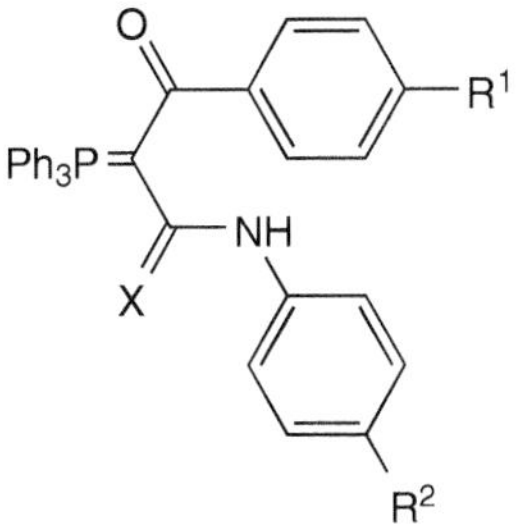

91	X	R^1	R^2	91	X	R^1	R^2	91	X	R^1	R^2
a	O	H	H	f	S	H	H	l	O	NO$_2$	H
b	O	H	NO$_2$	g	S	H	NO$_2$	m	O	Cl	H
c	O	H	Cl	h	S	H	Cl	n	O	Me	H
d	O	H	Me	i	S	H	Me	o	O	OMe	H
e	O	H	OMe	k	S	H	OMe	p	S	Me	H
								q	S	OMe	H

Scheme 7.54

This combination was chosen so as to examine the effects of electron-donating and electron-withdrawing substituents in the para positions on the reaction rate.

Two parent compounds **91a** and **91f** were found to react completely at 500 °C and 10^{-2} torr with the loss of ArCO or ArNCS (Scheme 7.55).

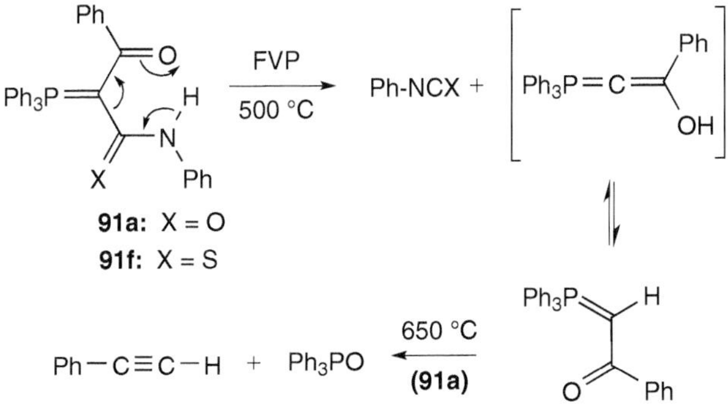

Scheme 7.55

For ylide **91a**, the electrocyclic process shown in Scheme 7.55 leads to PhNCO and the enol form of ylide, the latter undergoing a secondary reaction with further extrusion of Ph$_3$PO to give phenylacetylene at 650 °C. This secondary reaction was reduced to a good extent by lowering the temperature to 500 °C.

Table 7.9 Rate constants and relative rates at 127 °C in the gas phase ($R^1 = H$).

91	R^2	X = O $10^4 k$ s^{-1}	91	X = S $10^2 k$ s^{-1}	$k_{rel} = k_{(X = S)}/k_{(X = O)}$
a	H	3.69	f	1.34	36
b	4-NO$_2$	5.41	g	1.08	20
c	4-Cl	6.11	h	1.34	22
d	4-Me	4.32	i	2.35	54
e	4-OMe	7.12	k	2.32	33

Rate constants calculated at 127 °C are shown in Table 7.9 for varied substituents on the ArCO ring (**91a–k**) and Table 7.10 for varied substituents on the ArNH ring (**91l–q**).

In all cases, the thiocarbonyl ylides are more reactive than the corresponding carbonyl ylides. This is attributed to the existence of the ylide in its enolate form **92** in which the C—C bond to be broken has a double-bond character (Scheme 7.56). This enolate form is likely to be less significant for the thiocarbonyl ylides, a fact also supported by the higher values of J_{P-C}, thus leading to more rapid fragmentation.

Replacing the ArCO group in phosphorus ylides **91** by a CN group produced cyano(arylcarbonyl) phosphorus ylides **93** and their thio analogues [50] (Scheme 7.57).

Ylide **93a** at 500 °C eliminates Ph$_3$PO to give aniline. Ylide **93f** behaved similarly (Scheme 7.58). Various cyano-ynamines generated were found to undergo isomerization to ketenimines [50].

Table 7.10 Rate constants and relative rates at 127 °C in the gas phase ($R^2 = H$).

91	R^1	X = O $10^4 k$ s^{-1}	91	X = S $10^2 k$ s^{-1}	$k_{rel} = k_{(X = S)}/k_{(X = O)}$
a	H	3.69	f	1.34	36
l	4-NO$_2$	0.90	—	—	—
m	4-Cl	3.42	—	—	—
n	4-Me	16.42	p	3.10	19
o	4-OMe	16.18	q	4.53	28

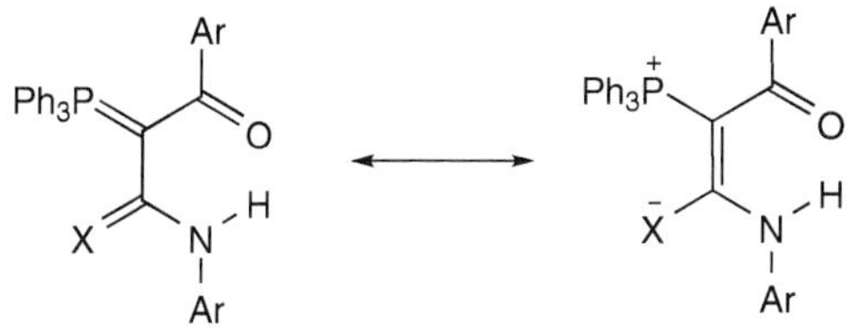

Scheme 7.56

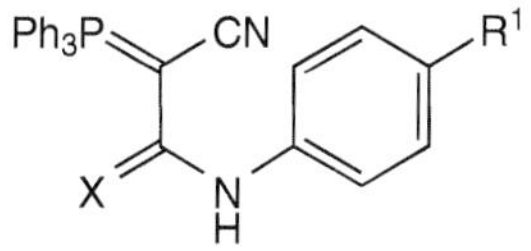

93	X	R^1	93	X	R^1
a	O	H	f	S	H
b	O	NO$_2$	g	S	NO$_2$
c	O	Cl	h	S	Cl
d	O	Me	I	S	Me
e	O	OMe	k	S	OMe

Scheme 7.57

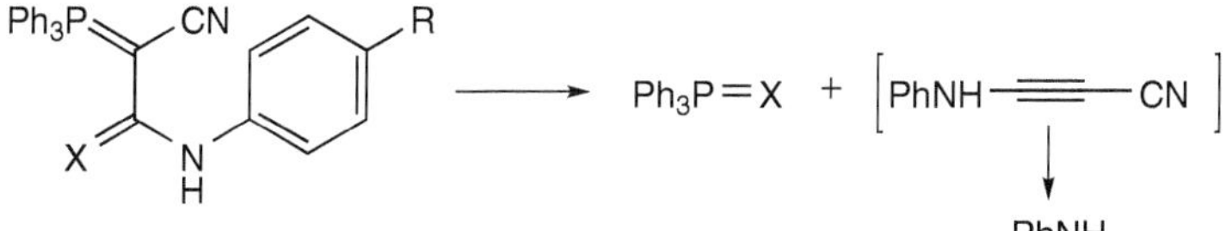

Scheme 7.58

Kinetic data presented in Table 7.11 show greater reactivity of the thio-carbonyl ylides **93f–k** over their carbonyl analogues **93a–e**, a feature also noted for ylides **91**; this is attributed both to the C=S group being more labile than C=O and to the greater nucleophilicity of sulfur versus oxygen.

We have also replaced ArCO in **91** with an MeCO group to prepare [(acetyl) (arylcarbamoyl)] **94a–e** and the thio analogues **95a–e** [51] (Scheme 7.59).

FVP of **94a** at 500 °C gave phenyl isocyanate (37%), Ph$_3$PO (37%), and ylide **96** (27%). Pyrolysis of **95a** at 500 °C gave phenyl isothiocyanate (45%) and Ph$_3$PO (92%) (Scheme 7.60).

Although the C=S bond is not involved in the six-membered transition state, the thiocarbonyl ylides **95** are consistently more reactive than their carbonyl analogues **94** (Table 7.12), showing relative rate differences of 4.6–42. This is attributed to the more favorable enolate form **97** of **94** than form **98** of **95** (Scheme 7.61), which is ascribed to the higher electronegativity of oxygen relative to sulfur.

Table 7.11 Rate constants and relative rates at 127 °C for gas phase pyrolysis of ylide **93**.

R	k(93a–e)10^6 k s^{-1}	k(93f–k)10^4 k s^{-1}	$k_{rel} = k_{(S)}/k_{(O)}$
H	4.01	1.54	38
NO$_2$	2.22	0.95	43
Cl	1.33	0.87	65
Me	1.66	0.92	55
OMe	–	0.74	–

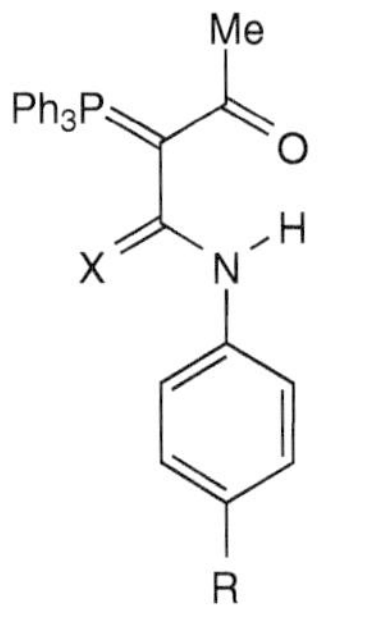

94/95	X	R
94a	O	H
94b	O	NO$_2$
94c	O	Cl
94d	O	Me
94e	O	OMe
95a	S	H
95b	S	NO$_2$
95c	S	Cl
95d	S	Me
95e	S	OMe

94a–e / 95a–e
X = O / S

Scheme 7.59

94a, X = O
95a, X = S

96

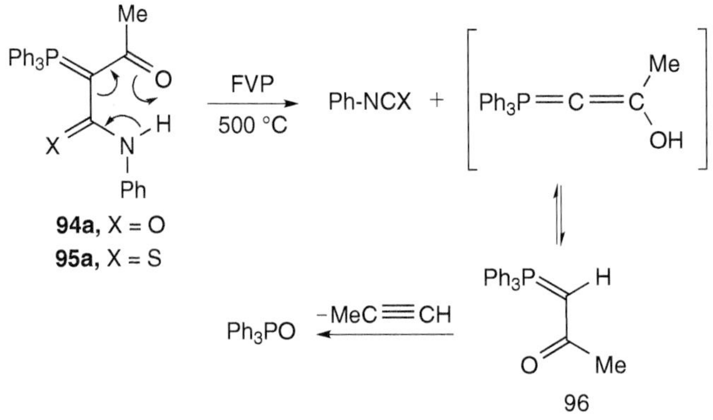

Scheme 7.60

Table 7.12 Rate constants and relative rates for **94** and **95** at 127 °C in the gas phase.

R	$X = O 10^4\ k\ s^{-1}$	$X = S 10^3\ k\ s^{-1}$	$k_{rel} = k_{(X\,=\,S)}/k_{(X\,=\,O)}$
H	3.95	1.81	4.6
4-NO$_2$	2.41	6.77	28
4-Cl	2.06	8.68	42
4-Me	4.88	9.03	19
4-OMe	6.79	12.07	18

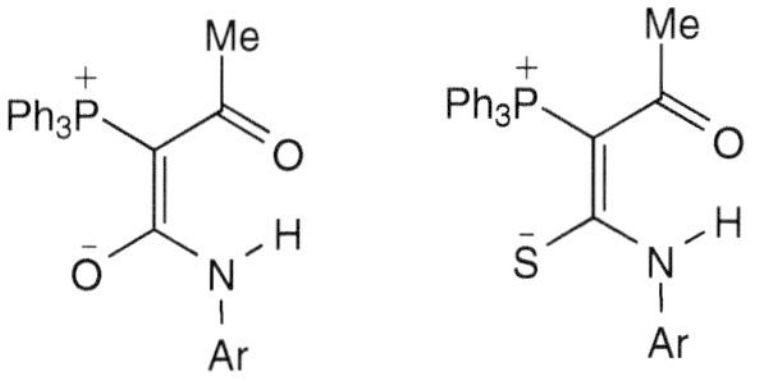

Scheme 7.61

To correlate reactivity with structure, a kinetic analysis of gas-phase pyrolysis was performed on triphenylphosphonium benzoylmethylides **99a–f** and carbonyl-stabilized arsonium ylides **100a–d** [52–54] (Scheme 7.62).

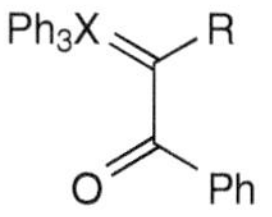

99 / 100	X	R
99a	P	H
99b	P	COPh
99c	P	Ph
99d	P	CO$_2$Et
99e	P	SPh
99f	P	Et
100a	As	H
100b	As	COPh
100c	As	Ph
100d	As	CO$_2$Et

Scheme 7.62

Triphenylphosphonium ylides **99a–f** gave only triphenylphosphine oxide and substituted phenylalkyne, as shown in Scheme 7.63. The structural rationale for extrusion of Ph$_3$PO from stabilized ylides has been ascribed to the presence of the enolate structure decomposing via a four-membered transition state.

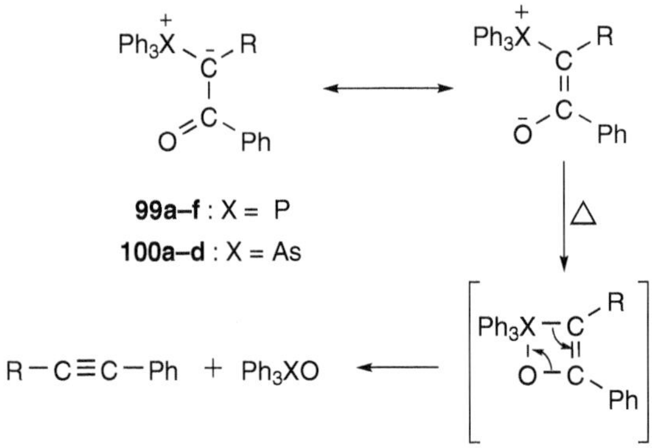

99a–f : X = P
100a–d : X = As

Scheme 7.63

On the other hand, ylide **100a**, which lacks an additional stabilizing substituent, pyrolyzed to give triphenylarsine and acetophenone through a carbene-reactive intermediate (Scheme 7.64). Capture of hydrogen by this carbene resulted in the formation of acetophenone.

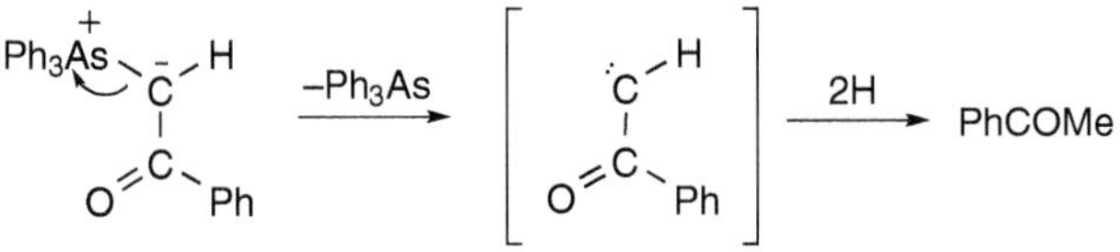

100a

Scheme 7.64

When pyrolysis of ylide **100a** was conducted in the presence of cyclohexene as a radical trap, no acetophenone was obtained, and triphenylarsine was the only product.

The trend in rates of reaction (Table 7.13) of the stabilized arsonium ylides **100b**, **100d** is the same as for their analogous phosphonium ylides **99b**, **99d**; the mechanism of pyrolysis (Scheme 7.63) is the same for the two sets of compounds (four-membered transition state). The reason for the higher reactivity of the phosphonium ylides (Table 7.14) relates to the much easier extrusion of triphenylphosphine oxide than of triphenylarsine oxide, as arsenic is a relatively poor oxygen scavenger [53–57]. The highest rate of reaction ($k = 4.4 \times 10^2$) in the arsenic series is that of ylide **100a** (R=H). This is because **100a** has no additional stabilizing substituent: this feature accounts for the formation of the reactive carbene intermediate following the extrusion of triphenylarsine rather than triphenylarsine oxide, and also accounts for the observed high rate of reaction and accordingly higher rate ratio (Table 7.13). The small contribution from the carbene pathway to the overall mechanism

Table 7.13 Rate constants ($k\,s^{-1}$) at 227 °C for pyrolysis of benzoyl-stabilized ylides **99a–f** and **100a–d**.

99	R	$10^4\,k\,s^{-1}$	100	R	$10^4\,k\,s^{-1}$
a	H	3.44	a	H	4.40×10^2
b	COPh	9.48×10^3	b	COPh	3.55
c	Ph	76.3	c	Ph	96.9
d	CO_2Et	21.4	d	CO_2Et	1.85
e	SPh	18.0	—	—	—
f	Et	18.6	—	—	—

Table 7.14 Ratio constants ($k\,s^{-1}$) at 227 °C of ylides **99a–d** and **100a–d**.

R	Ylide pair	(k_P/k_{As})	(k_{As}/k_P)
H	**100a/99a**		1.28×10^2
COPh	**99b/100b**	2.67×10^3	
Ph	**100c/99c**		1.27
CO_2Et	**99d/100d**	11.6	

of pyrolysis of ylide **100c** appears to be the reason this arsonium ylide is slightly more reactive than its phosphorus analogue **99c**.

The magnitude of change in reactivity with change in the structure of the stabilizing ketone/thione moiety was assessed by investigating the FVP of an interesting series of ketone/thione-stabilized triphenyl phosphonium methylides **101a–g**, **102a–d**, **103a,b**, and **104** [58] (Scheme 7.65).

The pyrolysate from each ylide consists of Ph_3PO or Ph_3PS, and an alkyne or its rearrangement product [58] (Scheme 7.66).

A mechanism that provides an adequate account for the observed products is shown in Scheme 7.67. It involves conformational pre-organization of the ylide, and a four-membered TS leading to the characterized products [42, 59, 60].

The rate constants $k\,s^{-1}$ calculated at 227 °C for the 14 ketone/thione-stabilized triphenylphosphonium methylides are given along with their structures in Scheme 7.68.

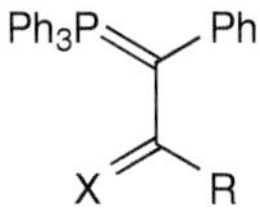

101a–g

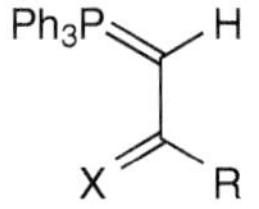

102a–d

101	X	R
a	O	CH_2PH
b	O	CH=CHPh
c	O	C≡CPh
d	O	NHPh
e	S	NHPh
f	S	$CH_2(CH_2)_4CH_3$
g	O	$CH_2(CH_2)_4CH_3$

102	X	R
a	O	$C(CH_3)_3$
b	O	CH_3
c	S	$C(CH_3)_3$
d	S	CH_3

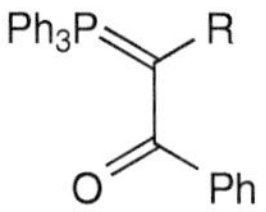

103a,b

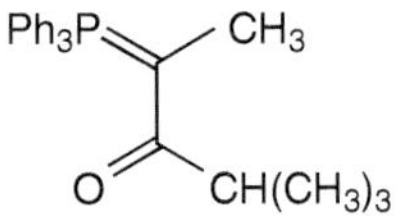

104

103	R
a	Ph
b	H

Scheme 7.65

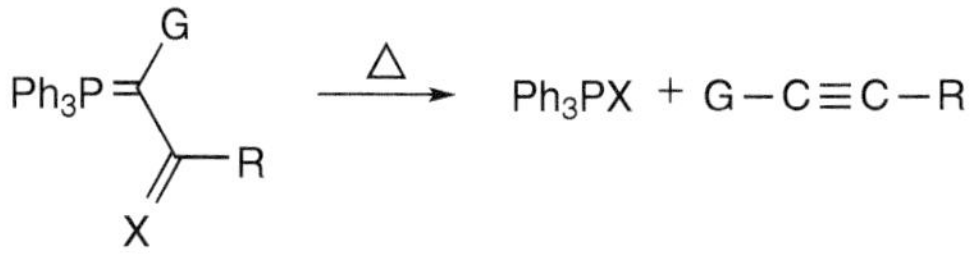

101; G = Ph
102; G = H

Scheme 7.66

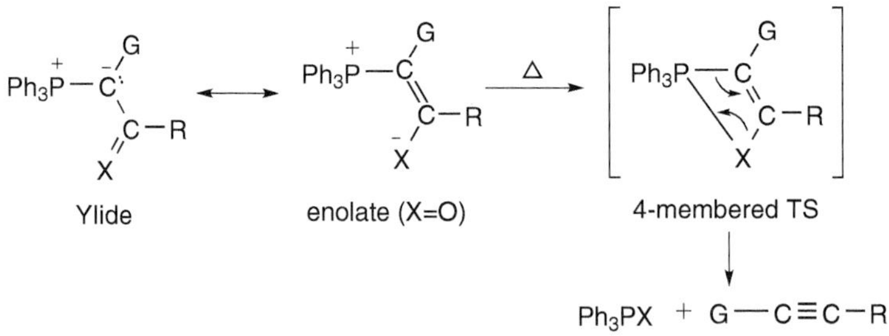
Ylide
enolate (X=O)
4-membered TS
Ph_3PX + G—C≡C—R

Scheme 7.67

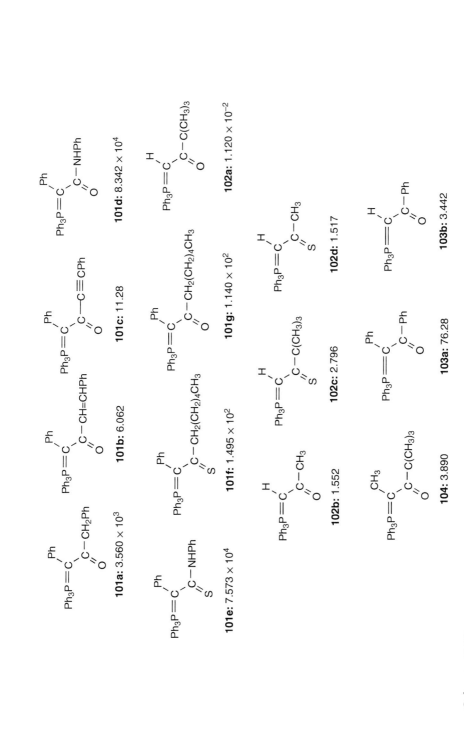

101a: 3.560×10^3 **101b:** 6.062 **101c:** 11.28 **101d:** 8.342×10^4

101e: 7.573×10^4 **101f:** 1.495×10^2 **101g:** 1.140×10^2 **102a:** 1.120×10^{-2}

102b: 1.552 **102c:** 2.796 **102d:** 1.517

104: 3.890 **103a:** 76.28 **103b:** 3.442

Scheme 7.68

The interpretation and rationalization of these results follow the mechanism suggested for the thermal gas-phase elimination of these ylides (Scheme 7.67). Substitution of Ph and CH_3 at the ylide carbon of **103a** and **104** has increased their thermal reactivity over the monosubstituted ylides **103b** and **102a** (Scheme 7.69).

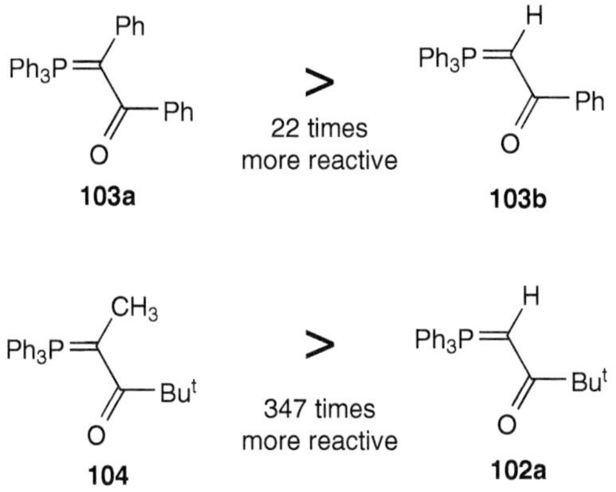

Scheme 7.69

All the monosubstituted ylides display lower thermal reactivity than the fully substituted triphenylphosphonium methylides as a result of conjugative and hyperconjugative effects in promoting ylide reactivity. C—H hyperconjugation of the methylene moiety inserted between the carbonyl and phenyl groups of ylide **101a** increased the reactivity of ylide **101a** over **103a** by a factor of 46.7. A more dramatic rate enhancement is observed when conjugation involves the delocalization of lone pairs of electrons: ylides **101d** and **101e** are ca. 103-fold more reactive than **103a**, and are also 23.4- and 21.3-fold more reactive than **101a**. These overall higher thermal reactivities of ylides **101d** and **101e** are associated with the lone pair of electrons on (N) being available for conjugative delocalization, leading to a substantial increase in negative charge at the oxide/sulfide centers of the ylides (Scheme 7.67).

Such studies have demonstrated for the first time the accessibility of kinetic investigations in gas-phase elimination reactions of stabilized triphenylphosphonium methylides [42, 50, 51] and have provided valuable support and insight into the mechanisms of pyrolysis and a quantitative basis for structure/molecular reactivity correlations [61].

References

1 Al-Bashir, R.F., Al-Awadi, N.A., and El-Dusouqui, O.M.E. (2011). *J. Phys. Org. Chem.* 24: 311–319.
2 Gilheany, D.G. (1994). *Chem. Rev.* 94: 1339–1374.
3 Al-Bashir, R.F., Al-Awadi, N.A., and El-Dusouqui, O.M.E. (2005). *Can. J. Chem.* 83: 1543–1553.
4 Trippett, S. and Walker, D.M. (1959). *J. Chem. Soc.*: 3874–3876.
5 Blanchard, M.L., Strzeleck, H., Martin, G.J. et al. (1976). *Bull. Soc. Chim. Fr.*: 2677–2679.
6 Strzeleck, H. (1966). *Ann. Chim (Paris)*: 201–220.
7 Shevchuk, M.I., Zabrodskaya, I.S., and Dombrovskil, A.V. (1969). *Zh. Obshch. Khim.* 39: 1282–1287.
8 Bestmann, H.J. and Pfohl, S. (1969). *Angew. Chem. Int. Ed. Engl.* 8: 762–763.
9 Toloctiko, A.F., Megera, I.V., Zykov, L.V., and Shevchuk, M.I. (1975). *Zh. Obshch. Khim.* 45: 2150–2154.
10 Bollens, E., Szonyi, F., Trabelsi, H., and Cambon, A. (1994). *J. Fluor. Chem.* 67: 177–181.
11 Bestmann, H.J., Schade, G., and Schmid, G. (1980). *Angew. Chem. Int. Ed. Engl.* 19: 822–824.
12 Gough, S.T.D. and Trippett, S. (1962). *J. Chem. Soc.* 453: 2333–2337.
13 Braga, A.L. and Comasseto, J.V. (1989). *Synth. Commun.* 19: 2877–2883.
14 Aitken, R.A. and Atherton, J.I. (1985). *J. Chem. Soc. Chem. Commun.*: 1140–1141.
15 Akiyama, S., Nakasuji, K., and Nakagawa, M. (1971). *Bull. Chem. Soc. Jpn.* 44: 2231–2236.
16 Märkl, G. (1961). *Chem. Ber.* 94: 3005–3010.
17 Braga, A.L., Comasseto, J.V., and Petragnani, N. (1984). *Tetrahedron Lett.* 25 (11): 1111–1114.
18 Shen, Y. and Zheng, J. (1987). *J. Fluor. Chem.* 35: 513–521.
19 Kobayashi, Y., Yamashita, T., Takahashi, K. et al. (1982). *Tetrahedron Lett.* 23 (3): 343–344.
20 Shen, Y., Zheng, J., and Huang, Y.Z. (1988). *J. Fluor. Chem.* 41: 363–369.
21 Shen, Y., Wenbiao, C., and Huang, Y. (1987). *Synth. Commun.*: 626–628.

22 Chen, Y., Lin, Y., and Xin, Y. (1985). *Tetrahedron Lett.* 26 (42): 5137–5138.

23 Chen, Y. and Qiu, W. (1987). *Synth. Commun.*: 42–43.

24 Bestmann, H.J. (1965). *Angew. Chem. Int. Ed. Engl.* 4: 645–660.

25 Aitken, R.A. and Atherton, J.I. (1994). *J. Chem. Soc. Perkin Trans.* 1: 1281–1284.

26 Bestmann, H.J., Kumar, K., and Schaper, W. (1983). *Angew. Chem. Int. Ed. Engl.* 22: 167–168.

27 Bestmann, H.J., Kumar, K., and Kisielowski, L. (1983). *Chem Ber.* 116: 2378–2382.

28 Aitken, R.A., Horsburgh, C.E.R., McCreadie, J.G., and Seth, S. (1994). *J. Chem. Soc. Perkin Trans.* 1: 1727–1732.

29 Aitken, R.A. and Seth, S. (1994). *J. Chem. Soc. Perkin Trans.* 1: 2461–2466.

30 Aitken, R.A., Herion, H., Horsburgh, C.E.R. et al. (1996). *J. Chem. Soc. Perkin Trans.* 1: 485–489.

31 Aitken, R.A., Boeters, C., and Morrison, J.J. (1994). *J. Chem. Soc. Perkin Trans.* 1: 2473–2480.

32 Aitken, R.A., Boeters, C., and Morrison, J.J. (1997). *J. Chem. Soc. Perkin Trans.* 1: 2625–2634.

33 Aitken, R.A., Balkovich, M., Bestmann, H.J. et al. (1999). *Synlett* (8): 1235–1236.

34 Aitken, R.A., Herion, H., Janosi, A. et al. (1993). *Tetrahedron Lett.* 34 (35): 5621–5622.

35 Abell, A.D., Heinicke, G.W., and Massy-Westropp, R.A. (1985). *Synth. Commun.*: 764–766.

36 Shen, Y., Cen, W., and Huang, Y. (1985). *Synthesis*: 159–160.

37 Aitken, R.A., Herion, H., Janosi, A. et al. (1994). *J. Chem. Soc. Perkin Trans.* 1: 2467–2472.

38 (a) Blitzk, T., Sicker, D., and Wilde, H. (1995). *Synthesis, Short Paper*: 236–238. (b) Aitken, R.A. and Karodia, N. (1997). *Liebigs Ann.*: 779–783.

39 Aitken, R.A., Drysdale, M.J., Hill, L. et al. (1999). *Tetrahedron* 55: 11039–11050.

40 Aitken, R.A. and Drysdale, M.J. (1991). *J. Chem. Soc. Chem. Commun.*: 512–513.

41 Aitken, R.A., Drysdale, M.J., Ferguson, G., and Lough, A.J. (1998). *J. Chem. Soc. Perkin Trans.* 1: 875–880.

42 Al-Bashir, R.F., Al-Awadi, N.A., and El-Dusouqui, O.M.E. (2005). *Can. J. Chem.* 83: 1543–1553.

43 Dib, H.H., Al-Awadi, N.A., Ibrahim, Y.A., and El-Dusouqui, O.M.E. (2003). *Tetrahedron* 59: 9455–9464.

44 Aitken, R.A., Drysdale, M.J., Ford, A. et al. (1993). *Phosphorus Sulfur Silicon* 75: 31–34.

45 Aitken, R.A., Drysdale, M.J., and Ryan, B.M. (1994). *J. Chem. Soc. Chem. Commun.*: 805–806.

46 Aitken, R.A., Drysdale, M.J., and Ryan, B.M. (1998). *J. Chem. Soc. Perkin Trans.* 1: 3345–3348.

47 Aitken, R.A., Armstrong, J.M., Drysdale, M.J. et al. (1999). *J. Chem. Soc. Perkin Trans.* 1: 593–604.

48 Aitken, R.A., Drysdale, M.J., and Ryan, B.M. (1993). *J. Chem. Soc. Chem. Commun.*: 1699–1700.

49 Aitken, R.A., Al-Awadi, N.A., Dawson, G. et al. (2005). *Tetrahedron* 61: 129–135.

50 Aitken, R.A., Al-Awadi, N.A., Dawson, G. et al. (2006). *Int. J. Chem. Kinet.* 38: 496–502.

51 Aitken, R.A., Al-Awadi, N.A., Dawson, G. et al. (2007). *Int. J. Chem. Kinet.* 39: 6–16.

52 Al-Bashir, R.F., Al-Awadi, N.A., and El-Dusouqui, O.M.E. (2008). *Arkivoc* (viii): 228–242.

53 Lloyd, D. (1994). *The Chemistry of Organic Arsenic, Antimony and Bismuth Compounds*. Chichester: Wiley.

54 Pandolfo, L., Bertani, R., Facchin, G. et al. (1995). *Inorg. Chim. Acta* 237: 27–35.

55 Johnson, A.W. and Schubert, H. (1970). *J. Org. Chem.* 35: 2678–2680.

56 Shen, Y., Fan, Z., and Qiu, W. (1987). *J. Organomet. Chem.* 320: 21–25.

57 Walker, B.J. (1985). *Org. Phos. Chem.* 15: 218–259.

58 Al-Bashir, R.F., Al-Awadi, N.A., and El-Dusouqui, O.M.E. (2011). *J. Phys. Org. Chem.* 24: 311–319.

59 Gilheany, D.G. (1994). *Chem. Rev.* 94: 1339–1374. and references therein.

60 Aitken, R.A., Al-Awadi, N.A., Balkovich, M.E. et al. (2003). *Eur. J. Org. Chem.*: 840–847.

61 Aitken, R.A., Dawson, G., Keddie, N.S. et al. (2009). *Chem. Commun.*: 7381–7383.

Index

Gas-Phase Pyrolytic Reactions: Synthesis, Mechanisms, and Kinetics,
First Edition. Nouria A. Al-Awadi.
© 2020 John Wiley & Sons, Inc. Published 2020 by John Wiley & Sons, Inc.